W9-ARK-281

Matrix Computation
for Engineers
and Scientists

Matrix Computation for Engineers and Scientists

Alan Jennings
Queen's University, Belfast

A Wiley – Interscience Publication

JOHN WILEY & SONS
London · New York · Sydney · Toronto

Library of Congress Cataloging in Publication Data:

Jennings, Alan.
Matrix computation for engineers and scientists.

'A Wiley–Interscience publication.'
1. Matrices. I. Title.
TA347.D4J46   519.4   76–21079

ISBN 0 471 99421 9

Printed in the United States of America

To my father

who has always had a
keen interest in the
mathematical solution
of engineering problems

Preface

In the past the sheer labour of numerical processes restricted their development and usefulness. Digital computers have removed this restriction, but at the same time have provided a challenge to those who wish to harness their power. Much work has been put into the development of suitable numerical methods and the computer organizational problems associated with their implementation. Also, new fields of application for numerical techniques have been established.

From the first days of computing the significance of matrix methods has been realized and exploited. The reason for their importance is that they provide a concise and simple method of describing lengthy and otherwise complicated computations. Standard routines for matrix operations are available on virtually all computers, and, where these methods are employed, duplication of programming effort is minimized. Matrices now appear on the school mathematics syllabus and there is a more widespread knowledge of matrix algebra. However, a rudimentary knowledge of matrix algebra should not be considered a sufficient background for embarking upon the construction of computer programs involving matrix techniques, particularly where large matrices are involved. Programs so developed could be unnecessarily complicated, highly inefficient or incapable of producing accurate solutions. It is even possible to obtain more than one of these deficiencies in the same program. The development of computer methods (most certainly those involving matrices) is an art which requires a working knowledge of the possible mathematical formulations of the particular problem and also a working knowledge of the effective numerical procedures and the ways in which they may be implemented on a computer. It is unwise to develop a very efficient program if it is so complicated that it requires excessive programming effort (and hence program testing time) or has such a small range of application that it is hardly ever used. The right balance of simplicity, economy and versatility should be sought which most benefits the circumstances.

Chapter 1 is intended to act as a review of relevant matrix algebra and hand computational techniques. Also included in this chapter is a discussion of the matrix properties which are most useful in numerical computation. In Chapter 2 some selected applications are briefly introduced. These are included so that the reader can see whether his particular problems are related to any of the problems mentioned. They also illustrate certain features which regularly occur in the formulation of matrix computational techniques. For instance:

(a) Alternative methods may be available for the solution of any one problem (as with the electrical resistance network, sections 2.1 and 2.2).
(b) Matrices often have special properties which can be utilized, such as symmetry and sparsity.
(c) It may be necessary to repeat the solution of a set of linear equations with modified right-hand sides and sometimes with modified coefficients (as with the non-linear cable problem, section 2.12).

Chapter 3 describes those aspects of computer programming technique which are most relevant to matrix computation, the storage allocation being particularly important for sparse matrices. Multiplication is the main matrix operation discussed in this chapter. Here it is interesting to note that some forethought is needed to program even the multiplication of two matrices if they are large and/or sparse. Numerical techniques for solving linear equations are presented in Chapters 4, 5 and 6. The importance of sparse matrices in many applications has been taken into account, including the considerable effect on the choice of procedure and the computer implementation.

Chapter 7 briefly introduces some eigenvalue problems and Chapters 8, 9 and 10 describe numerical methods for eigensolution. Although these last four chapters may be considered to be separate from the first six, there is some advantage to be gained from including procedures for solving linear equations and obtaining eigenvalues in the same book. For one reason, most of the eigensolution procedures make use of the techniques for solving linear equations. For another reason, it is necessary to be familiar with eigenvalue properties in order to obtain a reasonably comprehensive understanding of methods of solving linear equations.

Three short appendices have been included to help the reader at various stages during the preparation of application programs. They take the form of questionnaire checklists on the topics of program layout, preparation and verification.

Corresponding ALGOL and FORTRAN versions of small program segments have been included in Chapters 2, 3, 4 and 5. These segments are not intended for direct computer use, but rather as illustrations of programming technique. They have been written in such a way that the ALGOL and FORTRAN versions have similar identifiers and structure. In general this means that the ALGOL versions, while being logically correct, are not as elegant as they might be. To obtain a full appreciation of the complete text it is therefore necessary to have some acquaintance with computer programming. From the mathematical standpoint the text is meant to be as self-sufficient as possible.

I hope that the particular methods given prominence in the text, and the discussion of them, are not only justified but also stand the test of time. I will be grateful for any comments on the topics covered.

ALAN JENNINGS
Department of Civil Engineering,
Queen's University, Belfast

Acknowledgements

It was while on sabbatical leave at the Royal Military College, Kingston, Ontario, Canada, that I contemplated writing this book and started an initial draft. While I was at Kingston, and more recently at Belfast, I have had much encouragement and helpful advice from many colleagues (and tolerance from others). My particular thanks are to Professor A. A. Wells, head of the department of civil engineering at Queen's University, and Professor J. S. Ellis, head of the department of civil engineering at the R.M.C., Kingston, for helping me to provide time to prepare the script. Special thanks are due to Dr. M. Clint of the department of computer science at Queen's University who has carefully read the script and given many excellent comments. I would also like to thank D. Meegan of the department of engineering mathematics at Queen's University for carrying out checks on the program segments, Miss R. Tubman for competently typing the script and Mrs V. Kernohan for carefully preparing the figures.

Contents

It is a capital mistake to theorise before one has data

Sir Arthur Conan Doyle

Chapter 1
Basic Algebraic and Numerical Concepts

1.1 WHAT IS A MATRIX?

A matrix can be described simply as a rectangular array of elements. Thus

$$\mathbf{A} = \begin{bmatrix} 2 & 0 & 5 & 1 & 0 \\ 1 & 3 & 1 & 3 & 1 \\ 3 & 2 & 4 & 6 & 0 \end{bmatrix} \tag{1.1}$$

is a matrix of order 3×5 as it has three rows and five columns. The elements of a matrix may take many forms. In matrix (1.1) they are all real non-negative integers; however, they could be real or complex numbers, or algebraic expressions, or, with the restrictions mentioned in section 1.15, matrices themselves or matrix expressions. The physical context of the various elements need not be the same; if one of the elements is a measure of distance, it does not follow that the other elements have also to be measures of distance. Hence matrices may come from a large variety of sources and take a variety of forms. Computation with matrices will involve matrices which have elements in numerical form. However, matrices with elements of algebraic form will be of significance in the theoretical discussion of properties and procedures.

Matrix (1.1) could represent the numbers of different coins held by three boys, the columns specifying the five coin denominations (i.e. 1 p, 2 p, 5 p, 10 p and 50 p) while the rows differentiate the boys. The interpretation of matrix A would therefore be according to Table 1.1. Whereas any table of information could be considered as a matrix by enclosing the data within square brackets, such consideration would be fruitless unless it can operate with some other matrices in

Table 1.1 Possible interpretation of matrix (1.1)

			Coins		
	1 p	2 p	5 p	10 p	50 p
Tom	2	0	5	1	0
Dick	1	3	1	3	1
Harry	3	2	4	6	0

such a way that the rules of matrix algebra are meaningful. Before describing the basic rules of matrix algebra it is necessary to be able to specify any element of a matrix algebraically. The usual method for this is to replace whatever labels the rows and columns have by numbers, say 1 to m for rows and 1 to n for columns, and then to refer to the element on row i and column j of matrix A as a_{ij}.

A matrix is square if $m = n$ and is rectangular if $m \neq n$.

1.2 THE MATRIX EQUATION

Probably the most fundamental aspect of matrix algebra is that matrices are equal only if they are identical, i.e. they are of the same order and have corresponding elements the same. The identity (1.1) is a valid matrix equation which implies $m \times n$ ordinary equations defining each element a_{ij}, e.g. $a_{34} = 6$. This property of being able to represent a multiplicity of ordinary equations by a single matrix equation is the main power of matrix methods. (This is in distinct contrast to determinants where the equation

$$\begin{vmatrix} a_{11} & a_{12} \\ a_{21} & a_{22} \end{vmatrix} = \begin{vmatrix} 5 & 1 \\ 1 & 1 \end{vmatrix} \tag{1.2}$$

does not define the elements a_{11}, a_{12}, a_{21} and a_{22} but only specifies a relationship between them.

If they are of the same order, two matrices may be added by adding corresponding elements. If Table 1.1 describes the state of Tom, Dick and Harry's finances at the beginning of the day and if their transactions during the day are represented by Table 1.2 which specifies a further matrix H, then the state of their finances at the end of the day is given by the matrix

$$\left.\begin{aligned} &\mathbf{G} = \mathbf{A} + \mathbf{H} \\ &\text{i.e.} \\ &\mathbf{G} = \begin{bmatrix} 2 & 0 & 5 & 1 & 0 \\ 1 & 3 & 1 & 3 & 1 \\ 3 & 2 & 4 & 6 & 0 \end{bmatrix} + \begin{bmatrix} -2 & 1 & 2 & -1 & 0 \\ 0 & 0 & 2 & 3 & -1 \\ -1 & 2 & 3 & -1 & 0 \end{bmatrix} = \begin{bmatrix} 0 & 1 & 7 & 0 & 0 \\ 1 & 3 & 3 & 6 & 0 \\ 2 & 4 & 7 & 5 & 0 \end{bmatrix} \end{aligned}\right\} \tag{1.3}$$

Matrix subtraction may be defined in a corresponding way to matrix addition. Scalar multiplication of a matrix is such that all the elements of the matrix are

Table 1.2 Transactions of Tom, Dick and Harry (negative terms imply expenditure)

	Coins				
	1 p	2 p	5 p	10 p	50 p
Tom	−2	1	2	−1	0
Dick	0	0	2	3	−1
Harry	−1	2	3	−1	0

multiplied by the scalar. From these definitions it follows that, for instance, the matrix equation

$$\mathbf{A} = \mathbf{B} + \mu\mathbf{C} - \mathbf{D} \tag{1.4}$$

where **A**, **B**, **C** and **D** are 3 × 2 matrices and μ is a scalar, is equivalent to six simple linear equations of the form

$$a_{ij} = b_{ij} + \mu c_{ij} - d_{ij} \tag{1.5}$$

1.3 MATRIX MULTIPLICATION

Two matrices may only be multiplied if the number of columns of the first equals the number of rows of the second, in which case they are said to be *conformable.* If matrix **A**, of order $m \times p$, is multiplied by matrix **B**, of order $p \times n$, the product

$$\mathbf{C} = \mathbf{AB} \tag{1.6}$$

is of order $m \times n$ with typical element

$$c_{ij} = \sum_{k=1}^{p} a_{ik} b_{kj} \tag{1.7}$$

With **A** as in equation (1.1) and

$$\mathbf{B} = \begin{bmatrix} 1 & 1 \\ 1 & 2 \\ 1 & 5 \\ 1 & 10 \\ 1 & 50 \end{bmatrix} \tag{1.8}$$

the product matrix **C** = **AB** is given by

$$\begin{bmatrix} c_{11} & c_{12} \\ c_{21} & c_{22} \\ c_{31} & c_{32} \end{bmatrix} = \begin{bmatrix} 2 & 0 & 5 & 1 & 0 \\ 1 & 3 & 1 & 3 & 1 \\ \boxed{3 \quad 2 \quad 4 \quad 6 \quad 0} \end{bmatrix} \begin{bmatrix} \boxed{1} & 1 \\ \boxed{1} & 2 \\ \boxed{1} & 5 \\ \boxed{1} & 10 \\ \boxed{1} & 50 \end{bmatrix} = \begin{bmatrix} 8 & 37 \\ 9 & 92 \\ \boxed{15} & 87 \end{bmatrix} \tag{1.9}$$

The rule that the element c_{ij} is obtained by scalar multiplication of row i of **A** by column j of **B** has been illustrated for c_{31} by including the relevant elements in boxes. The choice of the matrix **B** has been such that the matrix **C** yields, in its first column, the total number of coins held by each boy and, in its second column, the total value, in pence, of the coins held by each boy.

Except in special cases the matrix product **AB** is not equal to the matrix product **BA**, and hence the order of the matrices in a product may not, in general, be reversed (the product **BA** may not even be conformable). In view of this it is not adequate to say that **A** is multiplied by **B**; instead it is said that **A** is *postmultiplied* by **B** or **B** is *premultiplied* by **A**. Unless either **A** or **B** contain zero elements the total number of multiplications necessary to evaluate **C** from equation (1.6) is $m \times p \times n$, with almost as many addition operations. Matrix multiplication can therefore involve a great deal of computation when m, p and n are all large.

If the multiplication of two large matrices is to be performed by hand it is advisable to include a check procedure to avoid errors. This can be done by including an extra row of column sums in **A** and an extra column of row sums in **B**. The resulting matrix **C** will then contain both row and column sum checks which enable any incorrect element to be pinpointed. With this checking procedure equation (1.9) would appear as

$$\begin{array}{c} \\ \\ \\ \Sigma \end{array}\begin{bmatrix} 2 & 0 & 5 & 1 & 0 \\ 1 & 3 & 1 & 3 & 1 \\ 3 & 2 & 4 & 6 & 0 \\ \hline 6 & 5 & 10 & 10 & 1 \end{bmatrix} \begin{bmatrix} 1 & 1 & \vert & 2 \\ 1 & 2 & \vert & 3 \\ 1 & 5 & \vert & 6 \\ 1 & 10 & \vert & 11 \\ 1 & 50 & \vert & 51 \end{bmatrix}^{\Sigma} = \begin{array}{c} \\ \\ \\ \Sigma \end{array}\begin{bmatrix} 8 & 37 & \vert & 45 \\ 9 & 92 & \vert & 101 \\ 15 & 87 & \vert & 102 \\ \hline 32 & 216 & \vert & 248 \end{bmatrix}^{\Sigma} \tag{1.10}$$

Multiple products

If the product matrix **C** of order $m \times n$ (equation 1.6) is further premultiplied by a matrix **D** of order $r \times m$, the final product can be written as

$$\underset{r \times n}{[\mathrm{F}]} = \underset{r \times m}{[\mathrm{D}]} \left(\underset{m \times p}{[\mathrm{A}]} \quad \underset{p \times n}{[\mathrm{B}]} \right) \tag{1.11}$$

It can be verified that the same result for **F** is obtained if the product **DA** is evaluated and the result postmultiplied by **B**. For this reason brackets are left out of multiple products so that equation (1.11) is written as

$$\mathbf{F} = \mathbf{DAB} \tag{1.12}$$

It is important to note that whereas **D(AB)** has the same value as **DA(B)** the order of evaluation of the products may be very important in numerical computation. For example, if **D** and **A** are of order 100×100 and **B** is of order 100×1, the total number of multiplications for **(DA)B** is 1,010,000 and for **D(AB)** is 20,000. It is therefore going to be roughly fifty times faster to evaluate **R** by multiplying **AB** first. If this calculation were to be performed by hand, the operator would not get far with the multiplication of **AB** before he realized that his method was unnecessarily long-winded. However, if a computer is programmed to evaluate the multiple product the wrong way round this oversight may remain buried in the program without detection. Although such an oversight appears only to involve a penalty of extra computation time, it is likely that the accuracy of the computed results would be less due to the greater accumulation of rounding errors (see section 3.2).

With **A** and **B** as in equation (1.9) and $\mathbf{D} = [1 \quad 1 \quad 1]$, it follows that

$$\mathbf{F} = [1 \quad 1 \quad 1]\begin{bmatrix} 2 & 0 & 5 & 1 & 0 \\ 1 & 3 & 1 & 3 & 1 \\ 3 & 2 & 4 & 6 & 0 \end{bmatrix}\begin{bmatrix} 1 & 1 \\ 1 & 2 \\ 1 & 5 \\ 1 & 10 \\ 1 & 50 \end{bmatrix} = [32 \quad 216] \tag{1.13}$$

the first element of which, $f_{11} = 32$, signifies that there is a total of thirty-two coins held by all of the boys and the second element, $f_{12} = 216$, signifies that their total value is 216 p.

1.4 SOME SPECIAL MATRIX FORMS

Row and column matrices

A row matrix is normally called a row vector and a column matrix, a column vector. For convenience, a column vector is often written horizontally rather than vertically. If this is done, braces { } rather than square brackets may be used to contain the elements.

The null matrix

The symbol **0** is used for a matrix having all of its elements zero. An example of its use is the equation

$$\mathbf{A} - \mathbf{B} - \mu\mathbf{C} + \mathbf{D} = \mathbf{0} \tag{1.14}$$

which is just an alternative statement of equation (1.4).

Diagonal matrices

A square matrix is diagonal if non-zero elements only occur on the *leading diagonal*, i.e. $a_{ij} = 0$ for $i \neq j$. The importance of the diagonal matrix is that it can be used for row or column scaling. Premultiplying a matrix by a conformable diagonal matrix has the effect of scaling each row of the matrix by the corresponding element in the diagonal matrix:

$$\begin{bmatrix} a_{11} & & \\ & a_{22} & \\ & & a_{33} \end{bmatrix}\begin{bmatrix} b_{11} & b_{12} \\ b_{21} & b_{22} \\ b_{31} & b_{32} \end{bmatrix} = \begin{bmatrix} a_{11}b_{11} & a_{11}b_{12} \\ a_{22}b_{21} & a_{22}b_{22} \\ a_{33}b_{31} & a_{33}b_{32} \end{bmatrix} \tag{1.15}$$

In the specification of the diagonal matrix the zero off-diagonal elements have been left blank. An alternative notation for the diagonal matrix is $\lceil a_{11} \quad a_{22} \quad a_{33} \rfloor$. Postmultiplication of a matrix by a conformable diagonal matrix has the effect of scaling each column by the corresponding element in the diagonal matrix.

The unit matrix

The symbol I is used to represent a diagonal matrix having all of its diagonal elements equal to unity. Its order is assumed to be conformable to the matrix or matrices with which it is associated. From the property of diagonal matrices it is immediately seen that pre- or postmultiplication by the unit matrix leaves a matrix unaltered, i.e.

$$\mathbf{AI} = \mathbf{IA} = \mathbf{A} \tag{1.16}$$

Triangular matrices

A lower triangular matrix is a square matrix having all elements above the leading diagonal zero. Similarly, an upper triangular matrix has all elements below the leading diagonal zero. A property of triangular matrices is that the product of two like triangular matrices produces a third matrix of like form, i.e.

$$\begin{bmatrix} 1 & & \\ 2 & 1 & \\ -1 & 1 & -1 \end{bmatrix}\begin{bmatrix} 2 & & \\ 1 & -1 & \\ 4 & 1 & 1 \end{bmatrix} = \begin{bmatrix} 2 & & \\ 5 & -1 & \\ -5 & -2 & -1 \end{bmatrix} \tag{1.17}$$

Fully populated and sparse matrices

A matrix is fully populated if all of its elements are non-zero and is sparse if only a small proportion of its elements are non-zero.

1.5 THE MATRIX TRANSPOSE AND SYMMETRY

The transpose of a matrix is obtained by interchanging the roles of rows and columns. Thus row i becomes column i and column j becomes row j. If the matrix is rectangular the dimensions of the matrix will be reversed. The transpose of matrix (1.1) is

$$\mathbf{A}^T = \begin{bmatrix} 2 & 1 & 3 \\ 0 & 3 & 2 \\ 5 & 1 & 4 \\ 1 & 3 & 6 \\ 0 & 1 & 0 \end{bmatrix} \tag{1.18}$$

A square matrix is said to be symmetric if it is symmetric about the leading diagonal, i.e. $a_{ij} = a_{ji}$ for all values of i and j. A symmetric matrix must be equal to its own transpose. Symmetric matrices frequently arise in the analysis of conservative systems and least squares minimization, and the symmetric property can normally be utilized in numerical operations.

A *skew symmetric matrix* is such that $a_{ij} = -a_{ji}$; hence $\mathbf{A}^T = -\mathbf{A}$ and the leading diagonal elements a_{ii} must be zero. Any square matrix may be split into the sum of a symmetric and a skew symmetric matrix. Thus

$$\mathbf{A} = \tfrac{1}{2}(\mathbf{A} + \mathbf{A}^T) + \tfrac{1}{2}(\mathbf{A} - \mathbf{A}^T) \qquad (1.19)$$

where $\frac{1}{2}(\mathbf{A} + \mathbf{A}^T)$ is symmetric and $\frac{1}{2}(\mathbf{A} - \mathbf{A}^T)$ is skew symmetric.

When dealing with matrices which have complex numbers as elements the *Hermitian transpose* is an important concept. This is the same as the normal transpose except that the complex conjugate of each element is used.

Thus if

$$\mathbf{A} = \begin{bmatrix} 5+i & 2-i & 1 \\ 6i & 4 & 9-i \end{bmatrix} \qquad (1.20)$$

$$\mathbf{A}^H = \begin{bmatrix} 5-i & -6i \\ 2+i & 4 \\ 1 & 9+i \end{bmatrix} \qquad (1.21)$$

A square matrix having $\mathbf{A}^H = \mathbf{A}$ is called a *Hermitian matrix* and if it is written as $\mathbf{A} = \mathbf{C} + i\mathbf{D}$ then $\mathbf{C}$ must be symmetric and $\mathbf{D}$ skew symmetric.

An inner product

When two column vectors $\mathbf{x} = \{x_1 x_2 \ldots x_n\}$ and $\mathbf{y} = \{y_1 y_2 \ldots y_n\}$ are multiplied together by transposing the first, the resulting scalar quantity

$$\mathbf{x}^T\mathbf{y} = \sum_{i=1}^{n} x_i y_i \qquad (1.22)$$

is often called an inner product. The inner product of a column vector with itself, i.e.

$$\mathbf{x}^T\mathbf{x} = \sum_{i=1}^{n} x_i^2 \qquad (1.23)$$

must be positive provided that $\mathbf{x}$ is real. (For a complex vector $\mathbf{x}$ then $\mathbf{x}^H\mathbf{x}$ must be positive provided only that $\mathbf{x}$ is not null.)

1.6 THE DETERMINANT OF A MATRIX

A square matrix has a determinant which is given by the following recursive formula:

$$|\mathbf{A}| = a_{11}M_{11} - a_{12}M_{12} + a_{13}M_{13} - \cdots - (-1)^n a_{1n}M_{1n} \qquad (1.24)$$

where M_{11} is the determinant of the matrix with row 1 and column 1 missing, M_{12} is the determinant of the matrix with row 1 and column 2 missing, etc. (It must also be noted that the determinant of a 1×1 matrix just equals the particular element.)

Hence,

$$\left.\begin{aligned}
&\begin{vmatrix} a_{11} & a_{12} \\ a_{21} & a_{22} \end{vmatrix} = a_{11}a_{22} - a_{12}a_{21} \\
&\text{and} \\
&\begin{vmatrix} a_{11} & a_{12} & a_{13} \\ a_{21} & a_{22} & a_{23} \\ a_{31} & a_{32} & a_{33} \end{vmatrix} = a_{11}(a_{22}a_{33} - a_{23}a_{32}) - a_{12}(a_{21}a_{33} - a_{23}a_{31}) \\
&\qquad\qquad\qquad\qquad + a_{13}(a_{21}a_{32} - a_{22}a_{31})
\end{aligned}\right\} \quad (1.25)$$

By rearranging the terms in a different order it is possible to expand the determinant, not by the first row, as has been done in equation (1.24), but by any other row or column. There are several useful properties which will not be given here, but it is important to know that the determinant of a matrix is zero if a row/column is zero or equal to a linear combination of the other rows/columns. Thus, for instance,

$$\begin{vmatrix} 3 & 1 & 4 \\ 2 & 1 & 0 \\ 1 & 1 & -4 \end{vmatrix}$$

exhibits the linear relationships (row 3) = −(row 1) + 2(row 2), (col 3) = 4(col 1) − 8(col 2), and consequently has a zero determinant.

1.7 THE SOLUTION OF SIMULTANEOUS EQUATIONS

A set of n linear simultaneous equations

$$\left.\begin{aligned}
a_{11}x_1 + a_{12}x_2 + \cdots + a_{1n}x_n &= b_1 \\
a_{21}x_1 + a_{22}x_2 + \cdots + a_{2n}x_n &= b_2 \\
\cdots\cdots\cdots\cdots\cdots\cdots\cdots\cdots \\
a_{n1}x_1 + a_{n2}x_2 + \cdots + a_{nn}x_n &= b_n
\end{aligned}\right\} \quad (1.26)$$

may be written in matrix form as

$$\mathbf{A}\mathbf{x} = \mathbf{b}$$

where $\mathbf{A}$ is an $n \times n$ coefficient matrix having typical element a_{ij} and $\mathbf{x}$ and $\mathbf{b}$ are column vectors of the variables and right-hand sides respectively. The solution of such a set of equations for $\mathbf{x}$ is a key operation in the solution of a vast number of problems.

Cramer's rule gives the solution in determinantal form, e.g. for a set of three equations

$$\left.\begin{aligned}
&x_1 = \frac{1}{|\mathbf{A}|}\begin{vmatrix} b_1 & a_{12} & a_{13} \\ b_2 & a_{22} & a_{23} \\ b_3 & a_{32} & a_{33} \end{vmatrix}, \quad x_2 = \frac{1}{|\mathbf{A}|}\begin{vmatrix} a_{11} & b_1 & a_{13} \\ a_{21} & b_2 & a_{23} \\ a_{31} & b_3 & a_{33} \end{vmatrix}, \\
&x_3 = \frac{1}{|\mathbf{A}|}\begin{vmatrix} a_{11} & a_{12} & b_1 \\ a_{21} & a_{22} & b_2 \\ a_{31} & a_{32} & b_3 \end{vmatrix}
\end{aligned}\right\} \quad (1.27)$$

Whereas a solution of the equations by Cramer's rule is numerically uneconomical as compared with elimination methods, and also difficult to automate, it does establish that there is a unique solution for all equations provided that $|\mathbf{A}| \neq 0$. A square matrix whose determinant is zero is known as a *singular matrix*, and the solution of equations which have a singular or near singular coefficient matrix will require special consideration.

1.8 GAUSSIAN ELIMINATION AND PIVOTAL CONDENSATION

Elimination methods are the most important methods of solving simultaneous equations either by hand or computer, certainly when the number of equations is not very large. The most basic technique is usually attributed to Gauss.

Consider the equations

$$\left.\begin{aligned} 10x_1 + x_2 - 5x_3 &= 1 \\ -20x_1 + 3x_2 + 20x_3 &= 2 \\ 5x_1 + 3x_2 + 5x_3 &= 6 \end{aligned}\right\} \tag{1.28}$$

It is possible to replace this set of equations by another in which each or any is scaled or in which any number have been linearly combined. By adding 20/10 times the first equation to the second its x_1 coefficient is eliminated. Similarly, by subtracting 5/10 times the first equation from the third equation its x_1 coefficient is also eliminated. The modified equations are

$$\left.\begin{aligned} 10x_1 + x_2 - 5x_3 &= 1 \\ 5x_2 + 10x_3 &= 4 \\ 2.5x_2 + 7.5x_3 &= 5.5 \end{aligned}\right\} \tag{1.29}$$

Subtracting 2.5/5 times the new second equation from the new third equation eliminates its x_2 coefficient giving

$$\left.\begin{aligned} 10x_1 + x_2 - 5x_3 &= 1 \\ 5x_2 + 10x_3 &= 4 \\ 2.5x_3 &= 3.5 \end{aligned}\right\} \tag{1.30}$$

Thus the equations have been converted into a form in which the coefficient matrix is triangular, from which the values of the variables can be obtained in reverse order by backsubstitution, i.e.

$$\left.\begin{aligned} x_3 &= \frac{3.5}{2.5} = 1.4 \\ x_2 &= \frac{4 - 10x_3}{5} = -2 \\ x_1 &= \frac{1 - x_2 + 5x_3}{10} = 1 \end{aligned}\right\} \tag{1.31}$$

The process is completely automatic however many equations are present, provided that certain *pivotal* elements are non-zero. In equations (1.28), (1.29) and (1.30) the pivotal elements, shown in italics, appear as denominators in the scaling factors and also the backsubstitution. Except where it is known in advance that these pivots cannot be zero, it is necessary to modify the elimination procedure in such a way that non-zero pivots are always selected. Of all the elements which could be used as pivot, pivotal condensation employs the one of largest absolute magnitude at each stage of the reduction. This not only ensures a non-zero pivot (except where the left-hand side matrix is singular) but also makes the scaling factors less than or equal to unity in modulus.

If pivotal condensation is used to solve equation (1.28), the first pivot must be either the x_1 or the x_3 coefficient of the second equation. If the x_3 coefficient is chosen as pivot then elimination of the other x_3 coefficients yields

$$\left.\begin{aligned} 5x_1 + 1.75x_2 &= 1.5 \\ -20x_1 + 3x_2 + 20x_3 &= 2 \\ 10x_1 + 2.25x_2 &= 5.5 \end{aligned}\right\} \tag{1.32}$$

The next pivot must be selected from either the first or third equation of (1.28). Hence the x_1 coefficient in the third equation is chosen, yielding the reduced equations

$$\left.\begin{aligned} 0.625x_2 &= -1.25 \\ -20x_1 + 3x_2 + 20x_3 &= 2 \\ 10x_1 + 2.25x_2 &= 5.5 \end{aligned}\right\} \tag{1.33}$$

The backsubstitution can now be performed in the sequence x_2, x_1, x_3.

An alternative method of carrying out pivotal condensation is to choose the same pivots, but then move them into the leading positions so that the triangular form of the reduction is retained. To place the x_3 element of the second equation of (1.28) into the leading position the first and second equations are interchanged, and also x_1 and x_3, giving

$$\left.\begin{aligned} 20x_3 + 3x_2 - 20x_1 &= 2 \\ -5x_3 + x_2 + 10x_1 &= 1 \\ 5x_3 + 3x_2 + 5x_1 &= 6 \end{aligned}\right\} \tag{1.34}$$

which reduces to

$$\left.\begin{aligned} 20x_3 + 3x_2 - 20x_1 &= 2 \\ 1.75x_2 + 5x_1 &= 1.5 \\ 2.25x_2 + 10x_1 &= 5.5 \end{aligned}\right\} \tag{1.35}$$

Since the x_1 coefficient in the third equation will be the next pivot, the second and third equations are interchanged and also x_1 and x_2 to give

$$\left.\begin{aligned} 20x_3 - 20x_1 + \quad 3x_2 &= 2 \\ 10x_1 + 2.25x_2 &= 5.5 \\ 5x_1 + 1.75x_2 &= 1.5 \end{aligned}\right\} \tag{1.36}$$

and the x_1 coefficient in equation x_3 eliminated to give the triangular form for backsubstitution.

1.9 EQUATIONS WITH MULTIPLE RIGHT-HAND SIDES

The two sets of simultaneous equations

$$\left.\begin{aligned} a_{11}x_{11} + a_{12}x_{21} + a_{13}x_{31} &= b_{11} \\ a_{21}x_{11} + a_{22}x_{21} + a_{23}x_{31} &= b_{21} \\ a_{31}x_{11} + a_{32}x_{21} + a_{33}x_{31} &= b_{31} \\ \text{and} \qquad & \\ a_{11}x_{12} + a_{12}x_{22} + a_{13}x_{32} &= b_{12} \\ a_{21}x_{12} + a_{22}x_{22} + a_{23}x_{32} &= b_{22} \\ a_{31}x_{12} + a_{32}x_{22} + a_{33}x_{32} &= b_{32} \end{aligned}\right\} \tag{1.37}$$

have the same coefficient matrix but different right-hand vectors. As a result their different solution sets are distinguished as $\{x_{11} \quad x_{21} \quad x_{31}\}$ and $\{x_{12} \quad x_{22} \quad x_{32}\}$. They may be conveniently expressed as one matrix equation

$$\begin{bmatrix} a_{11} & a_{12} & a_{13} \\ a_{21} & a_{22} & a_{23} \\ a_{31} & a_{32} & a_{33} \end{bmatrix} \begin{bmatrix} x_{11} & x_{12} \\ x_{21} & x_{22} \\ x_{31} & x_{32} \end{bmatrix} = \begin{bmatrix} b_{11} & b_{12} \\ b_{21} & b_{22} \\ b_{31} & b_{32} \end{bmatrix} \tag{1.38}$$

In general the matrix equation

$$\mathbf{AX} = \mathbf{B} \tag{1.39}$$

may be used to represent m sets of n simultaneous equations each having $\mathbf{A}$ $(n \times n)$ as the coefficient matrix, where $\mathbf{B}$ is an $n \times m$ matrix of the right-hand vectors compounded by columns. It is economical to solve these sets of equations in conjunction because the operations on the coefficient matrix to reduce it to triangular form only need to be performed once, however many sets of equations are involved. Gaussian elimination can be considered to be a series of operations on equation (1.39) to reduce it to the form

$$\mathbf{UX} = \mathbf{Y} \tag{1.40}$$

where $\mathbf{U}$ is an upper triangular matrix and $\mathbf{Y}$ is a modified right-hand side matrix, and then a series of operations (corresponding to the backsubstitution) to obtain

Table 1.3 Gaussian elimination for a 3 × 3 set of equations with two right-hand sides

Stage	Row operations	Left-hand side	Right-hand side	Σ
Initial equations		$\begin{bmatrix} 10 & 1 & -5 \\ -20 & 3 & 20 \\ 5 & 3 & 5 \end{bmatrix}$	$\begin{bmatrix} 1 & 1 \\ 2 & 7 \\ 6 & 6 \end{bmatrix}$	8 12 25
Elimination of first column	row 1 unchanged (row 2) + 2(row 1) (row 3) − 0.5(row 1)	10 1 −5 5 10 2.5 7.5	1 1 4 9 5.5 5.5	8 28 21
Triangular form	row 1 unchanged row 2 unchanged (row 3) − 0.5(row 2)	10 1 −5 5 10 2.5	1 1 4 9 3.5 1	8 28 7
Backsubstitution performed in order row 3, row 2, row 1	0.1{(row 1) − (new row 2) + 5(new row 3)} 0.2(row 2) − 10(new row 3) 0.4(row 3)	1 1 1	$\begin{bmatrix} 1 & 0.2 \\ -2 & 1 \\ 1.4 & 0.4 \end{bmatrix}$	2.2 0 2.8

Table 1.4 Multiplication and division operations for Gaussian elimination (n fully populated equations with m right-hand sides)

Type of matrix			Large n
Unsymmetric	Reduction stage	$n(n-1)\left[\frac{n+1}{3}+\frac{m}{2}\right]$	$\rightarrow \frac{n^3}{3}+\frac{n^2 m}{2}$
	Total	$\frac{n(n-1)(n+1)}{3}+n^2 m$	$\rightarrow \frac{n^3}{3}+n^2 m$
Symmetric	Reduction stage	$\frac{n(n-1)}{2}\left[\frac{n+4}{3}+m\right]$	$\rightarrow \frac{n^3}{6}+\frac{n^2 m}{2}$
	Total	$\frac{n(n-1)(n+4)}{6}+n^2 m$	$\rightarrow \frac{n^3}{6}+n^2 m$

the form

$$\mathbf{IX} = \mathbf{X} \tag{1.41}$$

Consider equation (1.28) with the additional right-hand side $\{1 \quad 7 \quad 6\}$. The reduction to the form of equation (1.41) is shown in Table 1.3 as a series of row operations. In the table a summation column has also been included. If the same operations are performed on the summation column as on the other columns, this acts as a check on the arithmetic. The solution appears in the right-hand side columns when the left-hand side has been converted to the unit matrix.

The number of multiplication and division operations necessary to solve sets of simultaneous equations in which all of the coefficients are non-zero is shown in Table 1.4. In the case where matrix **A** is symmetric a saving can be made in the left-hand side operations if a pivotal strategy is not needed. The fact that the amount of computation is cubic in n produces a virtual barrier against the solution of large-order fully populated equations by elimination. For hand calculation this barrier is in the region of 10--20 equations, whereas for digital computers it is in the region of 100—2,000 equations depending on the type of computer.

1.10 TRANSFORMING MATRIX EQUATIONS

It is possible to take any valid matrix equation and to pre- or postmultiply it by a conformable matrix. Thus from

$$\mathbf{AX} = \mathbf{B} \tag{1.42}$$

can be derived

$$\mathbf{DAX} = \mathbf{DB} \tag{1.43}$$

provided only that the number of columns of **D** is equal to the number of rows in **A** and **B**. If **D** is a rectangular matrix the number of elemental equations derived in equation (1.43) will be different from the number originally in equation (1.42).

Where the number of elemental equations increases the resulting equations must be linearly dependent.

Consider equation (1.42) as two elemental equations

$$\begin{bmatrix} 3 & -1 & 1 \\ 4 & 2 & 1 \end{bmatrix} \begin{bmatrix} x_1 \\ x_2 \\ x_3 \end{bmatrix} = \begin{bmatrix} 3 \\ -2 \end{bmatrix} \tag{1.44}$$

With

$$\mathbf{D} = \begin{bmatrix} -1 & 1 \\ 1 & 1 \\ 0 & 1 \end{bmatrix} \tag{1.45}$$

equation (1.43) gives the three elemental equations

$$\begin{bmatrix} 1 & 3 & 0 \\ 7 & 1 & 2 \\ 4 & 2 & 1 \end{bmatrix} \begin{bmatrix} x_1 \\ x_2 \\ x_2 \end{bmatrix} = \begin{bmatrix} -5 \\ 1 \\ -2 \end{bmatrix} \tag{1.46}$$

This does not provide a means of determining $\{x_1 \quad x_2 \quad x_3\}$ because of linear dependence within the coefficient matrix. If an attempt to solve the equations is made it will be discovered that the coefficient matrix is singular.

Although a pre- or postmultiplying matrix can sometimes be cancelled from a matrix equation, it is not always valid to do so. For example,

$$[3 \quad 1] \begin{bmatrix} x_1 \\ x_2 \end{bmatrix} = [3 \quad 1] \begin{bmatrix} 2 \\ 0 \end{bmatrix} \tag{1.47}$$

does not imply $\{x_1 \quad x_2\} = \{2 \quad 0\}$ as a unique solution, since, for example, $\{1 \quad 3\}$ is also a solution. The cancellation cannot be carried out in this case because it produces more elemental equations than originally present. If the number of elemental equations is reduced or remains the same, cancellation is usually valid. An exception is that a square matrix which is singular may not be cancelled. For example,

$$\begin{bmatrix} 3 & 1 \\ 6 & 2 \end{bmatrix} \begin{bmatrix} x_1 \\ x_2 \end{bmatrix} = \begin{bmatrix} 3 & 1 \\ 6 & 2 \end{bmatrix} \begin{bmatrix} 2 \\ 0 \end{bmatrix} \tag{1.48}$$

says no more than equation (1.47) and cancellation of the singular premultiplying matrix is invalid.

However, where the original matrix equation is satisfied whatever the numerical value of the pre- or postmultiplying matrix, cancellation can always be carried out. Thus if, for any vector $\mathbf{y} = [y_1 \quad y_2 \quad y_3]$, the following equation is true:

$$[y_1 \quad y_2 \quad y_3] \begin{bmatrix} 5 & 1 & 3 \\ 2 & 1 & 4 \\ 1 & 1 & 0 \end{bmatrix} \begin{bmatrix} x_1 \\ x_2 \\ x_3 \end{bmatrix} = [y_1 \quad y_2 \quad y_3] \begin{bmatrix} 8 \\ 0 \\ 2 \end{bmatrix} \tag{1.49}$$

then replacing y by [1 0 0], [0 1 0] and [0 0 1] in turn yields the rows of

$$\begin{bmatrix} 5 & 1 & 3 \\ 2 & 1 & 4 \\ 1 & 1 & 0 \end{bmatrix} \begin{bmatrix} x_1 \\ x_2 \\ x_3 \end{bmatrix} = \begin{bmatrix} 8 \\ 0 \\ 2 \end{bmatrix} \tag{1.50}$$

(which is equivalent to cancelling **y**).

1.11 THE RANK OF A MATRIX

The number of linearly independent vectors constituting a matrix is called its *rank*.

The matrix

$$\begin{bmatrix} 1 & 4 & 0 & 2 \\ 1 & 0 & 1 & -1 \\ -3 & -4 & -2 & 0 \end{bmatrix} \tag{1.51}$$

exhibits the relationship

$$(\text{row } 3) = -(\text{row } 1) - 2(\text{row } 2) \tag{1.52}$$

and also the relationships

$$\left.\begin{aligned} &(\text{col } 3) = (\text{col } 1) - 0.25(\text{col } 2) \\ \text{and} \quad &(\text{col } 4) = -(\text{col } 1) + 0.75(\text{col } 2) \end{aligned}\right\} \tag{1.53}$$

Hence, whether viewed by rows or columns, there are only two linearly independent vectors and the matrix is therefore of rank 2.

A matrix product must have a rank less than or equal to the smallest rank of any of the constituent matrices, e.g.

$$\left.\begin{aligned} &\underset{\text{rank } 2}{\begin{bmatrix} 1 & 4 & 0 & 2 \\ 1 & 0 & 1 & -1 \\ -3 & -4 & -2 & 0 \end{bmatrix}} \underset{\text{rank } 1}{\begin{bmatrix} 2 & 1 & 1 & -1 \\ -4 & -2 & -2 & 2 \\ 0 & 0 & 0 & 0 \\ -2 & -1 & -1 & 1 \end{bmatrix}} = \underset{\text{rank } 1}{\begin{bmatrix} -18 & -9 & -9 & 9 \\ 4 & 2 & 2 & -2 \\ 10 & 5 & 5 & -5 \end{bmatrix}} \\ \text{and} \\ &\underset{\text{rank } 2}{\begin{bmatrix} 2 & 4 & 0 \\ 1 & 6 & 0 \\ -3 & -6 & 0 \end{bmatrix}} \underset{\text{rank } 2}{\begin{bmatrix} 2 & 0 \\ -1 & 0 \\ 2 & -1 \end{bmatrix}} = \underset{\text{rank } 1}{\begin{bmatrix} 0 & 0 \\ -4 & 0 \\ 0 & 0 \end{bmatrix}} \end{aligned}\right\} \tag{1.54}$$

a corollary being that, if a matrix has rank r, then any matrix factor of it must have dimensions greater than or equal to r. For instance, matrix (1.51) can be specified as

$$\begin{bmatrix} 1 & 4 & 0 & 2 \\ 1 & 0 & 1 & -1 \\ -3 & -4 & -2 & 0 \end{bmatrix} = \begin{bmatrix} 1 & 1 \\ 1 & 0 \\ -3 & -1 \end{bmatrix} \begin{bmatrix} 1 & 0 & 1 & -1 \\ 0 & 4 & -1 & 3 \end{bmatrix} \tag{1.55}$$

but cannot be specified as the product of matrices of order 3×1 and 1×4.

A *minor* is the determinant obtained from an equal number of rows and columns of a matrix. The minor taken from rows 2 and 3 and columns 2 and 4 of matrix (1.51) is

$$\begin{vmatrix} 0 & -1 \\ -4 & 0 \end{vmatrix} = -4 \tag{1.56}$$

If a matrix has rank r then there must be at least one non-zero minor of order r and no non-zero minors of order greater than r.

It is important to use physical properties of the particular problem to establish the rank of a matrix because rounding errors are likely to confuse any investigation of rank by numerical means.

1.12 THE MATRIX INVERSE

The inverse of a square matrix $\mathbf{A}$ is designated as $\mathbf{A}^{-1}$ and is defined such that

$$\mathbf{AA}^{-1} = \mathbf{I} \tag{1.57}$$

It will be recognized that $\mathbf{A}^{-1}$ is the generalized solution, $\mathbf{X}$, of a set of simultaneous equations in which the unit matrix has been adopted as a multiple set of right-hand vectors,

$$\mathbf{AX} = \mathbf{I} \tag{1.58}$$

Matrix $\mathbf{A}$ will only have an inverse if it is non-singular, for otherwise equation (1.58) could not be solved.

Postmultiplying equation (1.57) by $\mathbf{A}$ gives

$$\mathbf{AA}^{-1}\mathbf{A} = \mathbf{IA} = \mathbf{AI} \tag{1.59}$$

Since the premultiplying matrix is non-singular it may be cancelled to give

$$\mathbf{A}^{-1}\mathbf{A} = \mathbf{I} \tag{1.60}$$

Premultiplication of any set of simultaneous equations $\mathbf{Ax} = \mathbf{b}$ by $\mathbf{A}^{-1}$ yields

$$\mathbf{A}^{-1}\mathbf{Ax} = \mathbf{A}^{-1}\mathbf{b} \tag{1.61}$$

and hence, by using equation (1.60),

$$\mathbf{x} = \mathbf{A}^{-1}\mathbf{b} \tag{1.62}$$

For example, since

$$\mathbf{A} = \begin{bmatrix} 10 & 1 & -5 \\ -20 & 3 & 20 \\ 5 & 3 & 5 \end{bmatrix} \tag{1.63}$$

has the inverse

$$\mathbf{A}^{-1} = \begin{bmatrix} -0.36 & -0.16 & 0.28 \\ 1.6 & 0.6 & -0.8 \\ -0.6 & -0.2 & 0.4 \end{bmatrix} \tag{1.64}$$

the solution of equations (1.28) can be obtained by the matrix multiplication

$$\begin{bmatrix} x_1 \\ x_2 \\ x_3 \end{bmatrix} = \begin{bmatrix} -0.36 & -0.16 & 0.28 \\ 1.6 & 0.6 & -0.8 \\ -0.6 & -0.2 & 0.4 \end{bmatrix} \begin{bmatrix} 1 \\ 2 \\ 6 \end{bmatrix} = \begin{bmatrix} 1 \\ -2 \\ 1.4 \end{bmatrix} \tag{1.65}$$

The inverse of a matrix may be derived from its adjoint (which is not defined here). However, this method, being an extension of Cramer's rule for simultaneous equations, is not an economic or convenient computational procedure.

Elimination methods are almost always adopted for determining inverses. The Gaussian elimination for matrix (1.63) can be performed by substituting a 3 x 3 unit matrix into the right-hand side of Table 1.3 and carrying out the same operations as before. If necessary the process can be modified to include a pivotal strategy. For large n approximately n^3 multiplications are necessary for the inversion of a fully populated matrix (reducing to $n^3/2$ if symmetry can be utilized). As *Gauss–Jordan elimination* is similar in efficiency to Gaussian elimination for fully populated inversions, it is a viable alternative. In Gauss–Jordan elimination, matrix coefficients above the diagonal, as well as those below the diagonal, are reduced to zero during the reduction stage. This means that no backsubstitution is required. A Gauss–Jordan elimination for the inverse of matrix (1.63) is given in Table 1.5.

1.13 SIGNIFICANCE OF THE INVERSE

The matrix inverse should be thought of as a useful algebraic concept rather than as an aid to numerical computation. This may be appreciated by examining the solution of simultaneous equations, as illustrated by equation (1.65). Although the solution of these equations is rapid once the inverse has been found, the process of finding the inverse involves significantly more computation than is required for the direct solution of the original equations, and hence cannot be justified numerically. For large sets of fully populated equations with multiple right-hand vectors, the number of multiplications required to form the inverse and then perform the matrix multiplications required to obtain the solution is approximately $n^3 + n^2m$,

Table 1.5 Gauss–Jordan elimination for inversion of a 3 × 3 matrix

Stage	Row operations	Left-hand side			Right-hand side			Σ
Initial matrix		10	1	−5	1			7
		−20	3	20		1		4
		5	3	5			1	14
Elimination of first column	0.1(row 1)	1	0.1	−0.5	0.1			0.7
	(row 2) + 20(new row 1)		5	10	2	1		18
	(row 3) − 5(new row 1)		2.5	7.5	−0.5		1	10.5
Elimination of second column	(row 1) − 0.1(new row 2)	1		−0.7	0.06	−0.02		0.34
	0.2(row 2)		1	2	0.4	0.2		3.6
	(row 3) − 2.5(new row 2)			2.5	−1.5	−0.5	1	1.5
Elimination of third column	(row 1) + 0.7(new row 3)	1			−0.36	−0.16	0.28	0.76
	(row 2) − 2(new row 3)		1		1.6	0.6	−0.8	2.4
	0.4(row 3)			1	−0.6	−0.2	0.4	0.6

as opposed to $n^3/3 + n^2m$ if the equations are solved directly. One argument put forward for the use of the inverse is where different right-hand vectors are to be processed but where these are not all available at the same time (for instance, where the second right-hand vector can only be calculated when the solution to the first right-hand vector has been obtained). However, even this does not create a problem for direct elimination provided that the full sequence of row operations are recorded, for it is a simple process to add another column to the right-hand side of Table 1.3 and carry out the row operations on just the additional columns. It appears therefore that, unless the inverse itself is specifically required or numerical efficiency is of no consequence, computation of inverses should be avoided.

The argument against forming unnecessary inverses has been for fully populated matrices. For sparse matrices the arguments are even stronger because the inverse of a sparse matrix is almost invariably fully populated.

1.14 THE TRANSPOSE AND INVERSE IN MATRIX EXPRESSIONS

The following transpose and inversion properties may be verified by trying small numerical examples:

(a) $(\mathbf{A}^T)^T = \mathbf{A}$

(b) $(\mathbf{A}^{-1})^{-1} = \mathbf{A}$

(c) $(\mathbf{A}^{-1})^T = (\mathbf{A}^T)^{-1}$

(As a result of this property the simplified notation $\mathbf{A}^{-T}$ will be adopted for the transpose of the inverse of matrix **A**.)

(d) $$\left.\begin{array}{l}\text{If } \mathbf{D} = \mathbf{ABC}\\ \text{then } \mathbf{D}^T = \mathbf{C}^T\mathbf{B}^T\mathbf{A}^T.\end{array}\right\} \qquad (1.66)$$

(This is known as the *reversal rule for transposed products.*)

(e) $$\left.\begin{array}{l}\text{If } \mathbf{D} = \mathbf{ABC} \text{ and } \mathbf{A}, \mathbf{B} \text{ and } \mathbf{C} \text{ are square and non-singular}\\ \text{then } \mathbf{D}^{-1} = \mathbf{C}^{-1}\mathbf{B}^{-1}\mathbf{A}^{-1}.\end{array}\right\} \qquad (1.67)$$

(This is known as the *reversal rule for inverse products.*)

The reversal rule for transposed products may be used to show that the matrix

$$\mathbf{C} = \mathbf{A}^T\mathbf{A} \qquad (1.68)$$

is symmetric, which follows from

$$\mathbf{C}^T = (\mathbf{A}^T\mathbf{A})^T = \mathbf{A}^T\mathbf{A} = \mathbf{C} \qquad (1.69)$$

Also the matrix $\mathbf{C} = \mathbf{A}^T\mathbf{BA}$ is symmetric provided that **B** is symmetric, and from rule (c) it follows that the inverse of a symmetric matrix is also symmetric.

The reversal rule for inverse products may sometimes be used to simplify matrix expressions involving more than one inverse. Consider the matrix equation

$$\mathbf{z} = (\mathbf{A} - \mathbf{B}^{-1}\mathbf{C}\mathbf{B}^{-T})^{-1}\mathbf{y} \qquad (1.70)$$

where **A**, **B** and **C** are $n \times n$ matrices and **y** and **z** are $n \times 1$ column vectors. Since

$$\mathbf{A} = \mathbf{B}^{-1}\mathbf{B}\mathbf{A}\mathbf{B}^T\mathbf{B}^{-T}$$

then

$$\mathbf{z} = [\mathbf{B}^{-1}(\mathbf{B}\mathbf{A}\mathbf{B}^T - \mathbf{C})\mathbf{B}^{-T}]^{-1}\mathbf{y} \tag{1.71}$$

and the reversal rule for inverse products gives

$$\mathbf{z} = \mathbf{B}^T(\mathbf{B}\mathbf{A}\mathbf{B}^T - \mathbf{C})^{-1}\mathbf{B}\mathbf{y} \tag{1.72}$$

If the matrices **A**, **B** and **C** and the vector **y** have known numerical values and it is required to evaluate **z**, then a suitable sequence of steps is shown in Table 1.6. In this procedure the need to form the inverse is circumvented by the evaluation of an intermediate set of variables **x** such that

$$(\mathbf{B}\mathbf{A}\mathbf{B}^T - \mathbf{C})\mathbf{x} = \mathbf{B}\mathbf{y} \tag{1.73}$$

Table 1.6 Evaluation of $\mathbf{z} = \mathbf{B}^T(\mathbf{B}\mathbf{A}\mathbf{B}^T - \mathbf{C})^{-1}\mathbf{B}\mathbf{y}$ where **A**, **B** and **C** are $n \times n$

Operation	Approx. number of multiplications if **A**, **B**, **C** fully populated
Multiply **By**	n^2
Multiply **BA**	n^3
Multiply $\mathbf{BAB}^T$	n^3
Subtract $\mathbf{BAB}^T - \mathbf{C}$	—
Solve $(\mathbf{BAB}^T - \mathbf{C})\mathbf{x} = (\mathbf{By})$	$\frac{n^3}{3} + n^2$
Multiply $\mathbf{z} = \mathbf{B}^T\mathbf{x}$	n^2
	Total = $2\frac{1}{3}n^3 + 3n^2$

1.15 PARTITIONING OF MATRICES

There are many reasons why it might be useful to partition matrices by rows and/or columns or to compound matrices to form a larger matrix. If this is done the complete matrix is called a *supermatrix* and the partitions are called *submatrices.*

The partitioning of a matrix will lead to corresponding partitions in related matrices. For instance, if matrix (1.1) is partitioned by columns into a 3×3 submatrix $\mathbf{A}_p$ and a 3×2 submatrix $\mathbf{A}_q$ as follows:

$$\mathbf{A} = [\mathbf{A}_p \quad \mathbf{A}_q] = \left[\begin{array}{ccc|cc} 2 & 0 & 5 & 1 & 0 \\ 1 & 3 & 1 & 3 & 1 \\ 3 & 2 & 4 & 6 & 0 \end{array}\right] \tag{1.74}$$

then, with the interpretation of the matrix given in Table 1.1, the submatrix $\mathbf{A}_p$ contains the information regarding the coins of denominations 5 p or less while submatrix $\mathbf{A}_q$ contains the information regarding the coins of denomination greater than 5 p. A corresponding split in the matrix derived from Table 1.2 is

$$\mathbf{H} = [\mathbf{H}_p \quad \mathbf{H}_q] = \left[\begin{array}{ccc|cc} -2 & 1 & 2 & -1 & 0 \\ 0 & 0 & 2 & 3 & -1 \\ -1 & 2 & 3 & -1 & 0 \end{array}\right] \tag{1.75}$$

and hence the corresponding partitioned form of **G** (equation 1.3) found by matrix addition of **A** and **H** is

$$\mathbf{G} = [\mathbf{G}_p \quad \mathbf{G}_q] = [\mathbf{A}_p \quad \mathbf{A}_q] + [\mathbf{H}_p \quad \mathbf{H}_q] \tag{1.76}$$

It can easily be verified that

$$\left.\begin{aligned} \mathbf{G}_p &= \mathbf{A}_p + \mathbf{H}_p \\ \mathbf{G}_q &= \mathbf{A}_q + \mathbf{H}_q \end{aligned}\right\} \tag{1.77}$$

signifying that equation (1.76) can be expanded as if the submatrices are elements of the supermatrices.

Consider now the matrix multiplication of equation (1.9). Since the first three rows of matrix **B** refer to coins of denomination 5 p or less and the last two rows refer to coins of denomination greater than 5 p, this matrix should be partitioned by rows such that

$$\mathbf{B} = \begin{bmatrix} \mathbf{B}_p \\ \mathbf{B}_q \end{bmatrix} = \left[\begin{array}{cc} 1 & 1 \\ 1 & 2 \\ 1 & 5 \\ \hline 1 & 10 \\ 1 & 50 \end{array}\right] \tag{1.78}$$

The matrix **C** cannot be partitioned at all by this criterion and hence

$$\mathbf{C} = [\mathbf{A}_p \quad \mathbf{A}_q] \begin{bmatrix} \mathbf{B}_p \\ \mathbf{B}_q \end{bmatrix} \tag{1.79}$$

It can be verified that

$$\mathbf{C} = \mathbf{A}_p\mathbf{B}_p + \mathbf{A}_q\mathbf{B}_q \tag{1.80}$$

i.e. the submatrices may be treated as elements of the supermatrix provided that the order of multiplication in the submatrix products is retained.

If a set of linear simultaneous equations partitioned as follows:

$$\left[\begin{array}{cccccc|ccccc} \times&\times&\times&\times&\times&\times&\times&\times&\times&\times&\times\\ \times&\times&\times&\times&\times&\times&\times&\times&\times&\times&\times\\ \times&\times&\times&\times&\times&\times&\times&\times&\times&\times&\times\\ \times&\times&\times&\times&\times&\times&\times&\times&\times&\times&\times\\ \times&\times&\times&\times&\times&\times&\times&\times&\times&\times&\times\\ \times&\times&\times&\times&\times&\times&\times&\times&\times&\times&\times\\ \hline \times&\times&\times&\times&\times&\times&\times&\times&\times&\times&\times\\ \times&\times&\times&\times&\times&\times&\times&\times&\times&\times&\times\\ \times&\times&\times&\times&\times&\times&\times&\times&\times&\times&\times\\ \times&\times&\times&\times&\times&\times&\times&\times&\times&\times&\times\\ \times&\times&\times&\times&\times&\times&\times&\times&\times&\times&\times \end{array}\right] \left[\begin{array}{c}\times\\\times\\\times\\\times\\\times\\\times\\\hline\times\\\times\\\times\\\times\\\times\end{array}\right] \begin{array}{l} \\ \\ \\ p \text{ variables} \\ \\ \\ \\ \\ q \text{ variables} \\ \\ \\ \end{array} = \left[\begin{array}{c}\times\\\times\\\times\\\times\\\times\\\times\\\hline\times\\\times\\\times\\\times\\\times\end{array}\right] \begin{array}{l} \\ \\ \\ p \text{ equations} \\ \\ \\ \\ \\ q \text{ equations} \\ \\ \\ \end{array} \tag{1.81}$$

is represented by the supermatrix equation

$$\begin{bmatrix} \mathbf{A}_{pp} & \mathbf{A}_{pq} \\ \mathbf{A}_{qp} & \mathbf{A}_{qq} \end{bmatrix} \begin{bmatrix} \mathbf{x}_p \\ \mathbf{x}_q \end{bmatrix} = \begin{bmatrix} \mathbf{b}_p \\ \mathbf{b}_q \end{bmatrix} \tag{1.82}$$

the corresponding submatrix equations are

$$\left.\begin{aligned} \mathbf{A}_{pp}\mathbf{x}_p + \mathbf{A}_{pq}\mathbf{x}_q &= \mathbf{b}_p \\ \mathbf{A}_{qp}\mathbf{x}_p + \mathbf{A}_{qq}\mathbf{x}_q &= \mathbf{b}_q \end{aligned}\right\} \tag{1.83}$$

Since $\mathbf{A}$ is square, assuming that it is also non-singular, the first equation of (1.83) can be premultiplied by $\mathbf{A}_{pp}^{-1}$ to give

$$\mathbf{x}_p = \mathbf{A}_{pp}^{-1}\mathbf{b}_p - \mathbf{A}_{pp}^{-1}\mathbf{A}_{pq}\mathbf{x}_q \tag{1.84}$$

and, substituting in the second equation of (1.83),

$$(\mathbf{A}_{qq} - \mathbf{A}_{qp}\mathbf{A}_{pp}^{-1}\mathbf{A}_{pq})\mathbf{x}_q = \mathbf{b}_q - \mathbf{A}_{qp}\mathbf{A}_{pp}^{-1}\mathbf{b}_p \tag{1.85}$$

Solving this equation for $\mathbf{x}_q$ and substituting the solution into equation (1.84) provides a method of solving the simultaneous equations. The solution of sparse simultaneous equations by submatrix methods is discussed in Chapter 5.

A further illustration of matrix partitioning arises in the solution of complex simultaneous equations, e.g.

$$\begin{bmatrix} 5+i & 3-i \\ 6-2i & 8+4i \end{bmatrix} \begin{bmatrix} x_1 \\ x_2 \end{bmatrix} = \begin{bmatrix} 6 \\ 5-5i \end{bmatrix} \tag{1.86}$$

Such equations may be written in the form

$$(\mathbf{A}_r + i\mathbf{A}_i)(\mathbf{x}_r + i\mathbf{x}_i) = \mathbf{b}_r + i\mathbf{b}_i \tag{1.87}$$

where, for equation (1.86), the following would be true

$$\mathbf{A}_r = \begin{bmatrix} 5 & 3 \\ 6 & 8 \end{bmatrix}, \quad \mathbf{A}_i = \begin{bmatrix} 1 & -1 \\ -2 & 4 \end{bmatrix}, \quad \mathbf{b}_r = \begin{bmatrix} 6 \\ 5 \end{bmatrix}, \quad \mathbf{b}_i = \begin{bmatrix} 0 \\ -5 \end{bmatrix} \tag{1.88}$$

and x_r and x_i are column vectors constituting the real and imaginary parts of the solution vector. Expanding equation (1.87) gives

$$\begin{aligned} \mathbf{A}_r \mathbf{x}_r - \mathbf{A}_i \mathbf{x}_i &= \mathbf{b}_r \\ \mathbf{A}_i \mathbf{x}_r + \mathbf{A}_r \mathbf{x}_i &= \mathbf{b}_i \end{aligned} \tag{1.89}$$

which may be written in supermatrix form as

$$\begin{bmatrix} \mathbf{A}_r & -\mathbf{A}_i \\ \mathbf{A}_i & \mathbf{A}_r \end{bmatrix} \begin{bmatrix} \mathbf{x}_r \\ \mathbf{x}_i \end{bmatrix} = \begin{bmatrix} \mathbf{b}_r \\ \mathbf{b}_i \end{bmatrix} \tag{1.90}$$

showing that a set of complex simultaneous equations of order n may be converted into a set of real simultaneous equations of order $2n$.

1.16 THE EIGENVALUES OF A MATRIX

An eigenvalue and corresponding eigenvector of a matrix satisfy the property that the eigenvector multiplied by the matrix yields a vector proportional to itself. The constant of proportionality is known as the eigenvalue.

For instance, the matrix

$$\mathbf{A} = \begin{bmatrix} 16 & -24 & 18 \\ 3 & -2 & 0 \\ -9 & 18 & -17 \end{bmatrix} \tag{1.91}$$

exhibits the property

$$\begin{bmatrix} 16 & -24 & 18 \\ 3 & -2 & 0 \\ -9 & 18 & -17 \end{bmatrix} \begin{bmatrix} 2 \\ 1 \\ 0 \end{bmatrix} = 4 \begin{bmatrix} 2 \\ 1 \\ 0 \end{bmatrix} \tag{1.92}$$

showing that it has an eigenvalue equal to 4 with a corresponding eigenvector $\{2 \quad 1 \quad 0\}$. As the eigenvector has been premultiplied by the matrix it is known as a right eigenvector. The algebraic equation for the eigenvalue λ and corresponding right eigenvector $\mathbf{q}$ of a matrix $\mathbf{A}$ is given by

$$\mathbf{A}\mathbf{q} = \lambda \mathbf{q} \tag{1.93}$$

For this equation to be conformable $\mathbf{A}$ must be square. Hence only square matrices have eigenvalues.

A method of finding the eigenvalues of a matrix $\mathbf{A}$ can be illustrated with reference to matrix (1.91). Equation (1.93) gives

$$\begin{bmatrix} 16 & -24 & 18 \\ 3 & -2 & 0 \\ -9 & 18 & -17 \end{bmatrix} \begin{bmatrix} q_1 \\ q_2 \\ q_3 \end{bmatrix} = \lambda \begin{bmatrix} q_1 \\ q_2 \\ q_3 \end{bmatrix} \tag{1.94}$$

which can be written in the form

$$\begin{bmatrix} 16-\lambda & -24 & 18 \\ 3 & -2-\lambda & 0 \\ -9 & 18 & -17-\lambda \end{bmatrix} \begin{bmatrix} q_1 \\ q_2 \\ q_3 \end{bmatrix} = \begin{bmatrix} 0 \\ 0 \\ 0 \end{bmatrix} \tag{1.95}$$

Apart from the trivial solution $\mathbf{q} = \mathbf{0}$, Cramer's rule may be used to show that a solution is possible if, and only if,

$$\begin{vmatrix} 16-\lambda & -24 & 18 \\ 3 & -2-\lambda & 0 \\ -9 & 18 & -17-\lambda \end{vmatrix} = 0 \tag{1.96}$$

Expanding this determinant in terms of λ gives the *characteristic equation*

$$\lambda^3 + 3\lambda^2 - 36\lambda + 32 = 0 \tag{1.97}$$

This can be factorized into

$$(\lambda - 4)(\lambda - 1)(\lambda + 8) = 0 \tag{1.98}$$

indicating not only that $\lambda = 4$ is an eigenvalue of the matrix but that $\lambda = 1$ and $\lambda = -8$ are also.

The general form of equation (1.95), with $\mathbf{A}$ of order n, is

$$(\mathbf{A} - \lambda\mathbf{I})\mathbf{q} = \mathbf{0} \tag{1.99}$$

any non-trivial solution of which must satisfy

$$\begin{vmatrix} a_{11}-\lambda & a_{12} & \cdots & a_{1n} \\ a_{21} & a_{22}-\lambda & \cdots & a_{2n} \\ \cdots & \cdots & \cdots & \cdots \\ a_{n1} & a_{n2} & \cdots & a_{nn}-\lambda \end{vmatrix} = 0 \tag{1.100}$$

The characteristic equation must have the general form

$$\lambda^n + c_{n-1}\lambda^{n-1} + \cdots + c_1\lambda + c_0 = 0 \tag{1.101}$$

The characteristic equation method is not a good general procedure for the numerical determination of the eigenvalues of a matrix. For a large fully populated matrix the number of multiplications required to obtain the coefficients of the characteristic equation is roughly proportional to n^4. However, it is useful in establishing most of the important algebraic properties of eigenvalues given in section 1.17. More effective numerical methods of determining eigenvalues and eigenvectors will be considered in Chapters 8, 9 and 10.

1.17 SOME EIGENVALUE PROPERTIES

(a) The characteristic equation can be factorized into the form

$$(\lambda - \lambda_1)(\lambda - \lambda_2) \ldots (\lambda - \lambda_n) = 0 \tag{1.102}$$

showing that a matrix of order n has n eigenvalues. It should be noted that these eigenvalues will not necessarily all be distinct since it is possible for multiple roots to exist, e.g. $\lambda_1 = \lambda_2$.

(b) The sum of the diagonal elements of a matrix is called the *trace* of the matrix. From equations (1.100), (1.101) and (1.102),

$$\mathrm{tr}(\mathbf{A}) = a_{11} + a_{22} + \cdots + a_{nn} = -c_{n-1} = \lambda_1 + \lambda_2 + \cdots \lambda_n \tag{1.103}$$

Hence the sum of the eigenvalues of a matrix is equal to the trace of the matrix.

(c) Also, from equations (1.100), (1.101) and (1.102),

$$|\mathbf{A}| = (-1)^n c_0 = \lambda_1 \lambda_2 \ldots \lambda_n \tag{1.104}$$

Hence the product of the eigenvalues of a matrix equals the determinant of the matrix. It also follows that a singular matrix must have at least one zero eigenvalue.

(d) Since the determinant of a matrix is equal to the determinant of its transpose, it follows that a matrix has the same eigenvalues as its transpose.

(e) The characteristic equation obtained from a real matrix eigenvalue problem must have real coefficients. Hence each eigenvalue of a real matrix must be either real or one of a complex conjugate pair of eigenvalues.

(f) Consider A to be a real symmetric matrix. Premultiplying the standard matrix equation by $\mathbf{q}^H$ gives

$$\mathbf{q}^H \mathbf{A} \mathbf{q} = \lambda \mathbf{q}^H \mathbf{q} \tag{1.105}$$

forming the Hermitian transpose of this equation and making use of the property $\mathbf{A}^H = \mathbf{A}$ gives

$$\mathbf{q}^H \mathbf{A} \mathbf{q} = \lambda^* \mathbf{q}^H \mathbf{q} \tag{1.106}$$

where λ^* is the complex conjugate of λ. However, $\mathbf{q}^H \mathbf{q} \neq 0$, unless $\mathbf{q}$ is a null vector. Hence it follows from equations (1.105) and (1.106) that $\lambda^* = \lambda$ and λ must be real. Hence all the eigenvalues of a real symmetric matrix are real. It is also possible to show that the eigenvectors can be written in real form.

(g) The determinant of a triangular matrix is simply the product of the diagonal elements. Therefore, if A is triangular,

$$|\mathbf{A} - \lambda \mathbf{I}| = (a_{11} - \lambda)(a_{22} - \lambda) \ldots (a_{nn} - \lambda) = 0 \tag{1.107}$$

By comparing with equation (1.102) it follows that the eigenvalues of a triangular (and hence also a diagonal) matrix are equal to the diagonal elements.

(h) If rows and corresponding columns of a matrix are interchanged the eigenvalues remain the same, e.g. equation (1.94) may be written with rows 1 and 2 and variables q_1 and q_2 interchanged as

$$\begin{bmatrix} -2 & 3 & 0 \\ -24 & 16 & 18 \\ 18 & -9 & -17 \end{bmatrix} \begin{bmatrix} q_2 \\ q_1 \\ q_3 \end{bmatrix} = \lambda \begin{bmatrix} q_2 \\ q_1 \\ q_3 \end{bmatrix} \tag{1.108}$$

(i) Consider a 4 x 4 matrix **A** with an eigenvalue satisfying equation (1.93). Scaling the second elemental equation by f and the second element of the eigenvector also by f yields

$$\begin{bmatrix} a_{11} & a_{12}/f & a_{13} & a_{14} \\ fa_{21} & a_{22} & fa_{23} & fa_{24} \\ a_{31} & a_{32}/f & a_{33} & a_{34} \\ a_{41} & a_{42}/f & a_{43} & a_{44} \end{bmatrix} \begin{bmatrix} q_1 \\ fq_2 \\ q_3 \\ q_4 \end{bmatrix} = \lambda \begin{bmatrix} q_1 \\ fq_2 \\ q_3 \\ q_4 \end{bmatrix} \tag{1.109}$$

Because the modified matrix has the same eigenvalue as the original matrix it may be concluded that the eigenvalues of a matrix are unaltered if a row is scaled by f and the corresponding column is scaled by $1/f$.

1.18 EIGENVECTORS

Except for the special case of a defective matrix discussed in section 8.8, every eigenvalue of a matrix is associated with a separate right eigenvector satisfying equation (1.93). If a particular eigenvalue is known, equation (1.93) defines a set of n simultaneous equations having the n components of the corresponding eigenvector as unknowns. For example, consider finding the right eigenvector corresponding to $\lambda = 1$ for matrix (1.91):

$$\begin{bmatrix} 16-1 & -24 & 18 \\ 3 & -2-1 & 0 \\ -9 & 18 & -17-1 \end{bmatrix} \begin{bmatrix} q_1 \\ q_2 \\ q_3 \end{bmatrix} = \begin{bmatrix} 0 \\ 0 \\ 0 \end{bmatrix} \tag{1.110}$$

Elimination of the first column coefficients below the diagonal gives

$$\begin{bmatrix} 15 & -24 & 18 \\ & 1.8 & -3.6 \\ & 3.6 & -7.2 \end{bmatrix} \begin{bmatrix} q_1 \\ q_2 \\ q_3 \end{bmatrix} = \begin{bmatrix} 0 \\ 0 \\ 0 \end{bmatrix} \tag{1.111}$$

Elimination of the subdiagonal element in column 2 yields the result that q_3 is indeterminate. Backsubstitution gives $q_1 = q_2 = 2q_3$.

A more general examination of the equations reveals that, because $|\mathbf{A} - \lambda\mathbf{I}| = 0$, then $\mathbf{A} - \lambda\mathbf{I}$ must be singular and matrix equation (1.93) can never give a unique solution. However, because it is homogeneous (i.e. having zero right-hand side), it does not lead to any inconsistency in the solution. Instead, one equation is redundant and only the relative value of the variables can be determined. By inspection it is seen that, if any vector **q** satisfies $\mathbf{Aq} = \lambda\mathbf{q}$, then a scalar multiple of the vector must also satisfy it. From equation (1.111) it may be deduced that the required eigenvector is $\{q_1 \quad q_2 \quad q_3\} = \{2 \quad 2 \quad 1\}$ with an implicit understanding that it can be arbitrarily scaled.

For a large fully populated unsymmetric matrix it requires approximately $n^3/3$ multiplications to perform the elimination for each eigenvector. Hence the determination of eigenvectors should not be considered as a trivial numerical operation, even when the corresponding eigenvalues are known.

It is possible to combine the standard eigenvalue equations for all eigenvalues and corresponding right eigenvectors in the form

$$\left.\begin{aligned} &\begin{bmatrix} \mathbf{A} \end{bmatrix}\begin{bmatrix} \mathbf{q}_1 & \mathbf{q}_2 & \cdots & \mathbf{q}_n \end{bmatrix} = \begin{bmatrix} \mathbf{q}_1 & \mathbf{q}_2 & \cdots & \mathbf{q}_n \end{bmatrix}\begin{bmatrix} \lambda_1 & & & \\ & \lambda_2 & & \\ & & \ddots & \\ & & & \lambda_n \end{bmatrix} \\ &\text{i.e.} \\ &\mathbf{AQ} = \mathbf{Q\Lambda} \end{aligned}\right\} \quad (1.112)$$

where Λ is a diagonal matrix of the eigenvalues and **Q** is a square matrix containing all the right eigenvectors in corresponding order.

Left eigenvectors

Since the eigenvalues of **A** and $\mathbf{A}^T$ are identical, for every eigenvalue λ associated with an eigenvector **q** of **A** there is also an eigenvector **p** of $\mathbf{A}^T$ such that

$$\mathbf{A}^T\mathbf{p} = \lambda\mathbf{p} \quad (1.113)$$

Alternatively, the eigenvector **p** can be considered to be a left eigenvector of **A** by transposing equation (1.113) to give

$$\mathbf{p}^T\mathbf{A} = \lambda\mathbf{p}^T \quad (1.114)$$

Table 1.7 shows the full eigensolution of matrix (1.91). Eigenvalue properties for unsymmetic matrices will be discussed further in section 8.8.

As a symmetric matrix is its own transpose, its left and right eigenvectors

Table 1.7 Full eigensolution of matrix (1.91)

Corresponding left eigenvectors	Eigenvalues	Corresponding right eigenvectors
{7 –10 6}	4	{2 1 0}
{–1 2 –1}	1	{2 2 1}
{1 –2 2}	–8	{–2 1 4}

Table 1.8 Eigensolution of a 3 x 3 symmetric matrix

Matrix	Eigenvalues	Corresponding eigenvectors
$\begin{bmatrix} 3 & 1 & 5 \\ 1 & 3 & 5 \\ 5 & 5 & -1 \end{bmatrix}$	9 2 −6	{1 1 1} {1 −1 0} {−0.5 −0.5 1}

coincide and so need not be distinguished from each other. Table 1.8 shows a 3 x 3 symmetric matrix together with its eigenvalues and eigenvectors. The eigenvectors have been scaled so that the largest element in each vector is 1.

1.19 NORMS AND NORMALIZATION

It is sometimes useful to have a scalar measure of the magnitude of a vector. Such a measure is called a *norm* and for a vector **x** is written as $\| \mathbf{x} \|$. A commonly used norm is the magnitude of the largest element, e.g. for $\mathbf{x} = \{7 \quad -10 \quad 6\}$, $\| \mathbf{x} \| = 10$. The eigenvectors quoted in Table 1.8 have each been scaled so that their largest element is unity. This scaling process is called *normalization* since it makes the norm of each vector equal to 1. Normalization of a real eigenvector must produce a unique result, except for sign, and hence can be used as a basis for comparison of numerical results and trial solutions involving eigenvectors.

The above norm takes no account of the magnitude of the smaller elements of the vector. A useful alternative which is sensitive to the size of these elements is the Euclidean norm described algebraically by

$$\| \mathbf{x} \|_E = \left(\sum_{i=1}^{n} | x_i |^2 \right)^{1/2} \tag{1.115}$$

which has the property that

$$\| \mathbf{x} \|_E^2 = \mathbf{x}^H \mathbf{x} \tag{1.116}$$

For $\mathbf{x} = \{7 \ -10 \ 6\}$, $\| \mathbf{x} \|_E \simeq 13.6$.

A family of vector norms can be described by the relationship

$$\| \mathbf{x} \|_b = \left(\sum_{i=1}^{n} | x_i |^b \right)^{1/b} \tag{1.117}$$

for which the Euclidean norm corresponds to $b = 2$ and the norm based on the magnitude of the largest element corresponds to $b \to \infty$.

Several norms for matrices have also been defined, for instance the Euclidean norm of $\mathbf{A}(m \times n)$ is

$$\| \mathbf{A} \|_E = \left(\sum_{i=1}^{m} \sum_{j=1}^{n} | a_{ij} |^2 \right)^{1/2} \tag{1.118}$$

1.20 ORTHOGONALITY CONDITIONS FOR EIGENVECTORS OF SYMMETRIC MATRICES

If $\mathbf{q}_i$ and $\mathbf{q}_j$ are any two eigenvectors of a symmetric matrix A corresponding to distinct eigenvalues λ_i and λ_j, then

$$\left.\begin{aligned} \mathbf{A}\mathbf{q}_i &= \lambda_i \mathbf{q}_i \\ \mathbf{A}\mathbf{q}_j &= \lambda_j \mathbf{q}_j \end{aligned}\right\} \tag{1.119}$$

Transposing the second of these equations and taking account of the symmetry of **A**,

$$\mathbf{q}_j^T \mathbf{A} = \lambda_j \mathbf{q}_j^T \tag{1.120}$$

Premultiplying the first equation of (1.119) by $\mathbf{q}_j^T$ and postmultiplying equation (1.120) by $\mathbf{q}_i$ give

$$\left.\begin{aligned} \mathbf{q}_j^T \mathbf{A}\mathbf{q}_i &= \lambda_i \mathbf{q}_j^T \mathbf{q}_i \\ \mathbf{q}_j^T \mathbf{A}\mathbf{q}_i &= \lambda_j \mathbf{q}_j^T \mathbf{q}_i \end{aligned}\right\} \tag{1.121}$$

Since $\lambda_i \neq \lambda_j$, the only way in which these equations can be compatible is for

$$\mathbf{q}_i^T \mathbf{A}\mathbf{q}_i = \mathbf{q}_j^T \mathbf{q}_i = 0 \tag{1.122}$$

The condition $\mathbf{q}_j^T \mathbf{q}_i = 0$ is called the *orthogonality condition* for the eigenvectors. (It may be verified that the eigenvectors of Table 1.8 satisfy this condition for each of the three possible combinations of i and j for which $i \neq j$.)

If each eigenvector $\mathbf{q}_i$ is scaled so that its Euclidean norm is unity, then

$$\mathbf{q}_j^T \mathbf{q}_i = 1 \tag{1.123}$$

Initially discounting the possibility that the matrix has coincident eigenvalues, the orthogonality condition can be combined with equation (1.123) to yield

$$\begin{bmatrix} \mathbf{q}_1^T \\ \hline \mathbf{q}_2^T \\ \hline \\ \hline \mathbf{q}_n^T \end{bmatrix} \begin{bmatrix} \mathbf{q}_1 & \mathbf{q}_2 & \cdots & \mathbf{q}_n \end{bmatrix} = \begin{bmatrix} 1 & 0 & \cdots & 0 \\ 0 & 1 & \cdots & 0 \\ \cdot & \cdot & \cdot & \cdot \\ 0 & 0 & \cdots & 1 \end{bmatrix} \tag{1.124}$$

Designating the compounded eigenvector set $[\mathbf{q}_1 \; \mathbf{q}_2 \ldots \mathbf{q}_n]$ as **Q**, then

$$\mathbf{Q}^T\mathbf{Q} = \mathbf{I} \; (= \mathbf{Q}\mathbf{Q}^T) \tag{1.125}$$

Any real matrix **Q** satisfying this equation is known as an orthogonal matrix. For example, normalizing the eigenvectors in Table 1.8 so that their Euclidean norms are unity gives

$$\mathbf{Q} = \frac{1}{\sqrt{6}} \begin{bmatrix} \sqrt{2} & \sqrt{3} & -1 \\ \sqrt{2} & -\sqrt{3} & -1 \\ \sqrt{2} & 0 & 2 \end{bmatrix} \tag{1.126}$$

which can be shown to satisfy equations (1.125).

Comparing equations (1.60) and (1.125) it follows that the inverse of **Q** is equal to its transpose, and hence **Q** cannot be singular. If a symmetric matrix has coincident eigenvalues it is still possible to find a full set of eigenvectors which obey equation (1.125), the only difference being that the eigenvectors associated with the coincident eigenvalues are not unique.

Let **x** be a vector of order n such that

$$\mathbf{x} = c_1\mathbf{q}_1 + c_2\mathbf{q}_2 + \cdots + c_n\mathbf{q}_n \tag{1.127}$$

where $\mathbf{q}_1, \mathbf{q}_2, \ldots, \mathbf{q}_n$ are the eigenvectors of an $n \times n$ symmetric matrix. This equation may be expressed in the form

$$\mathbf{x} = \mathbf{Q}\mathbf{c} \tag{1.128}$$

where $\mathbf{c} = \{c_1 \quad c_2 \ldots c_n\}$. Premultiplying by $\mathbf{Q}^T$ and using equation (1.125) gives

$$\mathbf{c} = \mathbf{Q}^T\mathbf{x} \tag{1.129}$$

Since this equation defines the coefficients **c** for any arbitrary vector **x** it follows that any arbitrary vector can be expressed as a linear combination of the eigenvectors of a symmetric matrix.

From equations (1.112) and (1.125) it follows that **A** may be factorized to give

$$\mathbf{A} = \mathbf{Q}\mathbf{\Lambda}\mathbf{Q}^T \tag{1.130}$$

1.21 QUADRATIC FORMS AND POSITIVE DEFINITE MATRICES

The function

$$[x_1 \quad x_2 \quad \ldots \quad x_n]\begin{bmatrix} a_{11} & a_{12} & \cdots & a_{1n} \\ a_{21} & a_{22} & \cdots & a_{2n} \\ \cdot & \cdot & \cdot & \cdot \\ a_{n1} & a_{n2} & \cdots & a_{nn} \end{bmatrix}\begin{bmatrix} x_1 \\ x_2 \\ \cdot \\ x_n \end{bmatrix} = \sum_{i=1}^{n}\sum_{j=1}^{n} a_{ij}x_ix_j \tag{1.131}$$

can be used to represent any quadratic polynomial in the variables $x_1, x_2 \ldots x_n$ and is called a *quadratic form*. The quadratic form is unchanged if **A** is transposed and hence the quadratic form of a skew symmetric matrix must be zero. Therefore, if any unsymmetric matrix **A** is separated into symmetric and skew symmetric components according to equation (1.19), the quadratic form is only sensitive to the symmetric component. For instance, the quadratic form

$$[x_1 \quad x_2]\begin{bmatrix} 5 & -2 \\ 6 & -1 \end{bmatrix}\begin{bmatrix} x_1 \\ x_2 \end{bmatrix} = 5x_1^2 + 4x_1x_2 - x_2^2 \tag{1.132}$$

is identical to the quadratic form of the symmetric component of the matrix, i.e.

$$[x_1 \quad x_2]\begin{bmatrix} 5 & 2 \\ 2 & -1 \end{bmatrix}\begin{bmatrix} x_1 \\ x_2 \end{bmatrix}$$

A matrix is said to be *positive definite* if its quadratic form is positive for all real non-null vectors **x**. Symmetric positive definite matrices occur frequently in equations derived by minimization or energy principles and their properties can often be utilized in numerical processes. The following are important properties:

(a) Consider any eigenvalue λ and corresponding eigenvector $\mathbf{q}$ of a symmetric positive definite matrix. Premultiplying the eigenvalue equation by $\mathbf{q}^T$ gives

$$\mathbf{q}^T\mathbf{A}\mathbf{q} = \lambda\mathbf{q}^T\mathbf{q} \tag{1.133}$$

Since the left-hand side and also the inner product $\mathbf{q}^T\mathbf{q}$ are positive, λ must be positive. Hence all of the eigenvalues of $\mathbf{A}$ are positive.

(b) Consider a symmetric matrix $\mathbf{A}$ having only positive eigenvalues and let the quadratic form be such that

$$\mathbf{x}^T\mathbf{A}\mathbf{x} = \mu\mathbf{x}^T\mathbf{x} \tag{1.134}$$

Expressing the vector $\mathbf{x}$ as a linear combination of the eigenvectors of $\mathbf{A}$ according to equation (1.128) gives

$$\mathbf{x}^T\mathbf{A}\mathbf{x} - \mu\mathbf{x}^T\mathbf{x} = \mathbf{c}^T\mathbf{Q}^T\mathbf{A}\mathbf{Q}\mathbf{c} - \mu\mathbf{c}^T\mathbf{Q}^T\mathbf{Q}\mathbf{c} = 0 \tag{1.135}$$

But $\mathbf{A}\mathbf{Q} = \mathbf{Q}\boldsymbol{\Lambda}$ and $\mathbf{Q}^T\mathbf{Q} = \mathbf{I}$, hence

$$\mathbf{c}^T\boldsymbol{\Lambda}\,\mathbf{c} - \mu\mathbf{c}^T\mathbf{c} = \sum_{i=1}^{n} c_i^2(\lambda_i - \mu) = 0 \tag{1.136}$$

Since all of the terms $(\lambda_i - \mu)$ cannot have the same sign,

$$\lambda_1 \leqslant \mu \leqslant \lambda_n \tag{1.137}$$

where λ_n and λ_1 are the maximum and minimum eigenvalues of $\mathbf{A}$ respectively. This result restricts the magnitude of the quadratic form of a symmetric matrix. As a corollary it is seen that a symmetric matrix whose eigenvalues are all positive must itself be positive definite. Combining this result with property (a), both a necessary and sufficient condition for a symmetric matrix to be positive definite is that all of its eigenvalues are positive.

(c) The determinant of a symmetric positive definite matrix must be positive since

$$|\,\mathbf{A}\,| = \lambda_1\lambda_2\ldots\lambda_n \tag{1.138}$$

(d) Consider $\mathbf{x}$ to have some zero components, e.g. for a 4 x 4 symmetric positive definite matrix let $\mathbf{x} = \{x_1 \quad 0 \quad 0 \quad x_4\}$:

$$\left.\begin{aligned}\mathbf{x}^T\mathbf{A}\mathbf{x} &= [x_1 \quad 0 \quad 0 \quad x_4]\begin{bmatrix} a_{11} & a_{12} & a_{13} & a_{14} \\ a_{21} & a_{22} & a_{23} & a_{24} \\ a_{31} & a_{32} & a_{33} & a_{34} \\ a_{41} & a_{42} & a_{43} & a_{44}\end{bmatrix}\begin{bmatrix} x_1 \\ 0 \\ 0 \\ x_4\end{bmatrix} \\ &= [x_1 \quad x_4]\begin{bmatrix} a_{11} & a_{14} \\ a_{41} & a_{44}\end{bmatrix}\begin{bmatrix} x_1 \\ x_4\end{bmatrix} > 0\end{aligned}\right\} \tag{1.139}$$

which shows that

$$\begin{bmatrix} a_{11} & a_{14} \\ a_{41} & a_{44} \end{bmatrix}$$

must also be positive definite. By appropriate choice of zero components in $\mathbf{x}$ any number of rows and corresponding columns could have been omitted from $\mathbf{A}$. Hence any principal minor of a symmetric positive definite matrix (obtained by omitting any number of corresponding rows and columns) must also be positive definite.

A positive semidefinite (or non-negative definite) matrix is similar to a positive definite matrix except that it also admits the possibility of a zero quadratic form. Symmetric positive semidefinite matrices have similar properties to symmetric positive definite matrices except that zero eigenvalues and a zero determinant are admissible both for the matrix itself and for its principal minors. Table 1.9 illustrates the properties of symmetric positive definite and semidefinite matrices by considering an example of each.

Table 1.9 Properties of symmetric positive definite and semidefinite matrices

		Positive definite matrix	Eigen-values	Deter-minant	Positive semidefinite matrix	Eigen-values	Deter-minant
Full matrix		$\begin{bmatrix} 3 & 1 & -1 \\ 1 & 3 & -1 \\ -1 & -1 & 5 \end{bmatrix}$	6, 3, 2	36	$\begin{bmatrix} 4 & 2 & -4 \\ 2 & 1 & -2 \\ -4 & -2 & 4 \end{bmatrix}$	9, 0, 0	0
2 x 2 Principal minors	1, 2	$\begin{bmatrix} 3 & 1 \\ 1 & 3 \end{bmatrix}$	4, 2	8	$\begin{bmatrix} 4 & 2 \\ 2 & 1 \end{bmatrix}$	5, 0	0
	1, 3	$\begin{bmatrix} 3 & -1 \\ -1 & 5 \end{bmatrix}$	$4+\sqrt{2}$, $4-\sqrt{2}$	14	$\begin{bmatrix} 4 & -4 \\ -4 & 4 \end{bmatrix}$	8, 0	0
	2, 3	$\begin{bmatrix} 3 & -1 \\ -1 & 5 \end{bmatrix}$	$4+\sqrt{2}$, $4-\sqrt{2}$	14	$\begin{bmatrix} 1 & -2 \\ -2 & 4 \end{bmatrix}$	5, 0	0
1 x 1 Principal minors	1	[3]	3	3	[4]	4	4
	2	[3]	3	3	[1]	1	1
	3	[5]	5	5	[4]	4	4

An $n \times n$ symmetric matrix

$$\mathbf{A} = \mathbf{B}^T\mathbf{B} \tag{1.140}$$

must be positive semidefinite, for if $\mathbf{x}$ is any arbitrary real vector of order n and

$$\mathbf{y} = \mathbf{B}\mathbf{x} \tag{1.141}$$

then

$$\mathbf{x}^T\mathbf{A}\mathbf{x} = \mathbf{x}^T\mathbf{B}^T\mathbf{B}\mathbf{x} = \mathbf{y}^T\mathbf{y} \geqslant 0 \tag{1.142}$$

It can similarly be shown that if **F** is a symmetric positive semidefinite matrix then

$$\mathbf{A} = \mathbf{B}^T\mathbf{F}\mathbf{B} \tag{1.143}$$

must also be symmetric and positive semidefinite.

1.22 GERSCHGORIN DISCS

Gerschgorin's theorem is useful for providing bounds to the magnitudes of the eigenvalues of a matrix. Consider the eigenvalue equation $\mathbf{A}\mathbf{q} = \lambda\mathbf{q}$ in which the eigenvector has been normalized such that the largest element $q_k = 1$. This equation may be expressed as

$$\begin{bmatrix} \times & \times & \cdots & \times & \cdots & \times \\ \times & \times & \cdots & \times & \cdots & \times \\ \cdot & \cdot & \cdot & \cdot & \cdot & \cdot \\ a_{k1} & a_{k2} & \cdots & a_{kk} & \cdots & a_{kn} \\ \cdot & \cdot & \cdot & \cdot & \cdot & \cdot \\ \times & \times & \cdots & \times & \cdot & \times \end{bmatrix} \begin{bmatrix} q_1 \\ q_2 \\ \cdot \\ 1 \\ \cdot \\ q_n \end{bmatrix} = \lambda \begin{bmatrix} q_1 \\ q_2 \\ \cdot \\ 1 \\ \cdot \\ q_n \end{bmatrix} \tag{1.144}$$

the k-th elemental equation of which is

$$\lambda - a_{kk} = \sum_{j \neq k} a_{kj} q_j \tag{1.145}$$

Since $|q_j| \leqslant 1$ it follows that

$$|\lambda - a_{kk}| \leqslant \sum_{j \neq k} |a_{kj}| \tag{1.146}$$

This can be interpreted on an Argand diagram as a statement that λ must lie within a circle, centre a_{kk} and radius $\Sigma_{j \neq k} |a_{kj}|$. Since the position of the largest element in an eigenvector is normally unknown, it is only possible to say that every eigenvalue must lie within the union of the discs constructed from n rows of the matrix according to equation (1.146).

The three Gerschgorin discs of the matrix (1.91) have (centre, radius) given by (16, 42), (–2, 3) and (–17, 27). The actual eigenvalues 4, 1 and –8 lie well within the union of these discs shown in Figure 1.1(a). If the matrix is unsymmetric the left eigenvectors may be used to give an alternative set of discs based on the columns of the matrix. The union of column discs for matrix (1.91) shown in Figure 1.1(b) show a restricted intercept on the real axis. Unsymmetric scaling of a matrix of the sort shown in equation (1.109) can often be used to advantage in restricting the envelope of the discs. Thus factoring row 2 by 4 and column 2 by ¼ gives smaller disc envelopes for matrix (1.91), as shown in Figure 1.1(c) and (d).

In the case where a set of r discs do not intersect with the other discs, it can be shown that the union of these discs must contain just r of the eigenvalues.

For a real symmetric matrix all of the eigenvalues must be real, and hence only the intercepts of the discs on the real axis of the Argand diagram are of significance.

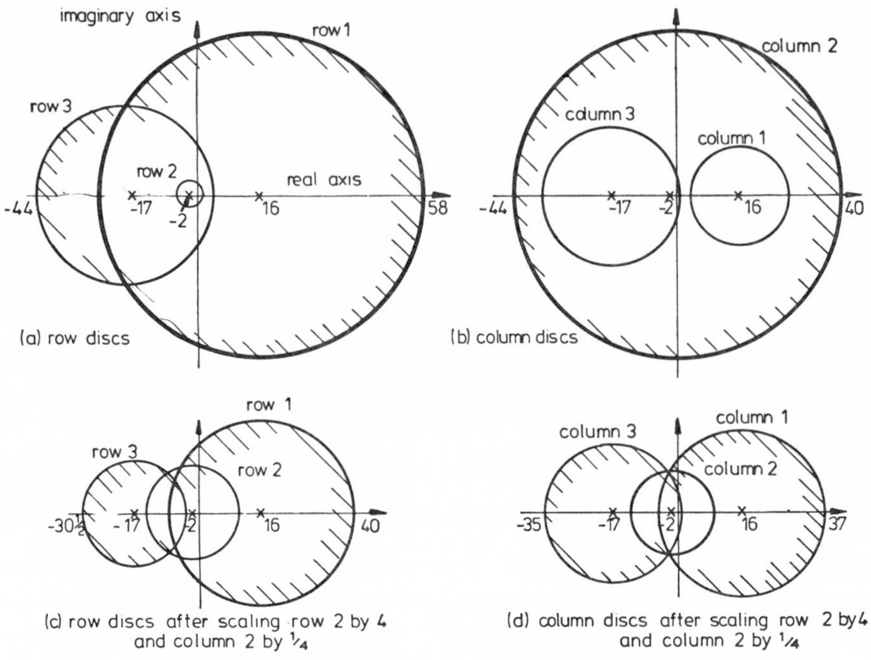

Figure 1.1 Gerschgorin discs of matrix (1.91)

Furthermore, it can be shown from equation (1.137) that

$$\lambda_1 \leqslant a_{ii} \leqslant \lambda_n \tag{1.147}$$

by adopting for **x** (equation 1.134) a vector which is null except for a unit term in position *i*. Therefore, for a real symmetric matrix

$$\left.\begin{aligned} &(a_{ii})_{\max} \leqslant \lambda_n \leqslant \left(a_{kk} + \sum_{j \neq k} |a_{jk}| \right)_{\max} \\ &\text{and} \\ &\left(a_{kk} - \sum_{j \neq k} |a_{jk}| \right)_{\min} \leqslant \lambda_1 \leqslant (a_{ii})_{\min} \end{aligned}\right\} \tag{1.148}$$

For the 3 x 3 symmetric matrix, Table 1.8, it is therefore possible to deduce by inspection that $3 \leqslant \lambda_n \leqslant 9$ and $-11 \leqslant \lambda_1 \leqslant -1$.

BIBLIOGRAPHY

Bickley, W. G., and Thompson, R. S. H. G. (1964). *Matrices, Their Meaning and Manipulation*, English Universities Press, London.

Forsythe, G. E. (1953). 'Solving linear equations can be interesting'. *Bull. Amer. Math. Soc.*, 59, 299–329.

Frazer, R. A., Duncan, W. J., and Collar, A. R. (1938). *Elementary Matrices and Some Applications to Dynamics and Differential Equations*, Cambridge University Press, Cambridge. Chap. 1.

Froberg, C. E. (1969). *Introduction to Numerical Analysis*, 2nd ed. Addison-Wesley, Reading, Massachusetts. Chap. 3.

Gere, J. M., and Weaver, W. (1965). *Matrix Algebra for Engineers*, Van Nostrand Reinhold, New York.

Hohn, F. E. (1973). *Elementary Matrix Algebra*, 3rd ed. Macmillan, New York.

Searle, S. R. (1966). *Matrix Algebra for the Biological Sciences (Including Applications in Statistics)*, Wiley, New York.

Steinberg, D. I. (1974). *Computational Matrix Algebra*, McGraw-Hill, New York.

Chapter 2
Some Matrix Problems

2.1 AN ELECTRICAL RESISTANCE NETWORK

This chapter provides a selection of examples in which matrix computation is relevant. All of the problems presented here have the solution of a set of linear simultaneous equations as a key part of the computation. Examples which involve the computation of eigenvalues have been left to Chapter 7.

As a simple example of a network problem consider the electrical resistance network shown in Figure 2.1. The battery in branch EA provides a constant voltage V across its terminals, and as a result current passes through all the branches of the circuit. The object of an analysis might be to determine how much current passes through each branch. The most basic formulation may be considered as that deriving immediately from Kirchhoff's laws, and this will be developed first.

Suppose that the voltage drop across branch AB in the direction A to B be v_{AB} and let the current passing along the branch from A to B be i_{AB}. Similar definitions may be adopted for each of the other branches, as shown in Figure 2.2.

Kirchhoff's voltage law states that the algebraic sum of the voltages round any closed circuit which is in equilibrium must be zero. Applying this law to the circuit ABE gives $v_{AB} + v_{BE} + v_{EA} = 0$. It is possible to write down four such circuit voltage equations as follows:

$$\left.\begin{array}{ll} \text{Circuit ABE:} & v_{AB} + v_{BE} + v_{EA} = 0 \\ \text{Circuit ADCB:} & v_{AD} - v_{CD} - v_{BC} - v_{AB} = 0 \\ \text{Circuit BCE:} & v_{BC} + v_{CE} - v_{BE} = 0 \\ \text{Circuit CDE:} & v_{CD} + v_{DE} - v_{CE} = 0 \end{array}\right\} \quad (2.1)$$

Other circuits can be found but, since their voltage equations are only linear combinations of those specified above, they do not provide any additional information. For instance, the equation for circuit ABCE can be obtained by adding the equations for circuits ABE and BCE.

Kirchhoff's current law states that the algebraic sum of the branch currents confluent at each node must be zero if the circuit is in equilibrium. Hence nodal

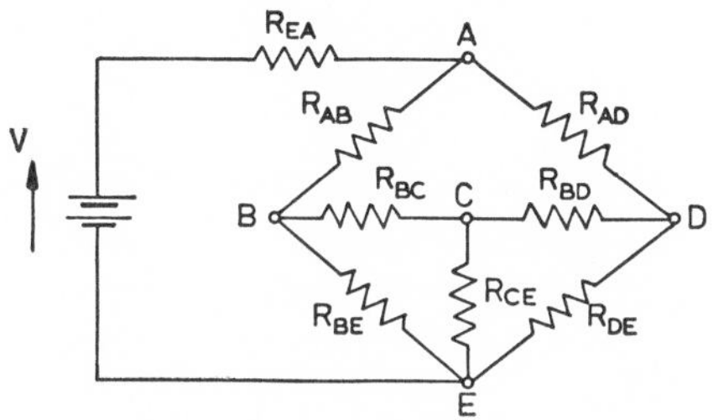

Figure 2.1 A Julie bridge

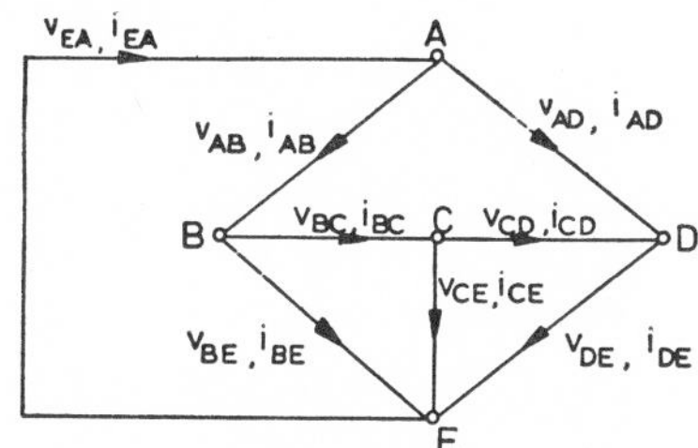

Figure 2.2 Branch voltage drops and currents for the Julie bridge

current equations for the Julie bridge are as follows:

$$\left.\begin{array}{ll} \text{Node A:} & i_{AB} + i_{AD} - i_{EA} = 0 \\ \text{Node B:} & -i_{AB} + i_{BC} + i_{BE} = 0 \\ \text{Node C:} & -i_{BC} + i_{CD} + i_{CE} = 0 \\ \text{Node D:} & -i_{AD} - i_{CD} + i_{DE} = 0 \end{array}\right\} \qquad (2.2)$$

The equation for node E can be obtained by summing the equations for the four other nodes. Hence it has not been specified.

Branch characteristics can be obtained from Ohm's law. In the branch AB this gives $v_{AB} = R_{AB}i_{AB}$, where R_{AB} is the resistance of the branch. Similar equations may be obtained for all the other branches except EA, where the presence of the voltage source results in the equation $v_{EA} = R_{EA}i_{EA} - V$.

The sixteen equations derived from Kirchhoff's laws and the branch characteristics may be written in matrix form as on the following page. If the branch resistances and the source voltage are known, equation (2.3) can be solved as a set of simultaneous equations to determine the branch voltages and currents. In this case the matrix of coefficients is unsymmetric and sparse. Although it is possible to alter the pattern of non-zero elements in this matrix by rearranging the order of the equations or the order of the variables, it is not possible, simply by doing this, to produce a symmetric matrix.

$$\left[\begin{array}{cccccccc|cccccccc}
1 & & & & 1 & & & 1 & & & & & & & & \\
-1 & 1 & -1 & -1 & & & & & & & & & & & & \\
 & & 1 & & -1 & 1 & & & & & & & & & & \\
 & & & 1 & & -1 & 1 & & & & & & & & & \\
\hline
 & & & & & & & & 1 & 1 & & & & & & -1 \\
 & & & & & & & & -1 & & 1 & & 1 & & & \\
 & & & & & & & & & & -1 & 1 & & 1 & & \\
 & & & & & & & & & -1 & & -1 & & & 1 & \\
\hline
1 & & & & & & & & -R_{AB} & & & & & & & \\
 & 1 & & & & & & & & -R_{AD} & & & & & & \\
 & & 1 & & & & & & & & -R_{BC} & & & & & \\
 & & & 1 & & & & & & & & -R_{CD} & & & & \\
 & & & & 1 & & & & & & & & -R_{BE} & & & \\
 & & & & & 1 & & & & & & & & -R_{CE} & & \\
 & & & & & & 1 & & & & & & & & -R_{DE} & \\
 & & & & & & & 1 & & & & & & & & -R_{EA}
\end{array}\right]
\left[\begin{array}{c}
v_{AB} \\ v_{AD} \\ v_{BC} \\ v_{CD} \\ \hline v_{BE} \\ v_{CE} \\ v_{DE} \\ v_{EA} \\ \hline i_{AB} \\ i_{AD} \\ i_{BC} \\ i_{CD} \\ i_{BE} \\ i_{CE} \\ i_{DE} \\ i_{EA}
\end{array}\right]
=
\left[\begin{array}{c}
0 \\ 0 \\ 0 \\ 0 \\ \hline 0 \\ 0 \\ 0 \\ 0 \\ \hline 0 \\ 0 \\ 0 \\ 0 \\ 0 \\ 0 \\ 0 \\ -V
\end{array}\right]
\tag{2.3}$$

2.2 ALTERNATIVE FORMS OF THE NETWORK EQUATIONS

It is possible to use nodal voltages or potentials instead of branch voltages. However, since only the relative voltages can be obtained from an analysis, the nodal voltages must be defined relative to a datum value. For the Julie bridge let the datum be the potential at E. Thus four voltages only are required which may be designated e_A, e_B, e_C and e_D (Figure 2.3). The branch voltages can be related to these nodal voltages by the voltage equilibrium equations

$$\left.\begin{aligned} v_{AB} &= e_A - e_B \\ v_{AD} &= e_A - e_D \\ v_{BC} &= e_B - e_C \\ v_{CD} &= e_C - e_D \\ v_{BE} &= e_B \\ v_{CE} &= e_C \\ v_{DE} &= e_D \\ v_{EA} &= -e_A \end{aligned}\right\} \quad (2.4)$$

Alternative current variables

It is possible to use loop current instead of branch currents. For instance, j_1 can describe a current which is continuous round the closed loop ABE. Similarly, loop currents j_2, j_3 and j_4 can be defined for loops ADCB, BCE and CDE respectively (Figure 2.3). The current in a particular branch must then be the sum of the loop currents passing through it, due attention being paid to the direction of the currents in the summation process, i.e.

$$\left.\begin{aligned} i_{AB} &= j_1 - j_2 \\ i_{AD} &= j_2 \\ i_{BC} &= -j_2 + j_3 \\ i_{CD} &= -j_2 + j_4 \\ i_{BE} &= j_1 - j_3 \\ i_{CE} &= j_3 - j_4 \\ i_{DE} &= j_4 \\ i_{EA} &= j_1 \end{aligned}\right\} \quad (2.5)$$

In loop analysis the loop currents are used as basic variables and the principal equations are the loop voltage equations. Substituting the branch characteristics for the Julie bridge into the loop voltage equations (2.1) give

$$\left.\begin{aligned} R_{AB}i_{AB} + R_{BE}i_{BE} + R_{EA}i_{EA} &= V \\ R_{AD}i_{AD} - R_{CD}i_{CD} - R_{BC}i_{BC} - R_{AB}i_{AB} &= 0 \\ R_{BC}i_{BC} + R_{CE}i_{CE} - R_{BE}i_{BE} &= 0 \\ R_{CD}i_{CD} + R_{DE}i_{DE} - R_{CE}i_{CE} &= 0 \end{aligned}\right\} \quad (2.6)$$

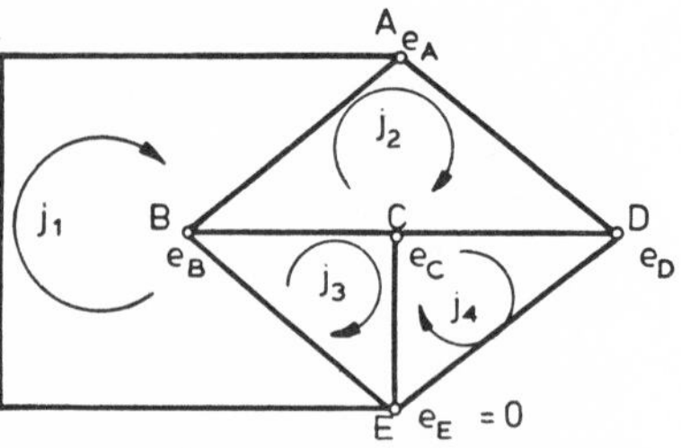

Figure 2.3 Nodal voltage and loop currents for the Julie bridge

and then substituting for the branch currents by means of equation (2.5) yields equations in the basic variables, namely

$$\begin{bmatrix} (R_{AB}+R_{BE}+R_{EA}) & -R_{AB} & -R_{BE} & \\ -R_{AB} & (R_{AB}+R_{AD}+R_{BC}+R_{CD}) & -R_{BC} & -R_{CD} \\ -R_{BE} & -R_{BC} & (R_{BC}+R_{BE}+R_{CE}) & -R_{CE} \\ & -R_{CD} & -R_{CE} & (R_{CD}+R_{CE}+R_{DE}) \end{bmatrix} \begin{bmatrix} j1 \\ j2 \\ j3 \\ j4 \end{bmatrix} = \begin{bmatrix} V \\ 0 \\ 0 \\ 0 \end{bmatrix} \quad (2.7)$$

Hence the number of simultaneous equations required to determine the loop currents has been reduced to four. Once these have been obtained it is a simple matter to substitute in equations (2.5) and in the branch characteristic equations to obtain the branch currents and voltages if they are required.

The coefficient matrix of equation (2.7) may be called the *loop resistance matrix* for the network. In general, loop resistance matrices are symmetric if the loops used for the loop currents correspond to the voltage loops and are also specified in the same order. The i-th diagonal element will be the sum of the resistances round the i-th loop, and the off-diagonal element in position (i, j) either will be minus the resistance of the branch common to loops i and j or, if there is not a common branch, will be zero. Hence this matrix can be constructed directly from a circuit diagram on which the resistances and loops have been specified.

Nodal analysis

In nodal analysis the nodal voltages are used as basic variables and the principal equations are the nodal current equations. Substituting the branch characteristics for the Julie bridge into the nodal current equations (2.2) gives the following equations:

$$\left.\begin{aligned} G_{AB}v_{AB} + G_{AD}v_{AD} - G_{EA}v_{EA} &= G_{EA}V \\ -G_{AB}v_{AB} + G_{BC}v_{BC} + G_{BE}v_{BE} &= 0 \\ -G_{BC}v_{BC} + G_{CD}v_{CD} + G_{CE}v_{CE} &= 0 \\ -G_{AD}v_{AD} - G_{CD}v_{CD} + G_{DE}v_{DE} &= 0 \end{aligned}\right\} \quad (2.8)$$

where, typically, $G_{AB} = 1/R_{AB}$ is the *conductance* of branch AB. Substituting for the branch voltages by means of equations (2.4) yields equations in the basic variables, namely

$$\begin{bmatrix} (G_{AB}+G_{AD}+G_{EA}) & -G_{AB} & & -G_{AD} \\ -G_{AB} & (G_{AB}+G_{BC}+G_{BE}) & -G_{BC} & \\ & -G_{BC} & (G_{BC}+G_{CD}+G_{CE}) & -G_{CD} \\ -G_{AD} & & -G_{CD} & (G_{AD}+G_{CD}+G_{DE}) \end{bmatrix} \begin{bmatrix} e_A \\ e_B \\ e_C \\ e_D \end{bmatrix} = \begin{bmatrix} G_{EA}V \\ 0 \\ 0 \\ 0 \end{bmatrix} \quad (2.9)$$

These simultaneous equations can be solved for the nodal voltages and, if the branch voltages and currents are required, they can be obtained by substituting the nodal voltages in equations (2.4) and the branch characteristic equations.

The coefficient matrix of equation (2.9) may be called the *node conductance matrix* for the network. In general, node conductance matrices are symmetric if the nodal voltage equations and the nodal currents are compatibly ordered. The i-th diagonal element will be the sum of the conductances of all the branches meeting at node i, and the off-diagonal element in position (i, j) either will be minus the conductance of the branch joining nodes i and j or, if they are not joined, will be zero. Hence this matrix can be constructed from a circuit diagram on which the resistances have been specified and the nodes numbered.

2.3 PROPERTIES OF ELECTRICAL RESISTANCE NETWORK EQUATIONS

Number of equations

For the Julie bridge both the loop and nodal analyses yield four equations for the basic variables. It is not generally true, however, that the number of equations will be the same for both methods. For instance, if a further branch AC is added to the circuit, then the number of loops will be increased to five, whereas the number of nodes will remain at four. If an analysis is to be performed by hand the method giving rise to the lesser number of equations is likely to be preferred, as this would normally give the most rapid solution. However, in the development of computer programs, the choice of method is likely to be determined by the ease of automation.

Table 2.1 Branch data for the network shown in Fig. 2.4

Branch no.	Node connections A	Node connections B	Conductance in mhos (= 1/ohms)	Voltage input
1	1	2	3.2	0
2	1	4	2.7	0
3	2	3	5.6	0
4	3	4	2.7	0
5	2	0	4.9	0
6	3	0	2.2	0
7	4	0	3.8	0
8	0	1	2.4	20.0

Automatic construction of the node conductance equations

Data describing the node connections, conductance and voltage input for each branch, as illustrated by Table 2.1, is all that is required to describe the network shown in Figure 2.4. A computer can be programmed to read in this table of data and construct the node conductance equations from the information it contains. Consider the construction of the node conductance matrix in a two-dimensional array store in either ALGOL or FORTRAN. Assume that computer stores have been declared as follows:

	ALGOL	FORTRAN
order of matrix, number of branches	**integer** *n,m*;	INTEGER N,M
node connections at ends A and B (e.g. columns 2, 3 of Table 2.1)	**integer array** *nodeA*, *nodeB* [1:*m*] ;	INTEGER NODEA(M), NODEB(M)
branch conductances (e.g. column 4 of Table 2.1)	**array** *conduc* [1:*m*] ;	REAL CONDUC(M)
node conductance matrix	**array** *A* [1:*n*,1:*n*] ;	REAL A(N,N)
working store	**integer** *i, j, k*; **real** *x*;	INTEGER I,J,K REAL X

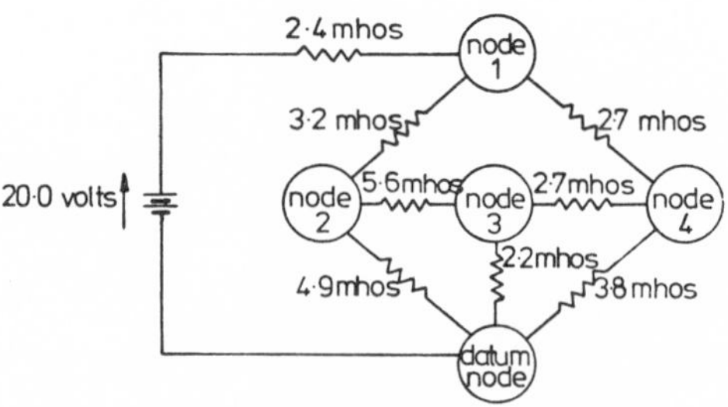

Figure 2.4 The network described by Table 2.1

and that N,M,NODEA, NODEB and CONDUC have been allocated their appropriate values. A program segment to form the node conductance matrix is as follows:

FORM NODE CONDUCTANCE MATRIX:

ALGOL

```
for i:=1 step 1 until n do
for j:=1 step 1 until n do A[i,j]:=0;
for k:=1 step 1 until m do
begin  i:=nodeA(k);
       j:=nodeB(k);
       x:=conduc(k);
       if i=0 then goto A0;
       A[i,i]:=A[i,i]+x;
       if j=0 then goto B0;
       A[j,i]:=A[i,j]:=-x;
    A0:A[j,j]:=A[j,j]+x;
B0:end forming node conductance matrix;
```

FORTRAN

```
  DO 1 I=1,N
  DO 1 J=1,N
1 A(I,J)=0.0
  DO 3 K=1,M
  I=NODEA(K)
  J=NODEB(K)
  X=CONDUC(K)
  IF(I.EQ.0)GO TO 2
  A(I,I)=A(I,I)+X
  IF(J.EQ.0)GO TO 3
  A(I,J)=-X
  A(J,I)=-X
2 A(J,J)=A(J,J)+X
3 CONTINUE
```

It will be noted that the datum node has been numbered zero in the input data, and in the program special provision has been made to omit the unwanted contributions to the node conductance matrix when a branch is connected to a datum node. It has been assumed that no two branches connect the same pair of nodes and also that no branch connects a node with itself. It is also possible to construct the right-hand vector automatically from branch data of the form shown in Table 2.1 and so produce a general program which constructs the node conductance equations for networks of any topology.

Component matrices

If equations (2.4) and (2.2) are written in matrix form, i.e.

$$\begin{bmatrix} v_{AB} \\ v_{AD} \\ v_{BC} \\ v_{CD} \\ v_{BE} \\ v_{CE} \\ v_{DE} \\ v_{EA} \end{bmatrix} = \begin{bmatrix} 1 & -1 & & \\ 1 & & & -1 \\ & 1 & -1 & \\ & & 1 & -1 \\ & 1 & & \\ & & 1 & \\ & & & 1 \\ -1 & & & \end{bmatrix} \begin{bmatrix} e_A \\ e_B \\ e_C \\ e_D \end{bmatrix} \tag{2.10}$$

and

$$\begin{bmatrix} 1 & 1 & & & & & & -1 \\ -1 & & 1 & & 1 & & & \\ & & -1 & 1 & & 1 & & \\ & -1 & & -1 & & & 1 & \end{bmatrix} \begin{bmatrix} i_{AB} \\ i_{AD} \\ i_{BC} \\ i_{CD} \\ i_{BE} \\ i_{CE} \\ i_{DE} \\ i_{EA} \end{bmatrix} = \begin{bmatrix} 0 \\ 0 \\ 0 \\ 0 \end{bmatrix} \tag{2.11}$$

it will be noted that there is a transpose relationship between the matrices of coefficients. If **v**, **e** and **i** are the vectors of branch voltages, nodal voltages and branch currents respectively, these equations can be written in the form

$$\mathbf{v} = \mathbf{A}\mathbf{e} \tag{2.12}$$

and

$$\mathbf{A}^T\mathbf{i} = \mathbf{0} \tag{2.13}$$

In matrix form the branch characteristic equations can be specified as

$$\mathbf{i} = \mathbf{G}(\mathbf{v} - \mathbf{v}_0) \tag{2.14}$$

where **G** is a diagonal matrix of branch conductances and $\mathbf{v}_0$ is a column vector of applied voltages. Substituting in equations (2.13) the values of **i** and **v** given by equations (2.14) and (2.12) yields the node conductance equations in the form

$$\mathbf{A}^T\mathbf{G}\mathbf{A}\mathbf{e} = \mathbf{A}^T\mathbf{G}\mathbf{v}_0 \tag{2.15}$$

in which $\mathbf{A}^T\mathbf{G}\mathbf{A}$ is the node conductance matrix and $\mathbf{A}^T\mathbf{G}\mathbf{v}_0$ is the right-hand vector. Hence the node conductance matrix may alternatively be constructed by a process of matrix multiplication involving the simpler matrix **A** and the diagonal matrix of branch conductances.

It is also possible to derive the loop resistance matrix as a matrix product by expressing the loop voltage and current equations (2.1) and (2.5) in matrix form and using a diagonal matrix of branch resistances to relate the branch voltages to the branch currents. There is a duality between the loop and nodal methods of analysis.

Form of equations

For the Julie bridge the node conductance matrix (equation 2.9) is clearly positive definite since its Gerschgorin discs lie entirely in the positive half-plane of the Argand diagram. It is possible to show that all node conductance matrices which have compatible ordering (and which are therefore symmetric) must be at least positive semidefinite by virtue of the fact that they have the same form as equation (1.143). Hence, if they are non-singular they must be positive definite. A singular node conductance matrix is obtained if the nodal voltages are indeterminate through the datum node being omitted from the circuit or through a part of the circuit being completely independent of the rest. Singular node conductance matrices are most likely to be encountered because of errors in the input of data.

A singular loop resistance matrix could be encountered if too many loops have been allocated for a particular circuit. For instance, in the Julie bridge, if an extra loop current j_5 is inserted round the outside of the circuit (through ADE) and a corresponding extra voltage equation included, then the loop resistance matrix will be singular. Except for such special cases all loop resistance matrices must be positive definite.

2.4 OTHER NETWORK PROBLEMS

A.C. electrical networks

Electrical network theory may be generalized from the resistance network of the previous section by the inclusion of inductance and capacitance properties. This allows the time-dependent behaviour of the network to be analysed. Numerical methods of determining the response of particular networks will depend on the time-dependent characteristics of the input voltages or currents (i.e. whether they can be represented by step functions, sinusoidal oscillations or other simple mathematical functions). In the important case of the response of a network to a sinusoidally varying input, node admittance equations may be used to obtain a solution for the nodal voltages. The difference between the node admittance parameters and the previous node conductance parameters is that, if a branch has inductance or capacitance, the associated parameters will have imaginary components. Hence the node admittance matrix will be complex. Normally, however, it will still be symmetric for consistently ordered equations.

Hydraulic networks

If a standpipe were to be erected at each junction (or node) of a hydraulic pipe network then water would rise to the position of the hydraulic head at that junction. The concept of hydraulic head is important because water in a pipe will flow from the end which has the higher hydraulic head to the end which has the lower hydraulic head. Hence the hydraulic head in the junctions and the rate of flow (or discharge) for the pipes of a hydraulic network are analogous to the nodal voltages and branch currents in an electrical resistance network. However, the analogy is not complete since the relationship between the discharge of a pipe and the difference between the hydraulic heads at its end is non-linear rather than linear. (The solution of non-linear equations is briefly discussed in section 2.12.)

Surveying network error analysis

Two types of surveying network error analysis are common. One is for the adjustment of level networks, which is very analogous to the electrical resistance network analysis. The other is the adjustment of triangulation networks. Both types of analysis will be discussed in section 2.6.

Analysis of framed structures

The analysis of framed structures can be considered as an extension of the network analysis principle by considering the joints and members of the frame to be nodes and branches respectively. In the stiffness method (which corresponds to a nodal analysis), the number of variables per joint may be anything up to six, i.e. three displacements in mutually perpendicular directions and also rotations about each of these directions as axis. Hence the number of equations may be quite large. However, if all of the displacement variables for each particular node are placed consecutively in the displacement vector, then the stiffness matrix can be divided into a submatrix form which has the same pattern of non-zero submatrices as the pattern of non-zero elements occurring in an electrical resistance network with the same node and branch configuration.

General properties of network equations

(a) Either loop or nodal analyses may be employed.

(b) The number of equations to be solved in a loop analysis is likely to be different from the number of equations to be solved in a nodal analysis.

(c) If networks of arbitrary geometry or topology are to be analysed, the automatic construction of nodal equations is normally more easily programmed than the automatic construction of loop equations.

(d) Frequently the coefficient matrix is symmetric and positive definite, although not always so. (An example of a network in which the equations cannot easily be put in symmetric form is an electrical network which contains an amplifier as one of its branches.)

(e) Where a large number of equations have to be solved the coefficient matrix will normally be sparse.

2.5 LEAST SQUARES FOR OVERDETERMINED EQUATIONS

In many problems the object is to obtain the best fit to a set of equations using insufficient variables to obtain an exact fit. A set of m linear equations involving n variables x_i, where $m > n$, may be described as being overdetermined. Taking the k-th equation as typical, then

$$a_{k1}x_1 + a_{k2}x_2 + \cdots + a_{kn}x_n = b_k \tag{2.16}$$

It will not normally be possible to satisfy all these equations simultaneously, and hence for any particular proposed solution one or more of the equations are likely to be in error.

Let e_k be the error in the k-th equation such that

$$a_{k1}x_1 + a_{k2}x_2 + \cdots + a_{kn}x_n = b_k + e_k \tag{2.17}$$

The most acceptable solution (or best fit) will not necessarily satisfy any of the equations exactly but will minimize an appropriate function of all the errors e_k. If the reliability of each equation is proportional to a weighting factor w_k, then a least squares fit finds the solution which minimizes $\Sigma_{k=1}^{m} w_k e_k^2$. Since this quantity can be altered by adjusting any of the variables then equations of the form

$$\frac{\partial}{\partial x_i} \Sigma w_k e_k^2 = 0 \tag{2.18}$$

must be satisfied for all the variables x_i. Substituting for e_k from equation (2.17) and differentiating with respect to x_i gives

$$\left(\sum_{k=1}^{m} w_k a_{ki} a_{k1}\right) x_1 + \left(\sum_{k=1}^{m} w_k a_{ki} a_{k2}\right) x_2 + \cdots + \left(\sum_{k=1}^{m} w_k a_{ki} a_{kn}\right) x_n$$

$$= \sum_{k=1}^{m} w_k a_{ki} b_i \tag{2.19}$$

Since there are n equations of this form a solution can be obtained.

Alternatively, these equations may be derived in matrix form by rewriting equations (2.17) as

$$\mathbf{Ax} = \mathbf{b} + \mathbf{e} \tag{2.20}$$

where $\mathbf{A}$ is an $m \times n$ matrix and $\mathbf{b}$ and $\mathbf{e}$ are column vectors of order m, and proceeding in the following way. If incremental changes in the variables are represented by the column vector $[\mathbf{dx}]$ then the corresponding incremental changes in the errors $[\mathbf{de}]$ satisfy

$$\mathbf{A}\,[\mathbf{dx}] = [\mathbf{de}] \tag{2.21}$$

However, from the required minimum condition for the sum of the weighted squares it follows that

$$\sum_{k=1}^{m} w_j e_j \mathrm{d}e_j = 0 \tag{2.22}$$

which can be expressed in matrix form as

$$[\mathbf{de}]^T \mathbf{We} = 0 \tag{2.23}$$

where $\mathbf{W}$ is a diagonal matrix of the weighting factors. Substituting for $\mathbf{e}$ and $[\mathbf{de}]$ using equations (2.20) and (2.21) gives

$$[\mathbf{dx}]^T \mathbf{A}^T \mathbf{W}(\mathbf{Ax} - \mathbf{b}) = 0 \tag{2.24}$$

Since this equation is valid for any vector [dx], it follows that

$$\mathbf{A}^T\mathbf{W}\mathbf{A}\mathbf{x} = \mathbf{A}^T\mathbf{W}\mathbf{b} \tag{2.25}$$

It can be verified that this is the matrix equivalent of equation (2.19). Furthermore, the coefficient matrix $\mathbf{A}^T\mathbf{W}\mathbf{A}$ is symmetric and probably positive definite (it must be positive semidefinite according to equation 1.143).

2.6 ERROR ADJUSTMENTS IN SURVEYING

An example of the use of the least squares method for linear equations comes in the error adjustment of level networks. If the vector $\mathbf{x}$ defines the altitude of various points above a datum, then a typical observation, say that point 1 lies 1.204 m above point 2, can be represented by an observational equation of the form

$$x_1 - x_2 = 1.204 \tag{2.26}$$

It is good surveying practice to make more observations than are strictly necessary, so producing an overdetermined set of equations. If a least squares error adjustment of the equations is performed, then all of the observations are taken into account in determining the most probable altitudes of the points.

As an example consider the network whose plan view is shown in Figure 2.5. The altitude of the points marked 1 to 4 are required relative to the datum point. Eight observations have been made which are marked along the lines of sight. The corresponding observation equations including possible errors are thus

$$\left.\begin{aligned} x_1 - x_2 &= 1.204 + e_1 \\ x_1 - x_4 &= 1.631 + e_2 \\ x_2 - x_3 &= 3.186 + e_3 \\ x_3 - x_4 &= -2.778 + e_4 \\ x_2 \qquad &= 1.735 + e_5 \\ x_3 \qquad &= -1.449 + e_6 \\ x_4 \qquad &= 1.321 + e_7 \\ -x_1 \qquad &= -2.947 + e_8 \end{aligned}\right\} \tag{2.27}$$

Writing these in the matrix form of equation (2.20) gives

$$\mathbf{A} = \begin{bmatrix} 1 & -1 & & \\ 1 & & & -1 \\ & 1 & -1 & \\ & & 1 & -1 \\ & 1 & & \\ & & 1 & \\ & & & 1 \\ -1 & & & \end{bmatrix} \quad \text{and} \quad \mathbf{b} = \begin{bmatrix} 1.204 \\ 1.631 \\ 3.186 \\ -2.778 \\ 1.735 \\ -1.449 \\ 1.321 \\ -2.947 \end{bmatrix} \tag{2.28}$$

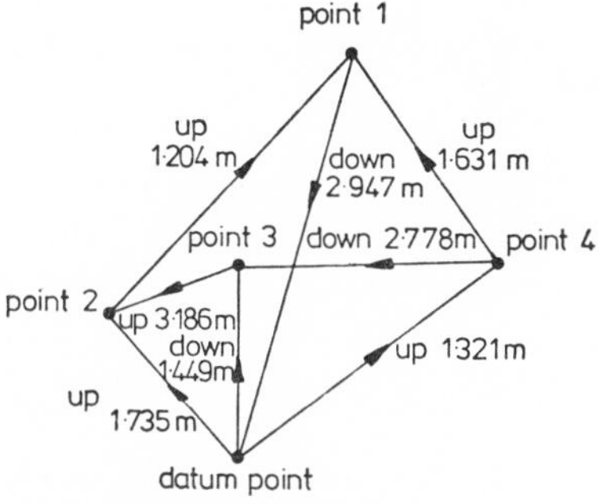

Figure 2.5 Observations for a level network

The reliability of observations made between distant points will be less than the reliability of observations made between near points. Weighting factors may be introduced to allow for this and any other factor affecting the relative reliability of the observations. Suppose that, for the problem under discussion, weighting factors are allocated as follows:

$$\mathbf{W} = \lceil 3.2 \quad 2.7 \quad 5.6 \quad 2.7 \quad 4.9 \quad 2.2 \quad 3.8 \quad 2.4 \rfloor \tag{2.29}$$

By substituting into equation (2.25) four simultaneous equations are obtained, namely

$$\left.\begin{array}{rrrrl} 8.3x_1 & -\ 3.2x_2 & & -2.7x_4 & = 15.3293 \\ -3.2x_1 & +\ 13.7x_2 & -\ 5.6x_3 & & = 22.4903 \\ & -5.6x_2 & +\ 10.5x_3 & -\ 2.7x_4 & = -28.5300 \\ -2.7x_1 & & -2.7x_3 & +\ 9.2x_4 & = 8.1167 \end{array}\right\} \tag{2.30}$$

These equations are called *observational normal equations* and their solution

$$x = \{2.9461 \quad 1.7365 \quad -1.4513 \quad 1.3209\} \tag{2.31}$$

defines the most probable altitudes for the points. Substituting the values computed for the x_i into the observation equations, the most probable errors in the observations are

$$e = \{0.0056 \quad -0.0058 \quad 0.0018 \quad 0.0058 \quad 0.0015 \quad -0.0023 \quad -0.0001 \quad 0.0009\} \tag{2.32}$$

(Note that these are not the actual errors which, of course, are unknown.)

This particular level network has been chosen as an example because it has the same topology as the electrical resistance network of section 2.2. Also, because the weighting factors correspond to the conductances, the coefficient matrix of the observational normal equations is identical to the node conductance matrix of the electrical network. To obtain the electrical network which is completely analogous to the level network it is necessary to insert applied voltages into each branch to correspond to the observations of the level network.

In trilateration and triangulation networks the horizontal positions of points are obtained by the measurement of horizontal distances and angles. To use a nodal analysis two variables per node are required, namely the Cartesian coordinates of each point relative to a suitable datum. If Fig. 2.5 represents a triangulation network and if the distance between points 1 and 2 is measured as 684.26 m, then the corresponding observation equation could be written as

$$(X_1 - X_2)^2 + (Y_1 - Y_2)^2 = 684.26^2 \tag{2.33}$$

where (X_1, Y_1) and (X_2, Y_2) are the coordinates of points 1 and 2. This equation is non-linear. However, if the redundant observations in the network are discarded and the position of the points estimated by methods of cooordinate geometry, then the variables can be redefined as changes to these estimated positions. If, for instance, in the first analysis points 1 and 2 are estimated to have coordinates (837.24 m, 589.29 m) and (252.10 m, 234.47 m) respectively, then with modifications (x_1, y_1) and (x_2, y_2) the observation equation including an error term becomes

$$(585.14 + x_1 - x_2)^2 + (354.82 + y_1 - y_2)^2 = (684.26 + e)^2 \tag{2.34}$$

Since the new variables represent small adjustments, second-order terms in the expansion of equation (2.34) may be neglected, giving

$$\frac{585.14}{684.26}(x_1 - x_2) + \frac{354.82}{684.26}(y_1 - y_2) = -0.0543 + e \tag{2.35}$$

which is a linear observation equation. Triangulation network observation equations can always be linearized in this way, and hence it is possible to adopt the least squares procedure to compute a solution.

In surveying network adjustments the loop analysis method is known as the method of correlatives.

2.7 CURVE FITTING

Consider the problem of finding the best quadratic polynomial to fit the six points (x, y) = (0.2, 1), (0.4, 2), (0.6, 3), (0.8, 5), (1.0, 5) and (1.2, 3) by a least squares minimization of adjustments to the y coordinates. If the polynomial has the form

$$y = c_0 + c_1 x + c_2 x^2 \tag{2.36}$$

the discrepancy between the value of the polynomial at the point k and the actual value is given by e_k, where

$$c_0 + c_1 x_k + c_2 x_k^2 = y_k + e_k \tag{2.37}$$

The complete set of six equations of this type may be written in the form $\mathbf{Ax} = \mathbf{b} + \mathbf{e}$ (equation 2.20), where

$$\mathbf{A} = \begin{bmatrix} 1 & x_1 & x_1^2 \\ 1 & x_2 & x_2^2 \\ 1 & x_3 & x_3^2 \\ 1 & x_4 & x_4^2 \\ 1 & x_5 & x_5^2 \\ 1 & x_6 & x_6^2 \end{bmatrix} = \begin{bmatrix} 1 & 0.2 & 0.04 \\ 1 & 0.4 & 0.16 \\ 1 & 0.6 & 0.36 \\ 1 & 0.8 & 0.64 \\ 1 & 1.0 & 1.00 \\ 1 & 1.2 & 1.44 \end{bmatrix}, \quad \mathbf{b} = \begin{bmatrix} y_1 \\ y_2 \\ y_3 \\ y_4 \\ y_5 \\ y_6 \end{bmatrix} = \begin{bmatrix} 1 \\ 2 \\ 3 \\ 5 \\ 5 \\ 3 \end{bmatrix} \tag{2.38}$$

and the coefficients of the quadratic polynomial $\{c_0 \quad c_1 \quad c_2\}$ are the variables **x**. If equal weighting factors are to be used then $\mathbf{W} = \mathbf{I}$, and substituting in equation (2.25) gives the normal equations

$$\begin{bmatrix} 6 & 4.2 & 3.64 \\ 4.2 & 3.64 & 3.528 \\ 3.64 & 3.528 & 3.64 \end{bmatrix} \begin{bmatrix} c_0 \\ c_1 \\ c_2 \end{bmatrix} = \begin{bmatrix} 19 \\ 15.4 \\ 13.96 \end{bmatrix} \tag{2.39}$$

Elimination of the subdiagonal elements gives

$$\begin{bmatrix} 6 & 4.2 & 3.64 \\ & 0.7 & 0.98 \\ & & 0.0597 \end{bmatrix} \begin{bmatrix} c_0 \\ c_1 \\ c_2 \end{bmatrix} = \begin{bmatrix} 19 \\ 2.1 \\ -0.5067 \end{bmatrix} \tag{2.40}$$

Hence by backsubstitution $\{c_0 \quad c_1 \quad c_2\} = \{-2.1000 \quad 14.8751 \quad -8.4822\}$. The solution is represented graphically in Figure 2.6 and the adjustments may be computed as $\mathbf{e} = \{-0.4643 \quad 0.4929 \quad 0.7715 \quad -0.6285 \quad -0.7071 \quad 0.5358\}$.

This method may be generalized by replacing equation (2.36) with

$$y = \Sigma c_i f_i \tag{2.41}$$

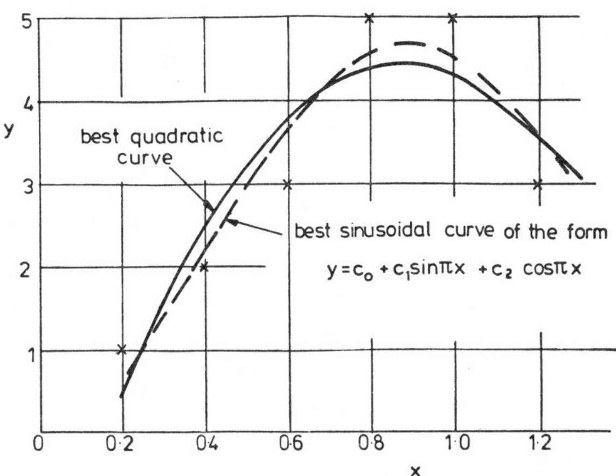

Figure 2.6 A curve fitting example

where the f_i are functions of one or more variables. A typical element a_{kj} of the matrix A is then the value of the j-th function at point k. If, instead of the quadratic curve, it had been required to find the best-fitting sinusoidal curve of the form

$$y = c_0 + c_1 \sin \pi x + c_2 \cos \pi x \tag{2.42}$$

functions (f_1, f_2, f_3) would be changed from $(1, x, x^2)$ to $(1, \sin \pi x, \cos \pi x)$ and hence equation $\mathbf{Ax} = \mathbf{b} + \mathbf{e}$ would remain unaltered except that matrix A would be replaced by

$$\mathbf{A} = \begin{bmatrix} 1 & \sin \pi x_1 & \cos \pi x_1 \\ 1 & \sin \pi x_2 & \cos \pi x_2 \\ 1 & \sin \pi x_3 & \cos \pi x_3 \\ 1 & \sin \pi x_4 & \cos \pi x_4 \\ 1 & \sin \pi x_5 & \cos \pi x_5 \\ 1 & \sin \pi x_6 & \cos \pi x_6 \end{bmatrix} = \begin{bmatrix} 1 & 0.5878 & 0.8090 \\ 1 & 0.9511 & 0.3090 \\ 1 & 0.9511 & -0.3090 \\ 1 & 0.5878 & -0.8090 \\ 1 & 0 & -1 \\ 1 & -0.5878 & -0.8090 \end{bmatrix} \tag{2.43}$$

The least squares solution of the modified equations is

$$\{c_0 \quad c_1 \quad c_2\} = \{2.0660 \quad 0.8886 \quad -2.4274\}$$

which gives rise to the supplementary curve shown in Figure 2.6.

In order to find the best plane

$$w = c_0 + c_1 u + c_2 v \tag{2.44}$$

to fit the points $(u, v, w) = (0, 0, 0), (1, 1, 1), (3, 0, 2), (1, 3, 2)$ and $(4, 2, 2)$ by least square adjustment of the w coordinates, functions (f_1, f_2, f_3) will be $(1, u, v)$, giving

$$\mathbf{A} = \begin{bmatrix} 1 & u_1 & v_1 \\ 1 & u_2 & v_2 \\ 1 & u_3 & v_3 \\ 1 & u_4 & v_4 \\ 1 & u_5 & v_5 \end{bmatrix} = \begin{bmatrix} 1 & 0 & 0 \\ 1 & 1 & 1 \\ 1 & 3 & 0 \\ 1 & 1 & 3 \\ 1 & 4 & 2 \end{bmatrix}, \quad \mathbf{b} = \begin{bmatrix} w_1 \\ w_2 \\ w_3 \\ w_4 \\ w_5 \end{bmatrix} = \begin{bmatrix} 0 \\ 1 \\ 2 \\ 2 \\ 2 \end{bmatrix} \tag{2.45}$$

The least square solution with equal weighting is

$$\{c_0 \quad c_1 \quad c_2\} = \{0.3167 \quad 0.3722 \quad 0.3500\},$$

which corresponds to discrepancies of

$$\mathbf{e} = \{0.3167 \quad 0.0389 \quad -0.5667 \quad -0.2611 \quad 0.5056\}$$

in the w coordinates of the five points.

In general, the more unknown coefficients that are used in equation (2.41), the lower will be the adjustments that are required to obtain a fit. However, there are dangers inherent in curve fitting that are likely to arise unless great care is taken, particularly where a large number of functions are used. These may be summarized as follows:

(a) Functions should not be chosen which give rise to linear dependence in the columns of A. This would occur in the last example if all the points (u_i, v_i) were collinear or if $f_4 = (u + v)$ was chosen as an additional function.

(b) It is possible to choose so many functions that the smoothing effect of the curve fit is lost. In this case the curve might behave erratically between the points being fitted, particularly near the extremities of the range. To use such a curve for extrapolation purposes could prove particularly disastrous.

(c) Even if the functions are chosen satisfactorily to give a unique and smooth curve, the normal equations may be ill conditioned, giving rise to either partial or complete loss of accuracy during solution. Ill-conditioning is discussed in more detail in Chapter 4 and an orthogonalization method of alleviating the effects of ill-conditioning is presented in section 4.19.

Provided that the functions chosen do not produce linear dependence in the matrix A, the normal equations will be symmetric and positive definite because the coefficient matrix is of the same form as equation (1.143).

2.8 A HEAT TRANSFER FIELD PROBLEM

Unlike problems of the network type, field problems do not give rise to finite sets of equations by a direct interpretation of the physical properties. Instead it is necessary to approximate the system to a discrete form, normally by choosing points or nodes at which to assign the basic variables. The errors involved in making this approximation will, in general, be smaller if more nodes are chosen. Hence the user must decide what particular idealization is likely to give him sufficient accuracy, bearing in mind that more accurate solutions tend to require much more computation.

Consider the heat transfer problem of Figure 2.7 in which the material surface AB is maintained at a high temperature T_H while the surface CDEF is maintained at a low temperature T_L, and the surfaces BC and FA are insulated. Estimates may be required for the steady state temperature distributions within the material and

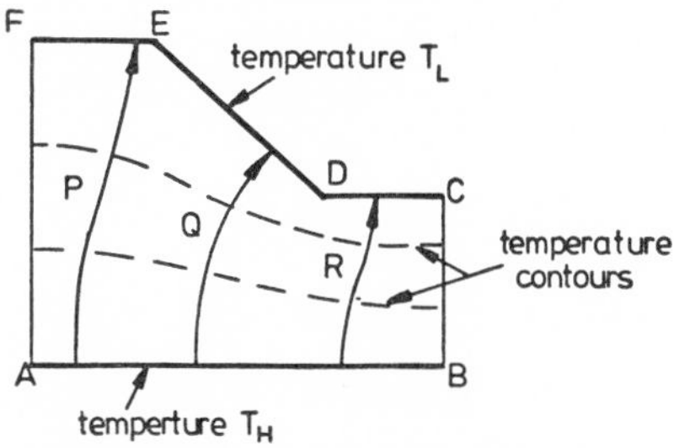

Figure 2.7 A heat transfer problem showing possible heat flow lines

also the rate of heat transfer, assuming that the geometry and boundary conditions are constant in the third direction.

There will be heat flow lines such as the lines P, Q and R in Figure 2.7 and also temperature contours which must be everywhere mutually orthogonal. If T is the temperature and q the heat flow per unit area, then the heat flow will be proportional to the conductivity of the material, k, and to the local temperature gradient. If s is measured along a line of heat flow then

$$q = -k \frac{\partial T}{\partial s} \tag{2.46}$$

Alternatively, the heat flow per unit area can be separated into components q_x and q_y in the x and y directions respectively. For these components it can be shown that

$$q_x = -k \frac{\partial T}{\partial x} \quad \text{and} \quad q_y = -k \frac{\partial T}{\partial y} \tag{2.47}$$

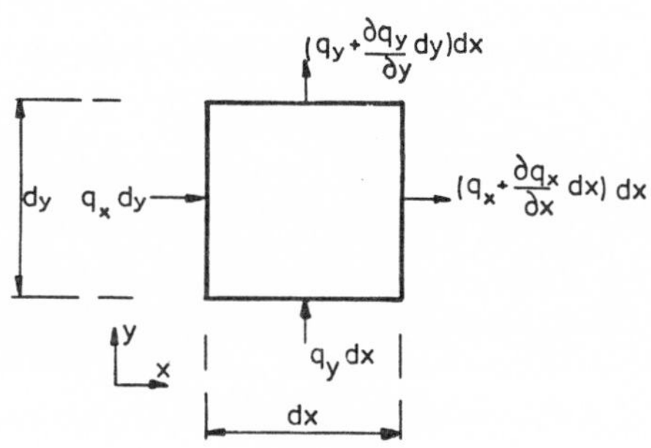

Figure 2.8 Heat flow across the boundaries of an element $dx \times dy \times 1$

Consider the element with volume $dx \times dy \times 1$ shown in Figure 2.8. In the equilibrium state the net outflow of heat must be zero, hence

$$\left(\frac{\partial q_x}{\partial x} + \frac{\partial q_y}{\partial y}\right) dx\, dy = 0 \tag{2.48}$$

Substituting for q_x and q_y from equations (2.47) gives

$$\frac{\partial^2 T}{\partial x^2} + \frac{\partial^2 T}{\partial y^2} = 0 \tag{2.49}$$

provided that the conductivity of the material is constant. This equation must be satisfied throughout the region ABCDEF, and, together with the boundary conditions, gives a complete mathematical statement of the problem. The boundary conditions for the problem being discussed are:

$$\left.\begin{aligned} &\text{on AB:} && T = T_H \\ &\text{on CDEF:} && T = T_L \\ \text{and}\quad &\text{on BC and FA:} && q_x = \frac{\partial T}{\partial x} = 0 \end{aligned}\right\} \tag{2.50}$$

Equation (2.49) is known as Laplace's equation. Other problems giving rise to equations of Laplace form are the analysis of shear stress in shafts subject to pure torsion and potential flow analysis in fluid mechanics. Even for Laplace's equation (which is one of the simpler forms of field equation) few analytical solutions exist, these being for simple geometrical configurations and boundary conditions. Where analytical solutions are not available it is necessary to resort to numerical techniques such as the finite difference or the finite element method to obtain an approximate solution.

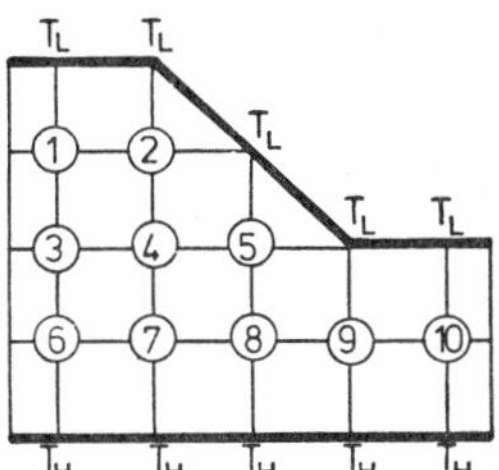

Figure 2.9 A finite difference mesh

2.9 THE FINITE DIFFERENCE METHOD

In the finite difference method a regular mesh is designated to cover the whole area of the field, and the field equation is approximated by a series of difference equations involving the magnitudes of the required variable at the mesh points. Figure 2.9 shows a square mesh suitable for an analysis of the heat transfer problem in which the temperatures of the points 1 to 10, designated by $T_1, T_2, \ldots, T_{10}$, are unknowns whose values are to be computed. For a square mesh the finite difference equation corresponding to Laplace's equation is

$$T_G + T_H - 4T_J + T_K + T_L = 0 \tag{2.51}$$

where the points G, H, K and L adjoin point J as shown in Figure 2.10. This equation, stating that the temperature at any point is equal to the average of the temperatures at the four neighbouring points, is likely to be less accurate when a large mesh size rather than a small one is chosen. Taking J to be each of the points 1 to 10 in turn yields ten equations; for instance, with

$$\begin{aligned} J = 4:&\quad T_2 + T_3 - 4T_4 + T_5 + T_7 = 0 \\ J = 5:&\quad T_4 - 4T_5 + T_8 = -2T_L \end{aligned}$$

The insulated boundaries BC and FA may be considered to act as mirrors, so that, for instance, with $J = 1$, the point immediately to the left of point 1 can be

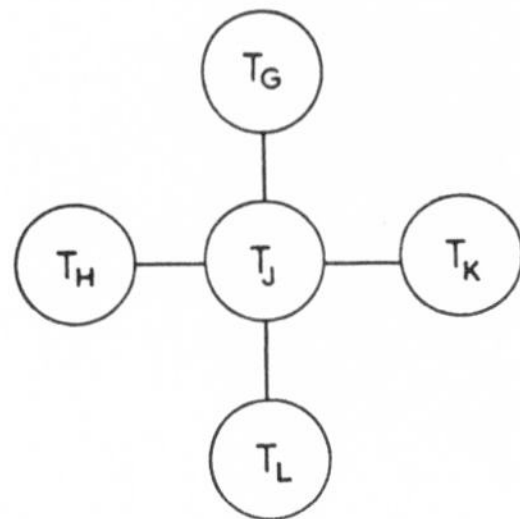

Figure 2.10 Finite difference linkage for Laplace's equation

assumed to have a temperature T_1, giving

$$-3T_1 + T_2 + T_3 = -T_L$$

The full set of equations may be written in matrix form as

$$\begin{bmatrix} 3 & -1 & -1 & & & & & & & \\ -1 & 4 & & -1 & & & & & & \\ -1 & & 3 & -1 & & -1 & & & & \\ & -1 & -1 & 4 & -1 & & -1 & & & \\ & & & -1 & 4 & & & -1 & & \\ & & -1 & & & 3 & -1 & & & \\ & & & -1 & & -1 & 4 & -1 & & \\ & & & & -1 & & -1 & 4 & -1 & \\ & & & & & & & -1 & 4 & -1 \\ & & & & & & & & -1 & 3 \end{bmatrix} \begin{bmatrix} T_1 \\ T_2 \\ T_3 \\ T_4 \\ T_5 \\ T_6 \\ T_7 \\ T_8 \\ T_9 \\ T_{10} \end{bmatrix} = \begin{bmatrix} T_L \\ 2T_L \\ \\ \\ 2T_L \\ T_H \\ T_H \\ T_H \\ T_L + T_H \\ T_L + T_H \end{bmatrix} \qquad (2.52)$$

The solution of this set of equations for the case $T_L = 0°$, $T_H = 100°$ yields the temperature distribution shown in Figure 2.11. The temperature contours have been drawn using linear interpolation along the mesh lines. The rate of heat flow may be estimated from the various temperature gradients.

By using a finite difference mesh of half the size (involving fifty-five temperature variables) the temperatures obtained at the ten points of the first difference scheme were {18.76° 13.77° 41.92° 37.09° 25.38° 69.66° 67.00° 61.38° 54.09° 51.18°} showing a maximum discrepancy of about 1° at points 5 and 9. For this particular problem, therefore, the crude ten-point idealization does appear to give a fairly accurate answer. However, for less regular geometrical configurations and boundary conditions a much finer mesh would be required to obtain a similar accuracy.

A set of finite difference equations obtained from Laplace's equation can always be written in symmetric form and, in addition, the coefficient matrix will be positive definite. In cases where the number of variables is large the equations will be sparse since there cannot be more than five non-zero coefficients in each

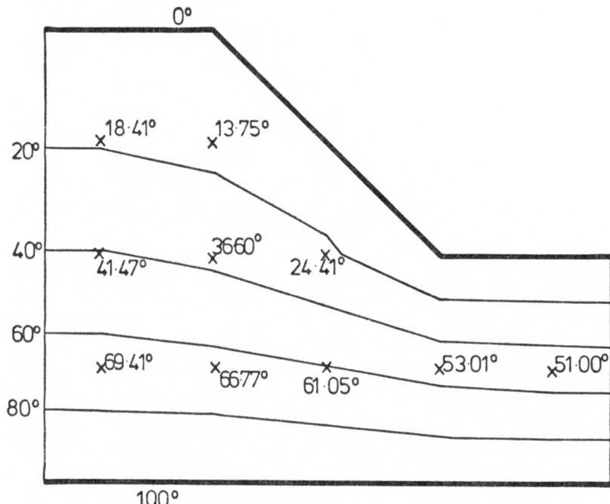

Figure 2.11 Finite difference solution for the heat transfer problem

equation if a square or rectangular mesh is used. Finite difference equations were normally solved by iterative (relaxation) techniques before computers were available.

2.10 THE FINITE ELEMENT METHOD

The finite difference method does not define the value of the variable uniquely elsewhere than at the mesh points. For instance, in the heat transfer problem, the seventh difference equation implies that the temperature variation is quadratic across points 6, 7 and 8, whereas the eighth difference equation implies a quadratic variation across points 7, 8 and 9. For any point on the mesh line between points 7 and 8 these two assumptions are likely to give a different value of temperature (Figure 2.12).

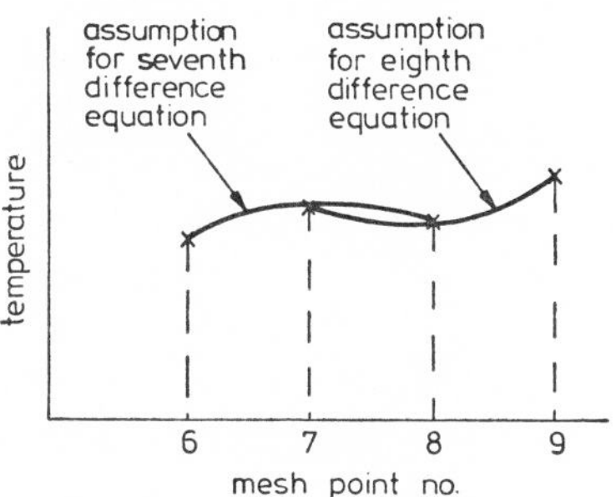

Figure 2.12 Finite difference idealization

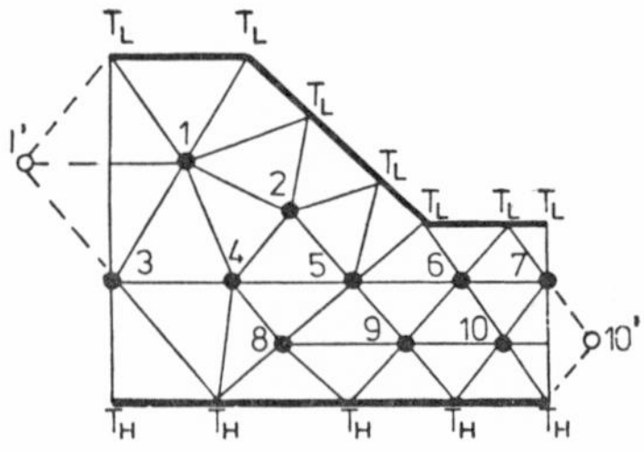

Figure 2.13 A finite element map

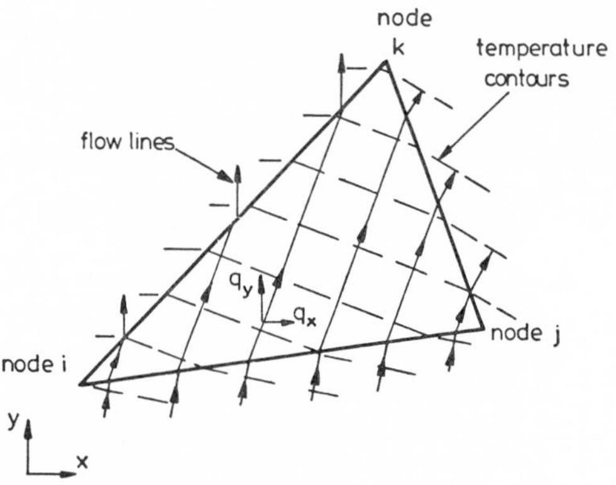

Figure 2.14 A triangular finite element with linear temperature variation

In the finite element method the region to be analysed is divided into a set of finite elements (e.g. triangles, rectangles, etc). The value of the variable at any point within a particular finite element is defined with respect to the value of the variables at certain points or nodes. For the simplest forms of finite element these points are the vertices of the element. Thus, in contrast with the finite difference method, the variable is uniquely defined over the whole field. The triangular element is the simplest element which gives versatility in representing arbitrary geometrical configurations, and will be described in relation to its possible use in the solution of the heat transfer problem. Figure 2.13 shows a finite element map which involves ten nodes at which temperature unknowns are to be found. By choosing the nodes closer together on the right of the region, a more accurate representation is likely to be obtained where the heat flows are largest.

Consider a triangular finite element (Figure 2.14) connecting any three nodes i, j and k whose coordinates are (x_i, y_i), (x_j, y_j) and (x_k, y_k) respectively, and assume that the temperature varies linearly over the element according to

$$T = c_1 + c_2 x + c_3 y \tag{2.53}$$

The temperatures at the three nodes may therefore be expressed as

$$\begin{bmatrix} T_i \\ T_j \\ T_k \end{bmatrix} = \begin{bmatrix} 1 & x_i & y_i \\ 1 & x_j & y_j \\ 1 & x_k & y_k \end{bmatrix} \begin{bmatrix} c_1 \\ c_2 \\ c_3 \end{bmatrix} \tag{2.54}$$

Inverting these equations and omitting the c_1 coefficient (which does not affect the temperature gradient and heat flow) gives

$$\begin{bmatrix} c_2 \\ c_3 \end{bmatrix} = \frac{1}{2\Delta} \begin{bmatrix} y_{jk} & y_{ki} & y_{ij} \\ x_{kj} & x_{ik} & x_{ji} \end{bmatrix} \begin{bmatrix} T_i \\ T_j \\ T_k \end{bmatrix} \tag{2.55}$$

where $y_{jk} = y_j - y_k$, $x_{kj} = x_k - x_j$, etc., and

$$\Delta = \frac{1}{2} \begin{vmatrix} 1 & x_i & y_i \\ 1 & x_j & y_j \\ 1 & x_k & y_k \end{vmatrix} \tag{2.56}$$

is the area of the triangle.

Since it has been assumed that T varies linearly over the whole element, the flow field will be parallel and uniform having components

$$\begin{aligned} q_x &= -kc_2 \\ q_y &= -kc_3 \end{aligned} \tag{2.57}$$

as shown in Figure 2.14.

Ideally the heat flowing to or from an edge of the triangle should be balanced by the heat flow in the neighbouring element. However, it is not possible to satisfy this condition along each line of the finite element map because there is an insufficient number of temperature variables. Instead, a flow balance can be obtained for each node by attributing to it one-half of the net loss or gain for each impinging line. For the triangular element connecting nodes i, j and k the heat loss along the edges (i, j), (j, k) and (k, i) are given, respectively, by

$$\left.\begin{aligned} Q_{ij} &= (y_j - y_i)q_x - (x_j - x_i)q_y \\ Q_{jk} &= (y_k - y_j)q_x - (x_k - x_j)q_y \\ \text{and} \quad Q_{ki} &= (y_i - y_k)q_x - (x_i - x_k)q_y \end{aligned}\right\} \tag{2.58}$$

Hence the heat loss due to flow within the element can be apportioned to the vertices according to $Q_i = ½(Q_{ki} + Q_{ij})$, $Q_j = ½(Q_{ij} + Q_{jk})$, $Q_k = ½(Q_{jk} + Q_{ki})$, giving

$$\begin{bmatrix} Q_i \\ Q_j \\ Q_k \end{bmatrix} = -\frac{k}{2} \begin{bmatrix} y_{jk} & x_{kj} \\ y_{ki} & x_{ik} \\ y_{ij} & x_{ji} \end{bmatrix} \begin{bmatrix} c_2 \\ c_3 \end{bmatrix} \tag{2.59}$$

Substituting for $\{c_2 \quad c_3\}$ from equation (2.55) gives

$$\begin{bmatrix} Q_i \\ Q_j \\ Q_k \end{bmatrix} = -\frac{k}{4\Delta} \begin{bmatrix} x_{jk}^2 + y_{jk}^2 & x_{jk}x_{ki} + y_{jk}y_{ki} & x_{jk}x_{ij} + y_{jk}y_{ij} \\ x_{ki}x_{jk} + y_{ki}y_{jk} & x_{ki}^2 + y_{ki}^2 & x_{ki}x_{ij} + y_{ki}y_{ij} \\ x_{ij}x_{jk} + y_{ij}y_{jk} & x_{ij}x_{ki} + y_{ij}y_{ki} & x_{ij}^2 + y_{ij}^2 \end{bmatrix} \begin{bmatrix} T_i \\ T_j \\ T_k \end{bmatrix} \tag{2.60}$$

Using the nodal coordinates it is possible to construct equations of the form of (2.60) for each finite element in Figure 2.13. Each of the temperatures T_i, T_j and T_k will either be one of the unknowns $T_1, \ldots, T_{10}$ or will be a fixed boundary temperature (point $1'$ in the mirror image position of point 1 is assumed to have a temperature T_1 and similarly T'_{10} is assumed to have a temperature T_{10}). For each of the node points 1 to 10 the heat loss attributable to the point from all the neighbouring elements may be summed and equated to zero (for the boundary nodes 3 and 7 only half of the contributions from elements $(1, 3, 1')$ and $(10, 7, 10')$ should be used).

The coefficient matrix of the resulting equations will be symmetric positive definite and have non-zero elements in the following pattern:

$$\begin{bmatrix} \times & \times & \times & \times & & & & & & \\ \times & \times & & \times & \times & & & & & \\ \times & & \times & \times & & & & & & \\ \times & \times & \times & \times & \times & & & \times & & \\ & \times & & \times & \times & \times & & \times & \times & \\ & & & & \times & \times & \times & & \times & \times \\ & & & & & \times & \times & & & \times \\ & & & \times & \times & & & \times & \times & \\ & & & & \times & \times & & \times & \times & \times \\ & & & & & \times & \times & & \times & \times \end{bmatrix} \begin{bmatrix} T_1 \\ T_2 \\ T_3 \\ T_4 \\ T_5 \\ T_6 \\ T_7 \\ T_8 \\ T_9 \\ T_{10} \end{bmatrix} = \begin{bmatrix} \times \\ \times \\ \times \\ \times \\ \times \\ \times \\ \times \\ \times \\ \times \\ \times \end{bmatrix} \tag{2.61}$$

The construction of these equations can be carried out on a computer in a similar way to the construction of the node conductance equations of an electrical resistance network (section 2.3) insofar as the equations may firstly be written down with zero coefficients throughout and then the contribution of each triangular element can be added into the appropriate locations. However, for each element, there will be up to nine contributions corresponding to the nine coefficients of equation (2.60). Where an element does not impinge on a fixed temperature boundary, all nine coefficients will contribute to the coefficient matrix, e.g. the coefficient arising from element (4, 5, 8) will modify coefficients (4, 4), (4, 5), (4, 8), (5, 4), (5, 5), (5, 8), (8, 4), (8, 5) and (8, 8) of the equations. Where the element has one of its nodes at a fixed temperature only four coefficients will add to the coefficient matrix, although two coefficients will modify the right-hand side vector. Thus, for example, the element linking nodes 2 and 5 with the boundary will affect coefficients (2, 2), (2, 5), (5, 2) and (5, 5) of the equations and will also modify rows 2 and 5 of the right-hand side vector. Where an element has two of its nodes at fixed temperatures, just one coefficient and one right-hand side element will be affected.

Finite elements with more nodes usually tend to give a more accurate solution, e.g. a triangular element which has nodes at the mid-side points as well as at the vertices will enable the linear temperature function (2.53) to be replaced by a quadratic function

$$T = c_1 + c_2x + c_3y + c_4x^2 + c_5xy + c_6y^2 \tag{2.62}$$

However, the calculation of the appropriate equations for such elements is more complicated than for the linear triangular element. In general the equations are obtained by minimizing an appropriate integral function. The integration, which has to be carried out over the whole area of the element, may have to be evaluated numerically in many cases. However, where simple mathematical functions have been chosen for the main variable (e.g. the quadratic function 2.62) numerical integration can usually be avoided.

Three-dimensional elements and elements with curved boundaries have also been derived.

2.11 A SOURCE AND SINK METHOD

Apart from the finite difference and finite element methods, other techniques may be adopted for the solution of field problems. A particular fluid mechanics problem which would be difficult to solve by either of these methods is the analysis of two-dimensional potential flow past a body of arbitrary shape, because the field under investigation extends to infinity. It is often possible to analyse such a problem by using flow singularities, i.e. sources and sinks, within the body. Consider first the classical case of a fluid source and sink pair in a uniform stream in which the source is situated upstream of the sink. The resulting flow pattern (Figure 2.15) has two distinct regions:

(a) the flow within a Rankine oval in which fluid emitted from the source flows into the sink, and

(b) the flow outside the Rankine oval in which the uniform stream is distorted because of the presence of the source and sink.

The second part of this flow pattern can be used to represent flow past a body of Rankine oval shape (the fact that the flow conditions within the oval cannot actually occur does not invalidate the solution outside the oval). This solution may be extended to give an approximation to the flow past symmetric bodies of more arbitrary shape by allocating a set of sources of strengths $m_1, m_2, \ldots, m_n$ acting at different points along the x axis and then finding the numerical values of $m_1, m_2, \ldots, m_n$ required to fit the particular profile at n suitable points. Figure 2.16 illustrates an example in which seven sources have been chosen together with

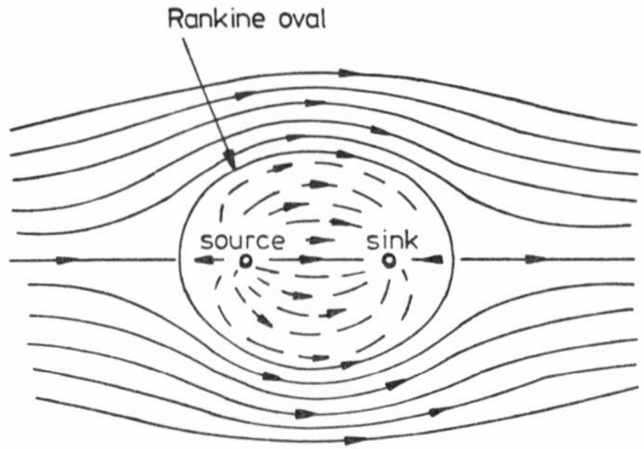

Figure 2.15 A source and sink pair in a uniform stream showing the resultant streamlines

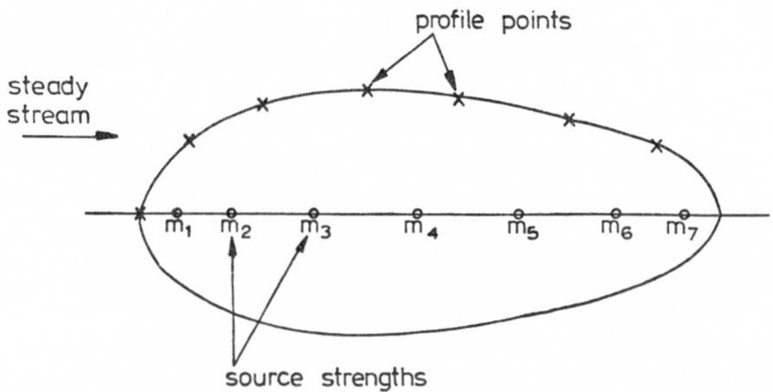

Figure 2.16 A point source distribution to represent the flow field round a symmetric body

seven points at which the body profile is to be matched with the actual profile. The *stream function* at any of the profile points can be derived as a linear combination of the source strengths and the steady stream velocity and can be equated to the known boundary value. This gives n linear simultaneous equations which may be solved to determine the source strengths (from which the approximate flow pattern may be predicted). The coefficient matrix of these simultaneous equations will be densely populated and unsymmetric. In the solution some of the sources should have negative strengths indicating that they are, in fact, sinks. The accuracy of the fit should be improved by increasing the number of sources and profile points, although the resulting equations are likely to be poorly conditioned. Accuracy may also be increased by:

(a) using more profile points than sources and using a least squares fit for the stream function at the profile points, or
(b) using line sources distributed along the centre line of the body instead of point sources.

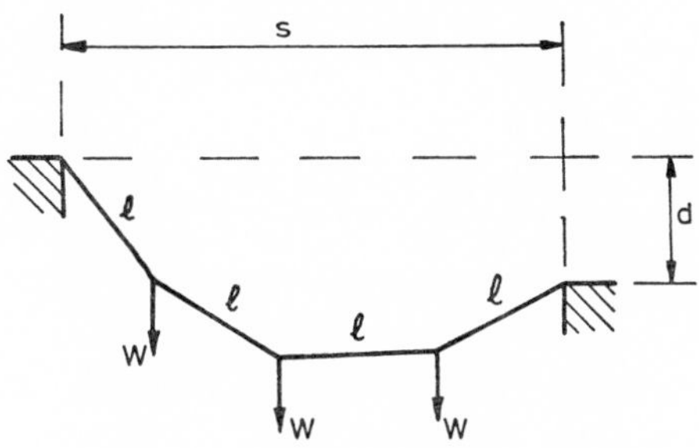

Figure 2.17 A freely hanging cable

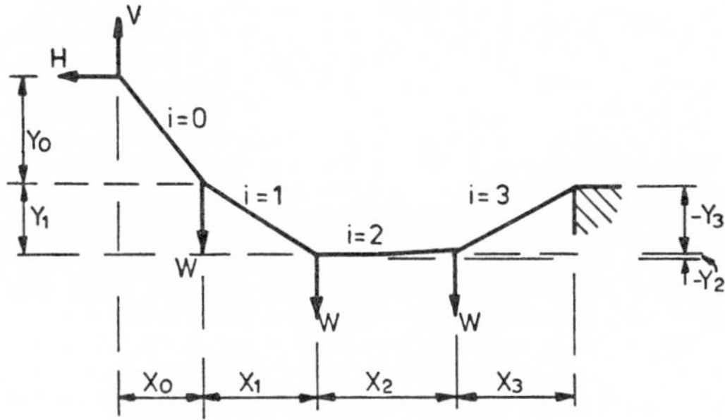

Figure 2.18 The cable problem with forces H and V defined

2.12 A NON-LINEAR CABLE ANALYSIS BY THE NEWTON–RAPHSON METHOD

Although matrices describe linear relationships, they may be used in the solution of non-linear as well as linear problems.

As an example of a non-linear problem with only two basic variables consider the cable shown in Figure 2.17 in which the horizontal and vertical components of span, s and d respectively, are known. It may be required to determine the equilibrium position of the cable. The cable is supporting three equal weights W at the nodes, and the segments of cable between the nodes, each of length ℓ, are assumed to be weightless (and therefore straight) and inextensible. This problem may be solved by taking as basic variables the horizontal and vertical components of cable tension, H and V, occurring at the left support and determining the geometry of the cable in terms of these variables (Figure 2.18).

Numbering the cable segments in the manner illustrated in Figure 2.18, the equilibrium of the cable nodes reveals that the horizontal component of tension in each segment is equal to H and that the corresponding vertical component of

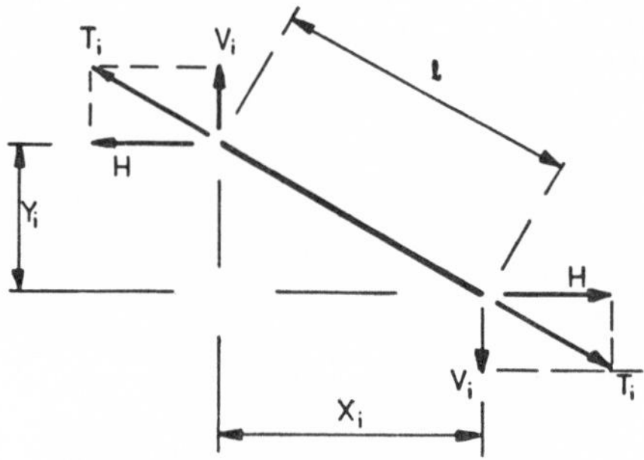

Figure 2.19 Equilibrium position for segment i

tension is given by

$$V_i = V - Wi \tag{2.63}$$

The resultant tension in each cable segment is given by

$$T_i = \sqrt{(H^2 + V_i^2)} \tag{2.64}$$

and the horizontal and vertical projections of segment i (Figure 2.19) are given by

$$\left.\begin{aligned} X_i &= \frac{H\ell}{T_i} \\ &\text{and} \\ Y_i &= \frac{V_i \ell}{T_i} \end{aligned}\right\} \tag{2.65}$$

respectively. However, for the cable to fit the necessary span conditions the following equations must be satisfied:

$$\sum_{i=0}^{3} X_i = s \quad \text{and} \quad \sum_{i=0}^{3} Y_i = d \tag{2.66}$$

Hence, by substituting for X_i, Y_i, T_i and V_i from equations (2.63) to (2.65),

$$\left.\begin{aligned} &H\ell \sum_{i=0}^{3} \left(\frac{1}{[H^2 + (V - Wi)^2]^{1/2}} \right) - s = 0 \\ &\text{and} \\ &\ell \sum_{i=0}^{3} \left(\frac{1}{[H^2 + (V - Wi)^2]^{1/2}} \right) - d = 0 \end{aligned}\right\} \tag{2.67}$$

These are two non-linear equations which define H and V but for which there is unlikely to be an analytical solution. However, it is possible to take trial values of H and V and determine values for V_i, T_i, X_i and Y_i, hence assessing the lack of fit in equations (2.66). A method of correcting these trial values is to assume that the load-deflection characteristics of the cable are linear over the range of the correction. Thus, if h and v are the assumed corrections to H and V, then for the linearized equations to be satisfied

$$\left.\begin{aligned}&\Sigma X_i + h\,\Sigma\frac{\partial X_i}{\partial H} + v\,\Sigma\frac{\partial X_i}{\partial V} - s = 0\\ &\text{and}\\ &\Sigma Y_i + h\,\Sigma\frac{\partial Y_i}{\partial H} + v\,\Sigma\frac{\partial Y_i}{\partial V} - d = 0\end{aligned}\right\} \tag{2.68}$$

The partial derivatives in these equations may be evaluated from

$$\left.\begin{aligned}&\frac{\partial X_i}{\partial H} = \frac{\ell(V - Wi)^2}{T_i^3}\\ &\frac{\partial X_i}{\partial V} = \frac{\partial Y_i}{\partial H} = \frac{-\ell H(V - Wi)}{T_i^3}\\ &\frac{\partial Y_i}{\partial V} = \frac{\ell H^2}{T_i^3}\end{aligned}\right\} \tag{2.69}$$

Since the linear assumption is inexact, the values $H + h$, $V + v$ may be used again as a revised set of trial forces. In this way an iterative scheme may be initiated.

Consider, for example, the case where $\ell = 10$ m, $s = 30$ m, $d = 12$ m and $W = 20$ kN . Substituting trial values of $H = 10$ kN and $V = 20$ kN into equation (2.68) gives

$$\left.\begin{aligned}&28.6143 + 1.0649h - 0.1789v - 30 = 0\\ &\text{and}\\ &8.9443 - 0.1789h + 1.7966v - 12 = 0\end{aligned}\right\} \tag{2.70}$$

The solution of these equations is $h = 1.6140$ and $v = 1.8616$. Hence the next trial values will be $H = 11.6140$, $V = 21.8616$. Table 2.2 shows the progress of the iterative process indicated above. Because the equations are fully revised at each iteration, convergence will be quadratic, i.e. when the errors are small they will be approximately squared at each iteration. In cases where the initial linear approximation is a good one it may be possible to obtain convergence without having to

Table 2.2 Convergence for the cable problem using the Newton–Raphson method

Iteration	H(kN)	V(kN)	Misclosure (m) Horizontal	Vertical
1	10	20	1.3857	3.0557
2	11.6140	21.8616	0.2489	0.1796
3	11.9474	22.0215	0.0061	0.0026
4	11.9556	22.0244	0.0000	0.0001

Table 2.3 Convergence for the cable problem using the modified Newton–Raphson method

Iteration	H(kN)	V(kN)	Misclosure (m) Horizontal	Vertical
1	10	20	1.3857	3.0557
2	11.6140	21.8616	0.2489	0.1796
3	11.8687	21.9869	0.0638	0.0387
4	11.9327	22.0148	0.0166	0.0097
5	11.9495	22.0219	0.0044	0.0025
6	11.9540	22.0238	0.0012	0.0007
7	11.9552	22.0243	0.0003	0.0002
8	11.9555	22.0244	0.0001	0.0000

re-evaluate the partial derivatives at each subsequent iteration. In this case the partial derivatives for the first iteration are used with each misclosure as it is calculated. The same initial values of H and V for the modified process give the convergence characteristics shown in Table 2.3.

Although this cable analysis has been described without the use of matrix notation, it is useful to adopt a matrix formulation for non-linear problems with more variables. A set of non-linear equations involving n variables $\mathbf{x} = \{x_1, x_2, \ldots, x_n\}$ can be written as

$$f(\mathbf{x}) = \mathbf{0} \tag{2.71}$$

Clearly, equations (2.67) can be seen to be in this form, with $\mathbf{x} = \{H, V\}$ and f defining the two non-linear functions. Suppose that $\mathbf{x}^{(k)}$ represents the trial values for the variables at iteration k and that

$$\mathbf{y}^{(k)} = f(\mathbf{x}^{(k)}) \tag{2.72}$$

Then the assumption that the system behaves linearly over the correction yields the matrix equation

$$\mathbf{A}^{(k)}(\mathbf{x}^{(k+1)} - \mathbf{x}^{(k)}) = -\mathbf{y}^{(k)} \tag{2.73}$$

where the typical element $a_{ij}^{(k)}$ of the matrix $\mathbf{A}^{(k)}$ corresponds to the partial derivative $\partial y_i/\partial x_j$ at $x_j^{(k)}$. In the cable analysis, equation (2.68) is in this form with

$$\mathbf{A}^{(k)} = \begin{bmatrix} \Sigma \dfrac{\partial X_i}{\partial H} & \Sigma \dfrac{\partial X_i}{\partial V} \\ \Sigma \dfrac{\partial Y_i}{\partial H} & \Sigma \dfrac{\partial Y_i}{\partial V} \end{bmatrix}, \quad \mathbf{x}^{(k+1)} - \mathbf{x}^{(k)} = \begin{bmatrix} h \\ v \end{bmatrix} \quad \text{and} \quad \mathbf{y}^{(k)} = \begin{bmatrix} \Sigma X_i - s \\ \Sigma Y_i - d \end{bmatrix} \tag{2.74}$$

The solution of equation (2.73) may be specified algebraically as

$$\mathbf{x}^{(k+1)} = \mathbf{x}^{(k)} - [\mathbf{A}^{(k)}]^{-1}\mathbf{y}^{(k)} \tag{2.75}$$

However, the numerical solution is most efficiently obtained by solving equation

(2.73) for $(\mathbf{x}^{(k+1)} - \mathbf{x}^{(k)})$ rather than by the inversion technique suggested by equation (2.75).

The modified method, in which the partial derivatives are not updated at each iteration, may be represented by the equation

$$\mathbf{A}^{(1)}(\mathbf{x}^{(k+1)} - \mathbf{x}^{(k)}) = -\mathbf{y}^{(k)} \tag{2.76}$$

While this method does not converge as rapidly as the basic method, it may be numerically more efficient since advantage can be taken of the fact that the coefficient matrix of the equations is the same at each iteration.

The cable problem is one in which the non-linearity does not cause great difficulty. Convergence is usually obtained from rough initial estimates for H and V. It is, however, important to appreciate that there is no solution when $\sqrt{(s^2 + d^2)} \geqslant 4\ell$, and that the position of link i is indeterminate when $H = 0$ and $V_i = 0$.

BIBLIOGRAPHY

Adams, J. A., and Rogers, D. F. (1973). *Computer-Aided Heat Transfer Analysis*, McGraw-Hill, New York.

Allen, D. N. de G. (1954). *Relaxation Methods in Engineering Science*, McGraw-Hill, New York.

Ames, W. F. (1969). *Numerical Methods for Partial Difference Equations*, Nelson, London. (Finite difference method).

Ashkenazi, V. (1967, 1968). 'Solution and error analysis of large geodetic networks'. *Survey Review*, 19, 166–173 and 194–206.

Batchelor, G. K. (1967). *An Introduction to Fluid Dynamics*, Cambridge University Press, Cambridge.

Brameller, A., Allan, R. N., and Hamam, Y. M. (1976). *Sparsity, Its Practical Application to Systems Analysis*, Pitman, London. (Discusses network equations).

Desai, C. S., and Abel, J. F. (1972). *Introduction to the Finite Element Method: a numerical method for engineering analysis*, Van Nostrand Reinhold, New York.

Fox, L. (Ed). (1962). *Numerical Solution of Ordinary and Partial Differential Equations*, Pergamon, Oxford.

Gallagher, R. H. (1975). *Finite Element Analysis: Fundamentals*, Prentice-Hall, Englewood Cliffs, New Jersey.

Guillemin, E. A. (1955). *Introductory Circuit Theory*, 2nd ed., Wiley, New York.

Jennings, A. (1962). 'The Free Cable', *The Engineer*, **214**, 1111–1112.

Pipes, L. A. (1963). *Matrix Methods for Engineering*, Prentice-Hall, Englewood Cliffs, New Jersey.

Rainsford, H. F. (1957). *Survey Adjustments and Least Squares*, Constable, London.

Southwell, R. V. (1946, 1956). *Relaxation Methods in Theoretical Physics*, Vols. 1 and 2. Clarendon Press, Oxford. (Finite difference applications).

Zienkiewicz, O. C. (1971). *The Finite Element Method in Engineering Science*, 2nd ed. McGraw-Hill, London.

Chapter 3
Storage Schemes and Matrix Multiplication

3.1 NUMERICAL OPERATIONS ON A COMPUTER

The use of a digital computer relies on intricate mechanical and electronic *hardware* and programmed *software* which interpret and execute the particular applications program.

It is well known that a computer usually operates with numbers in binary form (Table 3.1). However, since it is inconvenient to prepare numerical data in binary form, numbers are normally input in *binary-coded decimal* form. In binary-coded decimal form each decimal digit is represented by its binary equivalent, which can be specified as a pattern of holes on an input card or tape. A standard item of software converts input numbers from binary-coded decimal to binary form. Table 3.2 shows this conversion process for a positive integer. The operation is carried out by multiplying the binary equivalent of the highest decimal digit by 1010 (the binary equivalent of ten), adding it to the binary equivalent of the next highest decimal digit, and then repeating these operations till the required binary number is obtained. Two methods for converting a fractional number from binary-coded decimal form to binary form are shown in Tables 3.3 and 3.4. Procedures are also available for converting binary numbers into binary-coded decimal form for interpretation by decimal printing devices.

Table 3.1 Decimal/binary equivalents from 1 to 10

Decimal	Binary
1	0001
2	0010
3	0011
4	0100
5	0101
6	0110
7	0111
8	1000
9	1001
10	1010

Table 3.2 Conversion of an integer from decimal to binary

Decimal form	3	9	6	2
Binary-coded decimal form	0011	1001	0110	0010
Complete hundreds	0011×1010+1001=100111			
Complete tens	100111×1010+0110=110001100			
Binary form	110001100×1010+0010=111101111010			

Table 3.3 Conversion of a fractional number from decimal to binary form (working to an accuracy of ten binary digits)

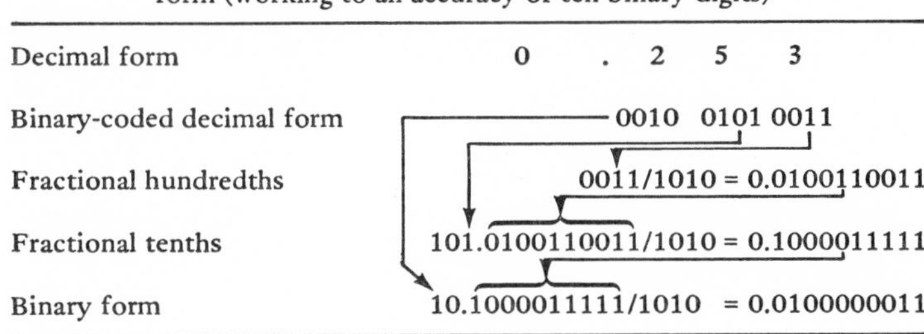

Decimal form	0 . 2 5 3
Binary-coded decimal form	0010 0101 0011
Fractional hundredths	0011/1010 = 0.0100110011
Fractional tenths	101.0100110011/1010 = 0.1000011111
Binary form	10.1000011111/1010 = 0.0100000011

Table 3.4 Conversion of a fractional number from decimal to binary form (alternative procedure)

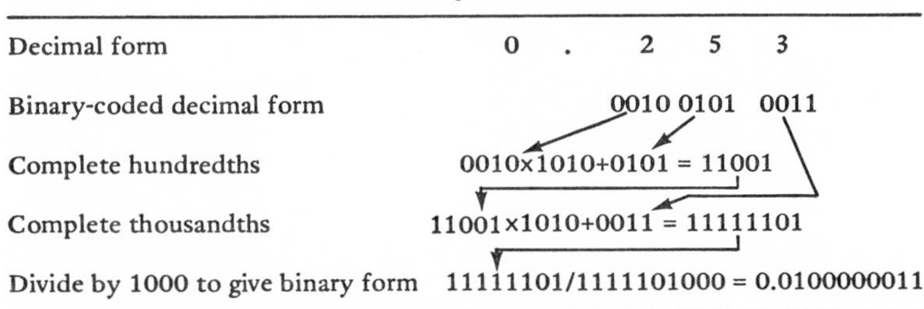

Decimal form	0 . 2 5 3
Binary-coded decimal form	0010 0101 0011
Complete hundredths	0010×1010+0101 = 11001
Complete thousandths	11001×1010+0011 = 11111101
Divide by 1000 to give binary form	11111101/1111101000 = 0.0100000011

In view of the availability of decimal/binary and binary/decimal conversion procedures it is unnecessary for users of computers or even applications programmers to be concerned with presenting or interpreting numbers in binary form.

Integer storage

In the computer, numbers are usually stored in words containing a fixed number of binary digits (or *bits*). An integer is normally stored in one such binary word. However, in order to be able to represent negative as well as positive integers the

Table 3.5 Integer storage using a 24-bit word

$-2^{23}=-8388608$	100000 000000 000000 000000
−8388607	100000 000000 000000 000001
	— — — — — — — —
−2	111111 111111 111111 111110
−1	111111 111111 111111 111111
0	000000 000000 000000 000000
1	000000 000000 000000 000001
2	000000 000000 000000 000010
	— — — — — — — —
8388607	011111 111111 111111 111111

most significant binary digit may be allocated a negative value. Hence a machine with a word length of 24 bits will permit the storage of integers i satisfying

$$-2^{23} \leqslant i < 2^{23} \tag{3.1}$$

as shown in Table 3.5. Since integers are normally used for indexing and counting operations the range limitations are not usually restrictive.

Fixed-point storage

In early computers the storage format for integers shown in Table 3.5 was extended to general real numbers by allocating a position for the binary point. This system has now been replaced by the more versatile floating-point storage format.

Floating-point storage

A number x to be stored in floating-point form is normally separated into the components a and b such that

$$x = a \times 2^b \tag{3.2}$$

Here b, called the *exponent*, is the smallest possible integer for which the *mantissa*, a, satisfies the restriction

$$-1 \leqslant a < 1 \tag{3.3}$$

If b is allocated 9 bits of storage, of which 1 bit is used for specifying the sign, then

$$-256 \leqslant b < 256 \tag{3.4}$$

and the possible range for $|x|$ is governed by

$$0.43 \times 10^{-77} \simeq 0.5 \times 2^{-256} < |x| < 1 \times 2^{255} \simeq 5.8 \times 10^{76} \tag{3.5}$$

Thus the overflow condition, in which a number has an exponent so large that it cannot be stored, will rarely be encountered in numerical operations. In the case of underflow, where a number has an exponent so small that it cannot be stored, it is set to zero by making $a = b = 0$. Frequently computers have facilities for indicating when underflow and overflow have taken place so that the user may take appropriate action.

Two word lengths are usually allocated for storing a single floating-point number. On the ICL 1906S computer 9 bits are reserved for the exponent in a 48-bit total. One bit is not used, leaving 38 bits available for the mantissa. This provides an accuracy of 37 significant binary figures for any number stored, which is equivalent to just over 11 significant decimal figures.

Double precision and complex storage

It is normally possible to store real numbers to a higher accuracy than the conventional floating-point format permits and also to store complex numbers. Both of these facilities require more words per number and the associated arithmetical operations take longer to perform.

3.2 ROUNDING ERRORS

When a number is to be stored in a computer it is not usually possible to store its value precisely and hence it is necessary to *round* it to the nearest number which can be stored exactly. A simple illustration is the rounding of the fractional part to 10 binary digits in the decimal/binary conversion of Table 3.4.

If a number has a precise value x and a rounded value $x + \epsilon x$ then the difference ϵx is the rounding error and $|\epsilon|$ may be called the rounding error level as it is the ratio of the magnitude of the rounding error to the magnitude of the number. If a floating-point number is stored with a mantissa of $t + 1$ bits then $|\epsilon x| \leqslant 2^{b-t-1}$. But since $|x| \geqslant 2^{b-1}$ it follows that

$$|\epsilon| \leqslant 2^{-t} \tag{3.6}$$

This shows that the maximum possible value of $|\epsilon|$, called the *relative precision*, is dependent only on the number of digits in the mantissa. There are usually sufficient bits in the mantissa for errors of this magnitude to be insignificant. However, in some numerical computations errors may accumulate or magnify as the calculation proceeds so that their effect is much more significant in the final answer, even to the extent of invalidating the results.

In order to appreciate how this error growth can occur consider arithmetical operations involving two numbers x_1 and x_2 which are represented in floating-point form as $x_1 + e_1x_1$ and $x_2 + e_2x_2$. Arithmetical operations are usually performed in a double-length register (i.e. with twice as many bits as in the mantissa of each operand) giving a double-length result which is consistent with the stored values of the operands $x_1(1 + e_1)$ and $x_2(1 + e_2)$. Thus for multiplication the double-length product would be

$$x_1(1 + e_1)x_2(1 + e_2) \simeq x_1x_2(1 + e_1 + e_2) \tag{3.7}$$

However, the result would then have to be rounded in order to enter it into a single-length store. If p is the value of the exact product x_1x_2, then designating the computed product as $p(1 + e_p)$ it can be shown that

$$e_p \simeq e_1 + e_2 + \epsilon \tag{3.8}$$

where ϵ is the rounding error satisfying the inequality (3.6). Alternatively, if an attempt is made to compute the quotient $q = x_1/x_2$ or the sum $s = x_1 + x_2$ the stored results, designated as $q(1 + e_q)$ or $s(1 + e_s)$, will be such that

$$e_q \simeq e_1 - e_2 + \epsilon \tag{3.9}$$

or

$$e_s \simeq e_1\left(\frac{x_1}{x_1 + x_2}\right) + e_2\left(\frac{x_2}{x_1 + x_2}\right) + \epsilon \tag{3.10}$$

respectively.

It is clear from an examination of these last three formulae that the error level in the stored result of an arithmetical operation may be higher than in the operands. However, for multiplication, division and addition (but not subtraction) the resulting error level cannot be much larger than the combined error levels in the operands.

On a computer, arithmetical expressions are converted into a series of machine instructions involving single arithmetical operations which are obeyed sequentially. For instance, a statement requesting the computation of y according to

$$y = x_1^2 + x_2^2 + \cdots + x_{10}^2 \tag{3.11}$$

would be translated by the compiler into a series of multiplications of the form

$$p_i = x_i \times x_i \tag{3.12}$$

interspersed by a series of additions which update the current sum, i.e.

$$s_i = p_i + s_{i-1} \tag{3.13}$$

Obviously there is a possibility that the error level gradually increases as the computations proceed.

Consider the case where values of x_i have been read into the computer and are therefore likely to have rounding errors, $x_i e_i$, due to decimal binary conversion such that $|e_i| < 2^{-t}$. If all the values of x_i have a similar magnitude it can be shown that the error level of the computed value of y (equation 3.11) must satisfy

$$|e_y| < 8.5 \times 2^{-t} \tag{3.14}$$

However, the probability that the error comes close to this limit is very remote. By making a statistical assessment it may be deduced that the probable error is about 1.03×2^{-t}. Since the probable error in the operands x_i is $\frac{1}{3} \times 2^{-t}$, the error level will probably be magnified by a factor of about 3 due to the arithmetical computation in this particular case.

The result of the error analysis for subtraction is given by equation (3.10) where x_1 and x_2 are assumed to have opposite signs. Since at least one of the factors $x_1/(x_1 + x_2)$ and $x_2/(x_1 + x_2)$ must have a modulus greater than unity the

Table 3.6 Error magnification due to cancellation — an example in decimal floating-point arithmetic

	Exact form	Rounded form
Subtraction of	0.5279362×10^2	0.5279×10^2
	0.5271691×10^2	0.5272×10^2
Unshifted difference	0.0007671×10^2	0.0007×10^2
Shifted difference	0.7671×10^{-1}	0.7000×10^{-1}

error level present in either one or both of the operands must be projected, in a magnified form, into the computed difference. This magnification will be most pronounced if the numbers being subtracted almost cancel each other. In cases of severe cancellation the computed result may have no accuracy at all.

The right-hand column of Table 3.6 gives the decimal floating-point subtractions of two almost equal four-figure decimal numbers. Because the result is shifted up, no rounding is required and hence the stored result is correct in relation to the stored operands. However, if the operands are themselves rounded versions of the numbers shown in the left-hand column, it is clear that the error level has been magnified by three decimal places through the subtraction. If, on the other hand, the four-figure operands had been computed in such a way that the lowest two decimal digits were inaccurate, then the computed result would be meaningless.

In the computation of an element of a matrix product according to the formula

$$c_{ij} = a_{i1}b_{1j} + a_{i2}b_{2j} + \cdots + a_{ip}b_{pj} \tag{3.15}$$

large error magnification will occur due to cancellation if the magnitude of c_{ij} is much less than any of the magnitudes of the component products of the form $a_{ik}b_{kj}$. However, if c_{ij} is much less significant than other elements of **C**, its low accuracy may not be detrimental to the overall accuracy of the computation. In matrix multiplication operations the loss of accuracy is usually insignificant. However, loss of accuracy is often a serious problem in the solution of linear equations and will be considered in Chapter 4.

3.3 ARRAY STORAGE FOR MATRICES

When programming in a high level language, vectors and matrices may be stored in a computer in the form of arrays, and access to particular elements will be by the use of subscripts and subscript expressions. If two-dimensional arrays are used to store matrices, simple matrix operations can be programmed easily. Thus if stores have been declared as follows:

	ALGOL	FORTRAN
matrix dimensions	**integer** n;	INTEGER N
square matrices **A,B,C**	**array** $A,B,C[1:n,1:n]$;	REAL A(N,N),B(N,N),C(N,N)
working stores	**integer** i,j; **real** x;	INTEGER I,J,LL
		REAL X

and, if the values of matrices A and B have been allocated, a program segment to implement the matrix addition

$$\mathbf{C} = \mathbf{A} + \mathbf{B} \tag{3.16}$$

will be:

MATRIX ADDITION

ALGOL	FORTRAN
for i :=1 **step** 1 **until** n **do**	DO 1 I=1,N
for j :=1 **step** 1 **until** n **do**	DO 1 J=1,N
$C[i,j]$:=$A[i,j]$+$B[i,j]$;	1 C(I,J)=A(I,J)+B(I,J)

Alternatively, a program segment to replace the matrix A by its own transpose will be:

$n \times n$ MATRIX TRANSPOSE

ALGOL	FORTRAN
for i :=2 **step** 1 **until** n **do**	DO 1 I=2,N
for j :=1 **step** 1 **until** $i-1$ **do**	LL=I−1
begin x :=$A[i,j]$;	DO 1 J=1,LL
$A[i,j]$:=$A[j,i]$;	X=A(I,J)
$A[j,i]$:=x	A(I,J)=A(J,I)
end $n \times n$ matrix transpose;	1 A(J,I)=X

Table 3.7 An example of array storage allocation

Case	For matrix A	For matrix B	For matrix C	Total storage requirement
1	5 x 3	3 x 4	4 x 2	35
2	50 x 10	10 x 20	20 x 10	900
3	200 x 10	10 x 400	400 x 5	8000
4	500 x 10	10 x 200	200 x 5	8000
Minimum blanket arrays	500 x 10	10 x 400	400 x 10	13000

In FORTRAN the array storage is allocated at the time that the program is compiled and hence the dimensions have to be specified in numerical form within the program. For example, if a program requires to allocate store for matrices **A**, **B** and **C** whose dimensions in four different cases to be analysed are as shown in Table 3.7, then the program will have to be ammended before each case is run in order to modify the array dimensions. A way of avoiding this is to allocate *blanket* arrays which have dimensions always greater than or equal to the corresponding matrix dimensions and, in writing the program, to allow for the fact that the matrices may only partially fill the arrays. Table 3.7 shows the minimum dimensions of the blanket arrays which are necessary to cater for the given cases. In ALGOL the array storage is allocated *dynamically*, i.e. during the execution of the program. This means that matrix dimensions may be read in as part of the data or even calculated from information within the computer, thus providing more flexibility for the programmer.

3.4 MATRIX MULTIPLICATION USING TWO-DIMENSIONAL ARRAYS

The most straightforward way of forming a matrix product

$$\underset{m \times n}{\mathbf{C}} = \underset{m \times l}{\mathbf{A}} \; \underset{l \times n}{\mathbf{B}} \tag{3.17}$$

using two-dimensional array storage is to form each element of the product matrix in turn according to the standard formula

$$c_{ij} = \sum_{k=1}^{l} a_{ik} b_{kj} \tag{3.18}$$

It is good programming technique to avoid unnecessary accessing of arrays, and hence a working store (say x) should be used to accumulate the element c_{ij}.

If the store allocations (assuming **A**, **B** and **C** to be real matrices) are:

	ALGOL	FORTRAN
matrix dimensions	**integer** m,l,n;	INTEGER M,L,N
matrices **A**, **B** and **C**	**array** $A[1:m,1:l]$, $B[1:l,1:n]$, $C[1:m,1:n]$;	REAL A(M,L),B(L,N), C(M,N)
working stores	**integer** i,j,k; **real** x;	INTEGER I,J,K REAL X

and the values of matrices A and **B** entered into their allocated stores, a program segment to perform the matrix multiplication is:

STANDARD MATRIX MULTIPLICATION

ALGOL

```
for i:=1 step 1 until m do
for j:=1 step 1 until n do
begin x:=0;
      for k:=1 step 1 until l do
      x:=A[i,k]*B[k,j]+x;
      C[i,j]:=x
end standard matrix multiplication;
```

FORTRAN

```
     DO 1 I=1,M
     DO 1 J=1,N
     X=0.0
     DO 2 K=1,L
   2 X=A(I,K)*B(K,J)+X
   1 C(I,J)=X
```

It is possible to reorganize the matrix multiplication procedure in such a way that the inner loop performs all the operations relevant to one element of A. Since a particular element a_{ik} may contribute to all of the elements on row i of the product matrix C, it is necessary to accumulate these elements simultaneously. Once the contributions from all the elements on row i of matrix A have been determined, then row i of matrix C will be complete. The multiplication sequence is illustrated schematically in Figure 3.1, a program segment for which is

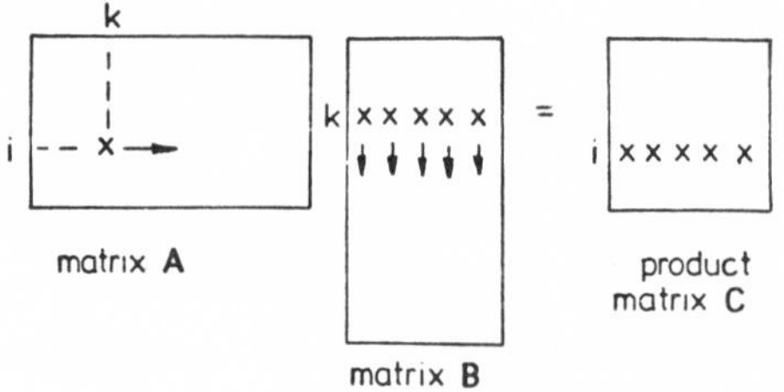

Figure 3.1 Scheme for simultaneous row matrix multiplication

as follows:

SIMULTANEOUS ROW MATRIX MULTIPLICATION

ALGOL	FORTRAN
for i:=1 **step** 1 **until** m **do**	DO 1 I=1,M
begin **for** j:=1 **step** 1 **until** n **do**	DO 2 J=1,N
$C[i,j]$:=0;	2 C(I,J)=0.0
for k:=1 **step** 1 **until** l **do**	DO 1 K=1,L
begin x:=$A[i,k]$;	X=A(I,K)
if $x \neq 0$ **then**	IF(X.EQ.0.0)GO TO 1
for j:=1 **step** 1 **until** n **do**	DO 3 J=1,N
$C[i,j]$:=$B[k,j] * x + C[i,j]$	C(I,J)=B(K,J)*X+C(I,J)
end row i	3 CONTINUE
end simultaneous row matrix multiplication;	1 CONTINUE

It will be noted that a test has been included to miss out multiplications associated with any element a_{ik} which is zero.

If the entire matrix multiplication can be performed within the available main store then there will be little difference in efficiency between the two multiplication strategies. The standard procedure will probably be preferred for fully populated matrices. If matrix **A** has a significant proportion of zero elements then the simultaneous row procedure could be more efficient than the standard procedure (even if the latter is modified to include a corresponding test to avoid zero multiplications), but the relative efficiency will depend on how the inner loops of the two schemes are implemented on the computer (see section 3.5). It is possible to devise a scheme for simultaneous column matrix multiplication as illustrated by Figure 3.2. This procedure may be beneficial in cases where matrix **B** has a significant proportion of zero elements. Yet another possible scheme for matrix multiplication is to accumulate all elements of the product matrix simultaneously. In this scheme the outer loop cycles counter k. For each value of k the full contributions due to column k of **A** and row k of **B** are added into the product matrix by cycling the i and j counters. This is illustrated in Figure 3.3.

In cases where block storage transfers between the main store and the backing store have to be carried out during a matrix multiplication, either explicitly or

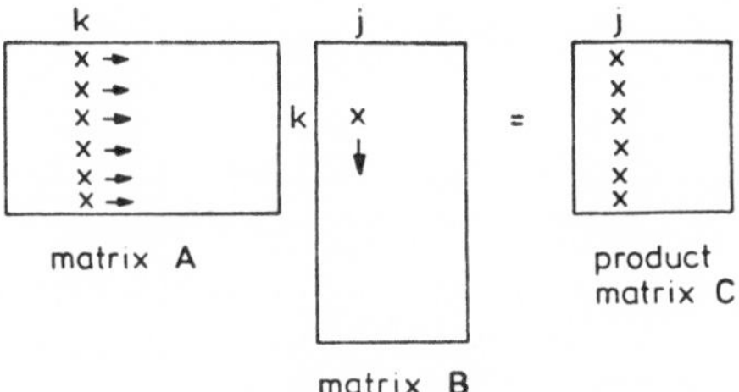

Figure 3.2 Scheme for simultaneous column matrix multiplication

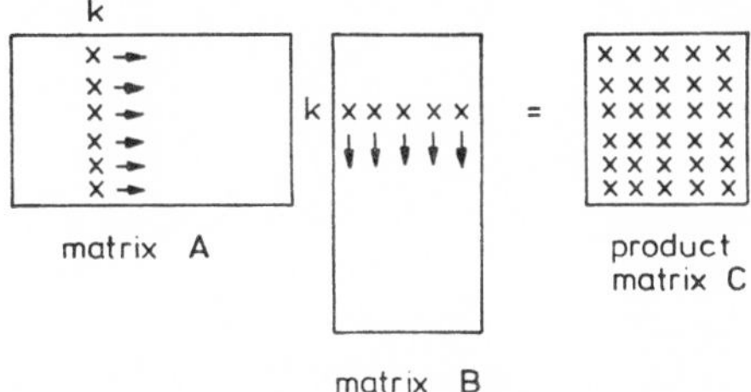

Figure 3.3 Scheme for simultaneous row and column matrix multiplication

implicitly, the choice of multiplication scheme may affect execution times very substantially (see section 3.7).

3.5 ON PROGRAM EFFICIENCY

A programmer has the choice of writing his program in a high level language such as ALGOL or FORTRAN or in a machine-oriented language, usually either machine code or a mnemonic version of machine code. If the program is written in a high level language the *compiler* produces an equivalent set of instructions in machine code, known as the *object program*, which the machine then executes. Applications programs are almost universally written in a high level language because they are easier to write, to test, to describe and to modify than programs written in a machine-oriented language. Furthermore, it is easy to implement an ALGOL or FORTRAN program on a computer with a different machine code to the computer for which it was originally written. However, it should be appreciated that an object program compiled from a high level language is unlikely to be the most efficient machine code program for the particular task, especially when the compiler is not an optimizing one.

Generally, each machine code instruction specifies an operation and also a storage location to which the operation refers. Thus a FORTRAN statement

```
A=B+C*D                                                  (3.19)
```

might be compiled as four machine code instructions which perform the following operations:

(a) Transfer the number in store C to a register in the arithmetic unit.
(b) Multiply the value of the register by the number in store D.
(c) Add the number in store B to the register.
(d) Transfer the number in the register to store A.

When arrays are declared, they are each allocated a consecutive sequence of stores. In FORTRAN a two-dimensional array is stored by columns. Thus if, at the time of compilation, the declaration REAL A(4,3) is encountered when, say, 593 is the next available storage location, stores will be allocated as follows:

593	594	595	596	597	598	599	600	601	602	603	604
A(1,1)	A(2,1)	A(3,1)	A(4,1)	A(1,2)	A(2,2)	A(3,2)	A(4,2)	A(1,3)	A(2,3)	A(3,3)	A(4,3)

(In ALGOL the store allocations will probably be similar, although it could be by rows instead of by columns.) When an assignment statement containing an array reference is compiled, the explicit value of the subscripts will not normally be available. For the 4 × 3 array shown above, A(I,J) will be located in store 588+I+4J. In general, for an $m \times n$ array the element (i, j) will be the $(i + mj - m)$-th member of the stored sequence. Thus for a FORTRAN statement

$$\text{A(I,J)=X} \tag{3.20}$$

the object program will need to precede the instructions to transfer the value of X to A(I,J) by machine code instructions to compute the address of the storage location A(I,J). The address computation will involve an integer multiplication and two additions, and will also contain a subtraction if m has not been subtracted during compilation.

In translating an assignment statement, a standard compiler does not examine neighbouring statements to decide if short-cuts may be made in the object program. Thus, if the assignment statement (3.19) is preceded by a statement to compute the value of B, the compiler does not optimize the sequence of machine codes for A=B+C*D to allow for the fact that B will already be present in an arithmetic register at the start of the sequence. Sometimes there will be scope for improvement within the compiled codes corresponding to a single assignment statement, e.g. the inner loop instruction within the FORTRAN simultaneous row matrix multiplication:

$$\text{C(I,J)=B(K,J)*X+C(I,J)} \tag{3.21}$$

The object code compiled from this statement will contain duplicate sets of codes to compute the address of C(I,J). A further way in which a machine code program could normally be written so that it is more efficient than the equivalent compiled object program is by making use of the systematic order in which array stores are usually called in matrix numerical processes. In particular, the integer multiplication to locate the storage addresses of a two-dimensional array element can nearly always be avoided.

At the expense of using more computer time for compilation, optimizing compilers save unnecessary duplication in the object program by inspecting the structure of the statement sequence being interpreted. For instance, an optimizing compiler might avoid duplications in finding the storage location C(I,J) when interpreting instruction (3.21). The effectiveness of an optimizing compiler depends on how many optimizing features are included and how well the original program is written.

Programs that involve large-order matrix operations are likely to spend much of the total computation time repeatedly executing just a few instructions within the inner loops of the matrix procedures. It is therefore possible to obtain moderate savings in execution time just by ensuring that these inner loops are either written in efficient machine code form or are compiled in a fairly efficient way. Of course, if well-tested and efficient library routines are available for the relevant matrix operations, these should be used.

3.6 ONE-DIMENSIONAL STORAGE FOR MATRIX OPERATIONS

A programmer is at liberty to store matrices in one-dimensional rather than two-dimensional arrays. An $m \times n$ matrix will require a one-dimensional array having mn elements and, if stored by rows, element (i, j) of the matrix will be stored in the $(ni + j - n)$-th location. Using this type of storage scheme it is often possible to produce a more efficient program than if two-dimensional array storage is used.

Consider the matrix addition, equation (3.16). With stores declared as follows:

	ALGOL	FORTRAN
length of arrays	**integer** *nn*;	INTEGER NN
matrices **A**,**B**,**C**	**array** *A*,*B*,*C*[1:*nn*];	REAL A(NN),B(NN),C(NN)
working store	**integer** *i*;	INTEGER I

a program segment for matrix addition is:

MATRIX ADDITION IN ONE-DIMENSIONAL STORE

ALGOL	FORTRAN
for *i*:=1 **step** 1 **until** *nn* **do**	DO 1 I=1,NN
C[*i*]:=*A*[*i*]+*B*[*i*];	1 C(I)=A(I)+B(I)

which is even simpler than the previous segment and also will produce a more efficient object code.

The standard matrix multiplication (equation 3.17) may be performed using the following storage declarations:

	ALGOL	FORTRAN
matrix dimensions	**integer** *m*,*l*,*n*;	INTEGER M,L,N,
length of arrays		ML,LN,MN
matrices **A**, **B** and **C**	**array** *A*[1:*m***l*],	REAL A(ML),
	B[1:*l***n*],*C*[1:*m***n*];	B(LN),C(MN)
working stores	**integer** *i*,*j*,*ka*,*kb*,*kc*,*kal*,*kau*;	INTEGER I,J,KA,KB,
		KC,KAL,KAU

The counters KA, KB and KC are used as subscripts when calling the arrays A, B and C respectively. If a row-wise storage scheme is used, the elements a_{ik} involved in one summation $\Sigma a_{ik} b_{kj}$ are found in consecutive addresses within array A, but the elements b_{kj} are spaced at intervals of n in array B. If integer stores KAL and KAU hold the appropriate lower and upper values of KA, the inner loop to compute the value of the element c_{ij} may be written as

ALGOL	FORTRAN
for *ka*:=*kal* **step** 1 **until** *kau* **do**	DO 2 KA=KAL,KAU
begin *x*:=*A* [*ka*] **B* [*kb*] +*x*;	X=A(KA)*B(KB)+X
kb:=*kb*+*n*	2 KB=KB+N
end forming *C* [*i,j*] ;	

Assignment statements to initiate counters, perform the required loops and store the results in the appropriate locations of array C may be included, so giving a full program segment to perform standard matrix multiplication in a one-dimensional store as:

MATRIX MULTIPLICATION IN ONE-DIMENSIONAL STORE

ALGOL

```
kc:=1;
kal:=0;
for i:=1 step 1 until m do
begin kau:=kal+l;
      kal:=kal+1;
      for j:=1 step 1 until n do
      begin kb:=j;
            x:=0;
            for ka:=kal step 1 until kau do
            begin x:=A[ka]*B[kb]+x;
                  kb:=kb+n
            end forming C[i,j];
            C[kc]:=x;
            kc:=kc+1
      end row of C;
end one-dimensional matrix multiplication;
```

FORTRAN

```
      KC=1
      KAL=0
      DO 1 I=1,M
      KAU=KAL+L
      KAL=KAL+1
      DO 1 J=1,N
      KB=J
      X=0.0
      DO 2 KA=KAL,KAU
      X=A(KA)*B(KB)+X
    2 KB=KB+N
      C(KC)=X
    1 KC=KC+1
```

In FORTRAN the EQUIVALENCE and COMMON statements may be used to enable a matrix stored as a two-dimensional array to be called as if it were a one-dimensional array stored by columns (rather than by rows as in the above example).

3.7 ON THE USE OF BACKING STORE

For computations involving large matrices it is likely that the main (i.e. fast access) store will not have sufficient capacity to hold all of the information required in a specific computation. In such circumstances at least some of the matrices will have to be held on backing store (magnetic tape, drums or discs) and transferred to and from the main store as needed. Because these items of equipment operate mechanically rather than electronically the transfer times are much greater than the execution times required for standard arithmetical operations. An individual transfer operation involves reading the information from a consecutive sequence of backing stores to specified locations in the main store, or conversely writing information from specified consecutive main stores into a consecutive sequence of backing stores. On computers which have a time-sharing facility, the control of the computer may be switched to other tasks while backing store transfers are proceeding. However, even in this case, there is a large penalty in computer time associated with transfer operations. If backing store has to be used during matrix operations, the number of transfers required could easily be the most important factor in assessing computing time.

As an illustration of how the number of transfers can be affected by the organization of a matrix operation, consider the evaluation of the matrix product $\mathbf{C} = \mathbf{AB}$ where $\mathbf{A}$ and $\mathbf{B}$ are fully populated 100×100 matrices. Suppose also that the main store has 1,500 available locations, which have been allocated in blocks of 500 to segments of each of the three matrices. The matrix multiplication procedure will transfer the first 500 elements of both $\mathbf{A}$ and $\mathbf{B}$ from the backing store and proceed as far as possible with the computation. In the case of the standard matrix multiplication scheme with matrices stored by columns, only five multiplications will be possible before another block of matrix $\mathbf{A}$ will need to be read into the main store. More than 200,000 transfers would be required to complete the matrix multiplication. In contrast, if simultaneous column multiplication were to be adopted using the same matrix storage scheme then only 2,040 transfers would be needed (and only 440 transfers if five columns were to be formed simultaneously).

If the available main store is sufficiently large to hold the whole of matrix $\mathbf{A}$ and at least one column each of matrices $\mathbf{B}$ and $\mathbf{C}$, then simultaneous column multiplication may be implemented with very few transfers. In the above example, if 11,000 storage locations are available in the main store, of which 10,000 are used to store matrix $\mathbf{A}$ and the remaining 1,000 are divided into two blocks of 500 to accommodate segments of matrices $\mathbf{B}$ and $\mathbf{C}$, then simultaneous column multiplication could be completed using a total of only forty transfer operations.

If matrix $\mathbf{B}$ is small and able to fit into the available main store, then simultaneous row multiplication should be considered, with matrices $\mathbf{A}$ and $\mathbf{C}$ stored by rows. Alternatively, if the product matrix $\mathbf{C}$ is small and able to fit into the available main store, then simultaneous row and column multiplication should be considered, with matrix $\mathbf{A}$ stored by columns and matrix $\mathbf{B}$ stored by rows.

In some computers a *paging system* automatically carries out storage transfers to and from the backing store as and when required. The programmer is relieved of the task of organizing and specifying the transfer operations and writes the program in a working store which is much larger than the available main store. However, as storage transfers are still carried out, it is still important to perform matrix operations in a way which is economical on the number of transfers. Hence, for instance, for the multiplication of large matrices stored by columns, the simultaneous column multiplication method will probably operate much more efficiently than the standard matrix multiplication method.

3.8 SPARSE STORAGE

Where matrices contain a significant proportion of zero elements it is possible to test whether each element is zero before carrying out arithmetical operations so that all trivial operations can be avoided. This technique is useful where the matrices are not large, i.e. having order less than 100. However, where larger sparse matrices are encountered it may well be desirable to avoid storing the zero elements. For matrices of order more than 1000 then it becomes virtually essential to store sparse matrices in some kind of sparse store. The rest of this chapter will be concerned with various sparse storage schemes and the resulting implications with regard to the execution of matrix operations.

If the definition of a sparse storage scheme is sufficiently broad to include any scheme which does not allocate a normal storage space to every element of a matrix, then there are a large variety of schemes which have been, or could be, implemented. Nearly all schemes make use of two storage components:

(a) A facility for storing either the non-zero elements or an area of the matrix which includes all of the non-zero elements. This usually takes the form of a one-dimensional array and will be called the *primary array*.

(b) A means of recognizing which elements of the matrix are stored in the primary array. This usually takes the form of one or more one-dimensional arrays of integer identifiers, which will be called the *secondary store.*

The more general schemes, which permit the storage of any sparse matrix, will be considered first, and the more specific schemes, such as band schemes, will be considered later.

3.9 BINARY IDENTIFICATION

A novel scheme which makes use of the binary nature of computer storage is to record the pattern of non-zero elements of the matrix as the binary digits of secondary array elements. The matrix

$$\mathbf{A} = \begin{bmatrix} 0 & 0 & 0 & 0 & 0 \\ 0 & 0 & 2.67 & 0 & 3.12 \\ -1.25 & 0.29 & 0 & 0 & 2.31 \end{bmatrix} \tag{3.22}$$

has a pattern of non-zero elements indicated by the binary sequence

$$\overbrace{0\ 0\ 0\ 0\ 0}^{\text{row 1}}\quad \overbrace{0\ 0\ 1\ 0\ 1}^{\text{row 2}}\quad \overbrace{1\ 1\ 0\ 0\ 1}^{\text{row 3}}$$

Hence this matrix could be stored by means of a primary array containing the five non-zero elements $\{2.67 \quad 3.12 \quad -1.25 \quad 0.29 \quad 2.31\}$ and a secondary store containing the binary sequence shown above. For this matrix the binary sequence could be held in a word with fifteen or more bits; however, for larger matrices a number of words would be required.

If an $m \times n$ matrix has r as the ratio of the number of non-zero elements to total elements and if two words each of γ bits are required to store each non-zero element, then the primary array will occupy $2mnr$ words and the secondary array will occupy approximately mn/γ words. Since $2mn$ words would be required to store the matrix in the conventional way, the storage compaction of the binary identification scheme may be expressed as the ratio c where

$$2mnc \simeq 2mnr + \frac{mn}{\gamma} \tag{3.23}$$

giving

$$c \simeq r + \frac{1}{2\gamma} \tag{3.24}$$

This storage scheme differs from other sparse schemes in that some storage space (a single bit in the secondary store) is allocated to every zero element. It is therefore less efficient for very sparse matrices than schemes which do not contain any storage allocation associated with zero elements. Moreover, the main drawback is the difficulty of implementing matrix operations with matrices stored in this way. Normally such implementations would produce much less efficient programs than could be achieved by using other sparse storage schemes.

3.10 RANDOM PACKING

Every non-zero element entered into the primary array may be identified by specifying its row and column numbers in the corresponding locations of two secondary arrays. Since each element is individually identified it is possible to store them in a random order. Thus matrix (3.22) could be represented by

$$\left.\begin{array}{ll} \text{Real array A} = \{0.29 & 3.12 \quad -1.25 \quad 2.67 \quad 2.31 \quad 0 \quad - \quad -\} \\ \text{Integer array IA} = \{\ 3 & 2 \quad\quad 3 \quad\quad 2 \quad\quad 3 \quad\quad 0 \quad - \quad -\} \\ \text{Integer array JA} = \{\ 2 & 5 \quad\quad 1 \quad\quad 3 \quad\quad 5 \quad\quad 0 \quad - \quad -\} \end{array}\right\} \tag{3.25}$$

One advantage of random packing is that extra non-zero elements can be added to the matrix by inserting them at the end of the list without disturbing the other items. It is often convenient to have a null entry in a secondary array to indicate termination of the list.

It is easy to construct the coefficient matrix for a network problem in this kind of store using a technique similar to that described in section 2.3. Because the diagonal elements are the only ones involving additions they may be accumulated in the first n storage locations. When off-diagonal elements are formed they may be added to the end of the list. Thus, if a node conductance matrix were formed from the branch data shown in Table 2.1, the resulting matrix would be stored as

$$\left.\begin{array}{lccccccccccccc} \text{Real array A} = \{ & 8.3 & 13.7 & 10.5 & 9.2 & -3.2 & -3.2 & -2.7 & -2.7 & -5.6 & -5.6 & -2.7 & -2.7 & 0\} \\ \text{Integer array IA} = \{ & 1 & 2 & 3 & 4 & 1 & 2 & 1 & 4 & 2 & 3 & 3 & 4 & 0\} \\ \text{Integer array JA} = \{ & 1 & 2 & 3 & 4 & 2 & 1 & 4 & 1 & 3 & 2 & 4 & 3 & 0\} \end{array}\right\} \tag{3.26}$$

One of the few operations which can easily be implemented with matrices stored in this way is the multiplication

$$\mathbf{y} = \mathbf{A}\mathbf{x} \tag{3.27}$$

where **A** is a sparse randomly packed matrix, and **x** and **y** are column vectors of orders n and m respectively, stored as standard one-dimensional arrays. ALGOL and FORTRAN versions of an algorithm to form this product are as follows:

POSTMULTIPLICATION OF A RANDOM MATRIX BY A VECTOR

ALGOL

```
for i:=1 step 1 until m do
y[i]:=0;
k:=1;
for i:=ia[k] while i≠0 do
begin  y[i]:=A[k]*x[ja[k]]+y[i];
       k:=k+1
end;
```

FORTRAN

```
     DO 1 I=1,M
   1 Y(I)=0.0
     K=1
   3 I=IA(K)
     IF(I.EQ.0)GO TO 2
     J=JA(K)
     Y(I)=A(K)*X(J)+Y(I)
     K=K+1
     GOTO 3
   2 - - -
```

A multiplication procedure of this form may be used in the conjugate gradient method for the solution of simultaneous equations (section 6.13), and hence network equations may be constructed and solved with the coefficient matrix in random packed form. If the network equations were always symmetric the storage requirement could be reduced by almost one-half by storing only one element of each off-diagonal pair. The multiplication procedure would have to be modified accordingly.

3.11 THE USE OF ADDRESS LINKS

Operations using randomly packed matrix storage may be extended considerably if links are specified to allow matrices to be scanned systematically. Consider the node conductance matrix (3.26) together with an extra integer array as follows:

Address	1	2	3	4	5	6	7	8	9	10	11	12	13
Real array A =	{8.3	13.7	10.5	9.2	−3.2	−3.2	−2.7	−2.7	−5.6	−5.6	−2.7	−2.7	0}
Integer array IA =	{1	2	3	4	1	2	1	4	2	3	3	4	0}
Integer array JA =	{1	2	3	4	2	1	4	1	3	2	4	3	0}
Integer array ILINK =	{5	9	11	0	7	2	6	12	10	3	8	4	0}

(3.28)

The array ILINK is constructed so that the non-zero elements of the matrix can be scanned in a row-wise sense. For example, if the first element a_{11} at address 1 is inspected, ILINK (1) points to the next non-zero element on row 1, i.e. a_{12} at address 5. Similarly, ILINK (5) points to a_{14} at address 7. Since there are no more non-zero elements on row 1, ILINK (7) points to the first non-zero element on row 2. Normally it is useful if the chain of elements can be entered at several points.

In the linked store (3.28) the addresses of the diagonal elements may be used as entry points. Thus, to find element a_{ij} it is possible to break into the chain either at address i, if $i < j$, or at address $i - 1$, if $j < i$, and continue along the chain until either element a_{ij} is reached or it is shown to be zero through being absent. In some cases where row address links are used it is possible to dispense with the row identifiers.

Apart from the forward row-wise linkage illustrated above, it is possible to specify reverse row-wise linkage, or forward or reverse column-wise linkage. The type of linkage (or linkages) required for a particular matrix will depend on the way (or ways) in which the matrix is to be scanned during matrix operations. However, address links should be used as little as possible because of the following disadvantages:

(a) A method is required to form the links.

(b) Extra storage space is required for the links unless they replace row or column identifiers.

(c) Extra computing time is used inspecting link addresses.

(d) Unless the matrix can be held entirely within the main store, frequent backing store transfers are likely to be necessary in matrix operations using the links.

3.12 SYSTEMATIC PACKING

If the elements of a sparse matrix have been read in or constructed in a systematic order or have been sorted into a systematic order there is no need to adopt both row and column indices for each element. For row-wise packing it is the row indices which may be dispensed with, except insofar as it is necessary to specify where each row begins.

The use of row addresses

The address of the first non-zero element in each row may be specified in a separate integer array. For example, matrix (3.22) could be represented by

$$\left.\begin{array}{lllllll} \text{Real array A} & = \{2.67 & 3.12 & -1.25 & 0.29 & 2.31\} \\ \text{Integer array JA} & = \{\ 3 & 5 & 1 & 2 & 5\ \} \\ \text{Integer array ISTART} & = \{\ 1 & 1 & 3 & 6\ \} \end{array}\right\} \tag{3.29}$$

The array of row addresses ISTART has been constructed so that the number of non-zero elements in row i is ISTART(I+1)–ISTART(I); hence for a matrix with m rows, ISTART will contain $m + 1$ entries.

The use of dummy elements

Either in place of row addresses, or as an adjunct to them, dummy elements may be included to indicate the start of each row and the end of the matrix. Several formats are possible for the dummy element and the corresponding entry in the column index array. For instance, a zero entry in the array JA could mark the presence of a dummy element and the dummy element itself could specify the row number (or be zero to indicate the end of the matrix). Hence matrix (3.22) would appear as

$$\left.\begin{array}{lllllllll}\text{Real array A} = \{ & \boxed{2} & 2.67 & 3.12 & \boxed{3} & -1.25 & 0.29 & 2.31 & \boxed{0}\ \} \\ \text{Integer array JA} = \{ & \boxed{0} & 3 & 5 & \boxed{0} & 1 & 2 & 5 & \boxed{0}\ \} \end{array}\right\} \tag{3.30}$$

Alternatively, the row number could be specified in the integer array and distinguished from column numbers by a change of sign. In this case the dummy element itself would not be used. Matrix (3.22) would appear as

$$\left.\begin{array}{lllllllll}\text{Real array A} = \{ & \boxed{\times} & 2.67 & 3.12 & \boxed{\times} & -1.25 & 0.29 & 2.31 & \boxed{\times}\ \} \\ \text{Integer array JA} = \{ & \boxed{-2} & 3 & 5 & \boxed{-3} & 1 & 2 & 5 & \boxed{0}\ \} \end{array}\right\} \tag{3.31}$$

In some cases (e.g. for the sparse matrix multiplication of section 3.14) it is easier to program matrix operations if the integer identifier for a dummy element is larger rather than smaller than the column indices. This may be achieved by making the identifier equal to the row number plus a constant, the constant being larger than the largest column number. In a similar way it may be convenient to use an even larger number to indicate the end of the matrix. Thus, matrix (3.22) would appear as

$$\left.\begin{array}{lllllllll}\text{Real array A} = \{ & \boxed{\times} & 2.67 & 3.12 & \boxed{\times} & -1.25 & 0.29 & 2.31 & \boxed{\times}\ \} \\ \text{Integer array JA} = \{ & \boxed{10002} & 3 & 5 & \boxed{10003} & 1 & 2 & 5 & \boxed{99999}\ \} \end{array}\right\} \tag{3.32}$$

A further alternative use for the dummy element is to specify in the column index position the number of elements in the next row. If a dummy element is included for each row, even if it is null, then there is no need to record the row number. Thus matrix (3.22) could be stored as

$$\left.\begin{array}{llllllll}\text{Real array A} = \{ & \boxed{\times\ \ \times} & 2.67 & 3.12 & \boxed{\times} & -1.25 & 0.29 & 2.31\ \} \\ \text{Integer array JA} = \{ & \boxed{0\ \ 2} & 3 & 5 & \boxed{3} & 1 & 2 & 5\ \} \end{array}\right\} \tag{3.33}$$

The number of rows in the matrix will need to be specified elsewhere.

In any of the dummy element schemes shown above except the first, the dummy elements in the real array may be omitted to save storage space. However, if this is done the addresses of the elements and their column indices will not coincide.

Semisystematic packing

With each of the above storage schemes it is possible to relax the constraint that the elements on a given row are in their correct sequence without altering the storage scheme. This could be described as a semisystematic packing.

3.13 SOME NOTES ON SPARSE PACKING SCHEMES

Zero elements

Although zero elements will normally be excluded from the packed matrix, all of the above storage schemes may be used when zero elements appear within the packed form.

DO loops

With the row address system packing, as illustrated in (3.29), the sequence of FORTRAN statements:

```
      KP=ISTART(I)
      KQ=ISTART(I+1)-1
      DO 1  K=KP,KQ
        - -
 1    - -
```

will normally set up a cycle to scan through all of the non-zero elements on row i. However FORTRAN compilers insist that DO loops are always executed at least once. Hence, if row i contains no stored elements, the DO loop will not operate correctly unless it is preceded by an IF statement which allows the DO loop to be bypassed in the case where no elements are present. Similarly, it may be necessary to take care with the use of DO loops with packed matrices of other types. The equivalent ALGOL statement:

for k:= *istart*[i] **step** 1 **until** *istart*[i+1]−1 **do** – – –;

will always operate correctly.

Compound identifiers

In the random packing scheme (3.26) it is possible to reduce the storage requirement by combining the two indices for each element so that they can be held in one integer store. A suitable compound identifier would be $\bar{n}i + j$ where $\bar{n}$ is an integer equal to or greater than the total number of columns in the matrix. In a similar way it is possible to avoid the use of dummy elements for systematic packing by using a compound identifier for the first element of each row. For example, matrix (3.22) could be represented by

$$\left.\begin{array}{llllllll} \text{Real array A} & = \{2.67 & 3.12 & -1.75 & 0.29 & 2.31 & \times & \} \\ \text{Integer array JA} & = \{2003 & 5 & 3001 & 2 & 5 & 999999\} & \end{array}\right\} \qquad (3.34)$$

However, unless compound identification yields necessary or highly desirable storage space savings, it should not be used because

(a) extra program will nearly always be required to interpret the compound identifiers and

(b) it must not be used for matrices whose orders are so large that overflow of the integer register would result.

Storage compaction

The most efficient of the packing schemes use one real store and one integer store per non-zero element. If 2γ bits and γ bits are required to store a real and an integer number respectively, and if r is the non-zero element ratio as defined in section 3.9, then the maximum storage compaction would be

$$c \simeq \frac{3r}{2} \tag{3.35}$$

By comparing with equation (3.24) it is seen that a matrix may be compacted into less storage space using an efficient packing scheme than using binary identification if $r < 1/\gamma$. However, it is more convenient to work with packing schemes which require more than the minimum amount of storage space. For random packing with separate row and column identifiers

$$c \simeq 2r \tag{3.36}$$

and for systematic packing with dummy elements

$$c \simeq \frac{3}{2}\left(r + \frac{1}{n}\right) \tag{3.37}$$

The use of mixed arrays

It is possible to use a single array to store both the non-zero elements of the matrix and the identifiers. For instance, the systematic packing scheme (3.31) could be amended to

$$\text{Real array A} = \{-2 \;\; \boxed{3 \;\; 2.67} \;\; \boxed{5 \;\; 3.12} \;\; -3 \;\; \boxed{1 \;\; -1.25} \;\; \boxed{2 \;\; 0.29} \;\; \boxed{5 \;\; 2.31} \;\; 0\} \tag{3.38}$$

where each non-zero element is preceded by its column number and each non-zero row is preceded by minus the row number. This type of storage scheme would be appropriate if the elements of the matrix were integers, and hence there would be no need to store integer identifiers in a real array.

3.14 OPERATIONS INVOLVING SYSTEMATICALLY PACKED MATRICES

Transposing a matrix

An $m \times n$ matrix stored in semisystematic form may be transposed into a separate store such that $\mathbf{B} = \mathbf{A}^T$. The following operations summarize how this may be performed with the matrix and its transpose stored in row-wise dummy element form. Also is given the intermediate stages in the operation if $\mathbf{A}$ is the 3×5 matrix (3.22) stored as in (3.32). An integer array designated ICOUNT, of order $n + 1$ is required for counters.

(a) The elements of the matrix are scanned, counting the number of elements in each column and accumulating the results in ICOUNT, ignoring for the time being the first storage space in ICOUNT. For the example ICOUNT would then contain {x 1 1 1 0 2}.

(b) ICOUNT may then be converted into a specification of the address locations of the dummy elements in the transposed form. Clearly the first entry will be 1. Provided that non-zero elements occur in every row of the transpose the conversion can be implemented in FORTRAN by executing

```
ICOUNT(I+1)=ICOUNT(I)+ICOUNT(I+1)+1
```

recursively from I=1 to n. However, this procedure will need to be modified if any row of the transpose could be null. For the example ICOUNT should then contain {1 3 5 7 7 10}.

(c) In ALGOL it may be possible to declare the primary and secondary arrays for the transpose at this stage to be precisely the correct length. In FORTRAN these arrays will have had to be declared earlier, probably necessitating an estimate of their lengths.

(d) The indices corresponding to the dummy elements may be entered into the secondary array JB. In the example JB will then contain

1	2	3	4	5	6	7	8	9	10
{10001	x	10002	x	10003	x	10005	x	x	99999}

(e) By adding 1 to each of the first n counters they will be ready to be used as pointers to indicate the address of the next free store for each of the rows. In the example ICOUNT becomes {2 4 6 8 8 x}.

(f) The elements of the matrix A should then be scanned again, placing each element into its correct position in the transpose. After placing each element the corresponding row pointer should be increased by one. In the example the first non-zero element in A is $a_{23} = 2.67$. The column number 3 identifies that the element should be placed in the transpose at address ICOUNT (3) = 6. After this has been done and ICOUNT (3) increased by one, the stores will be

	1	2	3	4	5	6	7	8	9	10
B = {	x	x	x	x	x	2.67	x	x	x	x }
JB = {	10001	x	10002	x	10003	2	10005	x	x	99999}
ICOUNT = {	2	4	7	8	8	x}				

On completion of the second scan the transpose will be fully formed. In the example the final result will be

	1	2	3	4	5	6	7	8	9	10
B = {	x	−1.25	x	0.29	x	2.67	x	3.12	2.31	x }
JB = {	10001	3	10002	3	10003	2	10005	2	3	99999}

Sorting a packed matrix into systematic form

The sorting of a randomly packed matrix without address links into row-wise systematically packed form may be carried out by two separate sorting operations. The first operation forms the transposed matrix with row-wise semisystematic packing in a separate store. The process is very similar to the matrix transposition described above, with the integer array counting the number of non-zero elements in each column of the original matrix. However, since the original matrix was in random packing the transpose will appear in semisystematic packed form. The second operation is simply the transposition exactly as described above. The final matrix may overwrite storage space used for the original matrix, provided that sufficient space is available for the dummy elements.

Matrix multiplication

The multiplication of matrices which are systematically packed is a complicated sparse matrix operation. Consider the matrix multiplication **C** = **AB**, where **A** and **B** are row-wise systematically packed matrices with dummy elements (as in 3.32) and the product **C** is to be stored in a similar manner.

The formation of row i of the product **C** only involves row i of matrix **A**, together with those rows of matrix **B** which correspond to the column numbers of non-zero elements in row i of matrix **A** (see Figure 3.4). If there are l such non-zero elements then, since l rows of **B** have to be accessed simultaneously it is necessary to carry pointers to the appropriate address locations in each of these rows. Initially, the pointers are set to the positions of the first non-zero elements in each of the l rows of **B**. (If a particular row of **B** is null the pointer must specify an address occupied by a dummy element.) The column number of the first non-zero element in row i of matrix **C** can then be obtained by scanning the column numbers of the elements of matrix **B** in the pointer positions and noting the least value. Once this column number is known the element itself may be obtained by means of a second scan of the elements in the pointer positions, using only those elements with the correct column numbers. Whenever an element of matrix **B** is used, the corresponding row pointer for **B** should be increased by one. By doing this the

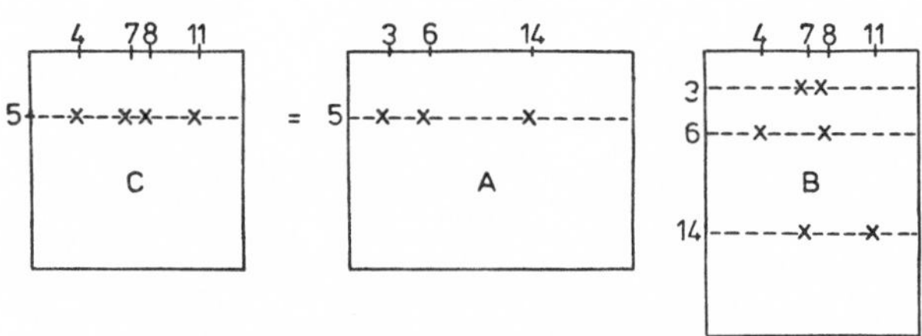

Figure 3.4 Sparse matrix multiplication procedure to form one row of product matrix

next scan of column numbers will reveal the column number of the next non-zero element in row i of matrix **C**. The process can be continued until all of the l rows of **B** are exhausted, by which time the row of matrix **C** will be fully formed.

Assume that the following declarations of storage have been made:

A,B,C	real primary arrays for matrices **A**, **B** and **C**
JA,JB,JC	integer secondary arrays for matrices **A**, **B** and **C**
I	row identifier = 10,000 + i
J	column number for current element of matrix **C**
JNL	next least column number
K	cycle counter 1 to L
KA,KB,KC	pointers for matrices **A**, **B** and **C**
KIB	integer array of pointers for L rows of matrix **B**
L	number of non-zero elements in row i of matrix **A**
X	accumulator for element of **C**
KAK	KA+K (an ALGOL equivalent is not required)

The pointers KA, KIB, KC indicate respectively:

KA	the position of the dummy element starting row i of matrix **A**
KIB	the positions of the next non-zero elements on L rows of matrix **B**
KC	the position of the next free location in matrix **C**.

An algorithm to construct row i of matrix **C** may be represented by:

ROW i OF SPARSE MATRIX MULTIPLICATION

ALGOL

```
        jc[kc]:=i;
        kc:=kc+1;
        jnl:=1;
  scan:j:=jnl;
        x:=0;
        jnl:=99999;
        for k:=1 step 1 until l do
        begin kb:=kib[k];
              if jb[kb]≠j then goto skip;
              x:=A[ka+k]*B[kb]+x;
              kb:=kb+1;
              kib[k]:=kb;
          skip: if jb[kb]<jnl then jnl:=jb[kb]
        end element loop;
        if x≠0 then
        begin C[kc]:=x;
              jc[kc]:=j;
              kc:=kc+1
        end writing element;
        if jnl<10000 then goto scan;
        if j=1 and x=0 then kc:=kc−1;
```

FORTRAN

```
     JC(KC)=I
     KC=KC+1
     JNL=1
   4 J=JNL
     X=0.0
     JNL=99999
     DO 1  K=1,L
     KB=KIB(K)
     IF(JB(KB).NE.J)GO TO 2
     KAK=KA+K
     X=A(KAK)*B(KB)+X
     KB=KB+1
     KIB(K)=KB
   2 IF(JB(KB).LT.JNL)JNL=JB(KB)
   1 CONTINUE
     IF(X.EQ.0.0)GO TO 3
     C(KC)=X
     JC(KC)=J
     KC=KC+1
   3 IF(JNL.LT.10000)GO TO 4
     IF(J.EQ.1.AND.X.EQ.0.0)KC=KC−1
```

Some notes on the matrix multiplication algorithm

(1) The scan used to form an element of C has been combined with the scan used to find the column number of the next element of C.

(2) For the first scan the column number has been set to 1 (but this may not yield an element of C).

(3) Because the row identifiers are larger than any possible row index, the test to find the next least column number is correct, even when some of the rows of **B** are exhausted.

(4) The most heavily utilized parts of the program have been enclosed in boxes.

(5) The number of passes through the most heavily utilized part of the program is approximately equal to the product of the number of non-zero elements in the product matrix **C** and the average number of non-zero elements per row of **A**.

(6) If the average number of non-zero elements per row of **A** is much larger than the average number of non-zero elements per column of **B**, then it may be more efficient to form $\mathbf{C}^T = \mathbf{B}^T\mathbf{A}^T$ than to form $\mathbf{C} = \mathbf{AB}$ if using row-wise packing schemes for the matrices. Alternatively, a column-wise packing scheme could be used.

(7) At the beginning of each new row i it is necessary to reallocate the KIB pointers. In order to do this easily it is useful to have a row address sequence for matrix **B** (see 3.29) in addition to the dummy elements.

(8) If the product matrix **C** is symmetric it is possible to modify the program to form only the lower triangle and leading diagonal by altering one statement (in the FORTRAN version IF(JNL.LT.10000)GO TO 4 should be altered to IF(JNL.LE.I)GO TO 4).

3.15 SERIES STORAGE OF SPARSE MATRICES

Consider the evaluation of

$$\mathbf{M} = \mathbf{A}^T\mathbf{B}\mathbf{A} \tag{3.39}$$

using systematically packed matrix storage throughout. The following three operations will be required:

$$\left.\begin{array}{lll} \text{(a)} & \text{sparse matrix multiplication} & \mathbf{C} = \mathbf{BA} \\ \text{(b)} & \text{sparse transpose} & \mathbf{A}^T \\ \text{(c)} & \text{sparse matrix multiplication} & \mathbf{M} = \mathbf{A}^T\mathbf{C} \end{array}\right\} \tag{3.40}$$

involving sparse matrix stores for **A**, **B**, **C**, $\mathbf{A}^T$ and **M**. However, instead of using five separate primary and secondary arrays it is possible to store all of the matrices within one primary and one secondary array, provided that the starting address for each matrix is specified. If the sparse representations of matrices **A**, **B**, **C**, $\mathbf{A}^T$ and **M** occupy 300, 100, 400, 280 and 200 locations respectively, then the occupation of the primary and secondary arrays can be arranged as illustrated in Figure 3.5. Note

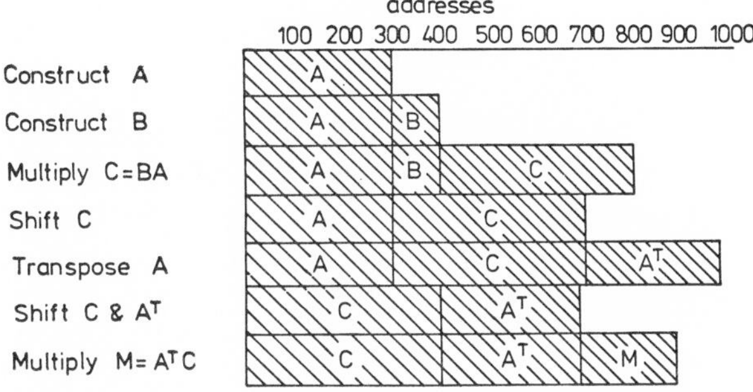

Figure 3.5 Example of the utilization of a sparse store for forming $M = A^TBA$

that when the matrices **B** and **A** are no longer required their storage space is re-used by shifting down the matrices stored above them. Although the arrays would have to be of order 1,280 to store all five matrices simultaneously, as a result of the reallocation of storage, arrays of order 980 are sufficient for the whole computation.

3.16 REGULAR PATTERN STORAGE SCHEMES

These storage schemes do not require a secondary array to define the positions of stored elements.

Triangular matrices

A lower triangular matrix of order n may be stored in a one-dimensional array having $n(n + 1)/2$ elements. If the matrix is stored by rows the storage sequence is given by

$$\text{Array L} = \{l_{11} \mid l_{21} \quad l_{22} \mid l_{31} \quad l_{32} \quad l_{33} \mid \ldots l_{nn}\} \tag{3.41}$$

where element l_{ij} is stored at address $(i/2)(i - 1) + j$.

Hessenberg matrices

An upper Hessenberg matrix is an upper triangular matrix with the addition that the principal subdiagonal elements may also be non-zero (see matrix 8.58). If an upper Hessenberg matrix **H** of order n is stored by rows, then a one-dimensional array having $(n/2)(n + 3) - 1$ elements will be required such that

$$\text{Array H} = \{h_{11} \ldots h_{1n} \mid h_{21} \ldots h_{2n} \mid h_{32} \ldots h_{3n} \mid h_{43} \ldots h_{4n} \mid \ldots h_{nn}\} \tag{3.42}$$

where element h_{ij} is stored at address $(i/2)(2n + 3 - i) - n + j - 1$.

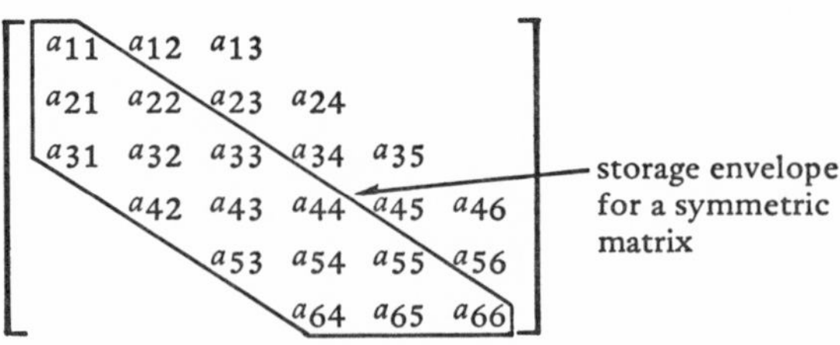

Figure 3.6 A diagonal band matrix

Diagonal band matrices

Figure 3.6 shows a diagonal band matrix A of order 6 and bandwidth 5. If the matrix is symmetric it may be completely represented by storing the elements with indices satisfying $j \leqslant i$ and hence can be stored by rows as

$$\text{Array A} = \{a_{11} \mid a_{21} \quad a_{22} \mid a_{31} \quad a_{32} \quad a_{33} \mid a_{42} \quad a_{43} \quad a_{44} \mid a_{53} \quad a_{54} \quad a_{55} \mid a_{64} \quad a_{65} \quad a_{66}\} \tag{3.43}$$

In general, if the matrix A is of order n and bandwidth $2b - 1$, the array will have $(b/2)(2n - b + 1)$ elements, with a_{ij} $(j \leqslant i)$ being stored in the location with address $(i/2)(i - 1) + j$ if $i \leqslant b$ or with address $(i - b/2)(b - 1) + j$ if $i \geqslant b$.

An alternative method of storing diagonal band matrices is to include dummy elements so that the same number of stores are allocated for each row. The matrix defined above could then be stored as

Array A =

$$\{\text{x} \quad \text{x} \quad a_{11} \mid \text{x} \quad a_{21} \quad a_{22} \mid a_{31} \quad a_{32} \quad a_{33} \mid a_{42} \quad a_{43} \quad a_{44} \mid a_{53} \quad a_{54} \quad a_{55} \mid a_{64} \quad a_{65} \quad a_{66}\} \tag{3.44}$$

In the general case a_{ij} $(j \leqslant i)$ will be stored at the location with address $ib - i + j$. Another alternative is to use a two-dimensional array of order $n \times b$ to store a symmetric band matrix. If dummy elements are included as above, then element a_{ij} $(j \leqslant i)$ would be located at the address with subscripts $(i, b - i + j)$.

Tridiagonal matrices

A tridiagonal matrix has a bandwidth of 3. It is usually stored as a linear array of order n of the diagonal elements together with one or two arrays of order $n - 1$ of the subdiagonal and superdiagonal elements (only one of these arrays being necessary if the matrix is symmetric).

Advantages of regular pattern storage schemes

Matrix operations can be programmed with virtually the same facility using these regular pattern storage schemes as they can be using standard matrix storage.

Furthermore, there is no storage requirement for secondary arrays and neither is there any computation time used in processing the secondary arrays. It is therefore advantageous to adopt these types of storage where possible. Even where some of the elements within the particular regular pattern storage are zero, it may be more economical and convenient to use a regular pattern storage than a random or systematic packing scheme.

3.17 VARIABLE BANDWIDTH STORAGE

Variable bandwidth storage is similar to diagonal band storage (array 3.43) except that the bandwidth may vary from row to row. Figure 3.7 shows a 6 x 6 symmetric matrix in which the number of elements to be stored for the six rows are 1, 1, 2, 1, 5 and 4 respectively. This matrix may be accommodated in the one-dimensional array:

$$\begin{array}{l}\quad\quad\quad\quad\;\; 1 \quad\; 2 \quad\;\; 3 \quad\;\; 4 \quad\;\; 5 \quad\;\; 6 \quad\;\; 7 \quad\;\; 8 \quad\;\; 9 \quad\; 10 \quad 11 \quad 12 \quad 13 \quad 14 \\ \text{Array A} = \{a_{11} \mid a_{22} \mid a_{32} \;\; a_{33} \mid a_{44} \mid a_{51} \;\; a_{52} \;\; a_{53} \;\; a_{54} \;\; a_{55} \mid a_{63} \;\; a_{64} \;\; a_{65} \;\; a_{66}\}\end{array} \tag{3.45}$$

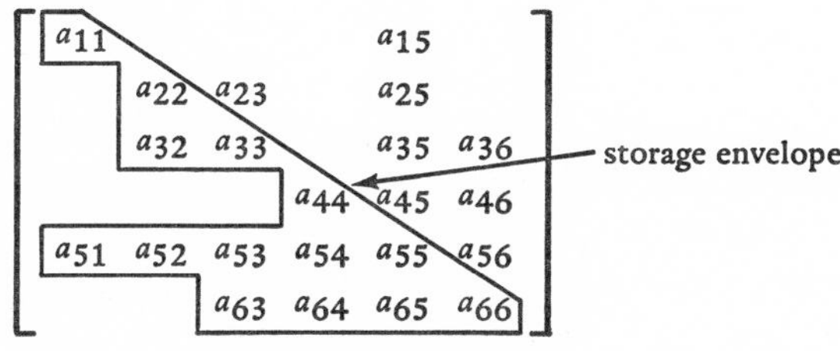

Figure 3.7 A symmetric variable bandwidth matrix

However, in order to interpret this array it is necessary to know the various row lengths or, more conveniently, the addresses of the diagonal elements. For the array (3.45) the addresses of the six diagonal elements may be specified as

$$\text{Integer array AD} = \{1 \quad 2 \quad 4 \quad 5 \quad 10 \quad 14\} \tag{3.46}$$

If element a_{ij} is contained within the stored sequence it will be located at address $\text{AD}(i) + j - i$. Also, the number of stored elements on row i of the matrix will be $\text{AD}(i) - \text{AD}(i-1)$ for all values of i greater than 1.

A variable bandwidth store may be used to represent either a symmetric matrix, as shown in Figure 3.7, or a lower triangular matrix. Since direct access can be obtained to any specified element it has some similarities with regular pattern storage schemes, but because row lengths can be individually controlled it is more versatile. It is particularly suitable for elimination methods of solving sparse symmetric positive definite simultaneous equations (Chapter 5).

3.18 SUBMATRIX STORAGE SCHEMES

Although submatrix storage can be used for full matrices, it is most useful for sparse matrices where only the non-null submatrices need to be stored. Since the submatrices have to be referenced within program loops during processing it will normally be essential that all of the non-null submatrices representing one matrix be stored within the same primary array.

Fixed pattern submatrix storage

The simplest type of submatrix storage is where non-null submatrices are all of the same order and form a regular pattern within the matrix. Suppose that Figure 3.6 represents a 60 x 60 symmetric matrix divided into submatrices a_{ij} of order 10 x 10, such that the specified diagonal band pattern of submatrices contain all of the non-zero elements of the matrix. The matrix can be stored as shown in array (3.43) except that each item a_{ij} corresponds to the sequence of 100 elements which specify the particular submatrix. Because of the regular structure of the matrix it is easy to compute the address of any submatrix element. This means that processing operations are not difficult to implement in submatrix form.

Submatrix packing schemes

Where the non-null submatrices do not form a regular pattern it is necessary to carry identification parameters for each submatrix. The submatrices may be packed in a one-dimensional primary array of sufficient length. The identification parameters may be stored in secondary arrays of length equal to the number of non-null submatrices. The following identification parameters may have to be stored for each submatrix:

(a) submatrix row and column indices,
(b) the primary array address of the first element in the submatrix, and
(c) address links.

In addition, if the submatrices are not all of the same order it will be necessary to specify separately the number of rows and columns corresponding to each submatrix row and column index.

Comments on submatrix storage schemes

There are some similarities between element and submatrix packing schemes. For instance, it is possible to pack submatrices either randomly or systematically, and also to extend the operations which can be performed with element packing to a corresponding submatrix packing scheme. Both the storage requirement for the identification parameters and the computer times to process these parameters will become less significant as the orders of the submatrix increase. This means that submatrix storage schemes may be efficient, particularly where matrices can be conveniently segmented into submatrices of large order. However, if the use of large submatrices entails having a large proportion of zero elements within the stored submatrices, the advantage gained by using large submatrices will be more than offset by the need to store and process the zero elements.

BIBLIOGRAPHY

Dijkstra, E. W. (1962). *A Primer of ALGOL 60 Programming*, Academic Press, London.

Jennings, A. (1968). 'A sparse matrix scheme for the computer analysis of structures'. *Int. J. of Computer Mathematics*, 2, 1–21.

Livesley, R. K. (1957). *An Introduction to Automatic Digital Computers*, Cambridge University Press, Cambridge.

McCracken, D. D. (1962). *A Guide to ALGOL Programming*, Wiley, New York.

McCracken, D. D. (1972). *A Guide to FORTRAN IV Programming*, 2nd ed. Wiley, New York.

Pollock, S. V. (1965). *A Guide to FORTRAN VI*, Columbia University Press, New York.

Tewarson, R. P. (1973). *Sparse Matrices*, Academic Press, New York.

Vowels, R. A. (1975). *ALGOL 60 and FORTRAN IV*, Wiley, London.

Wilkinson, J. H. (1963). *Rounding Errors in Algebraic Processes*, HMSO, London.

Chapter 4
Elimination Methods for Linear Equations

4.1 IMPLEMENTATION OF GAUSSIAN ELIMINATION

In Chapter 1 it was seen that Gaussian elimination could be carried out either with or without pivot selection and that it was a more satisfactory method for solving linear simultaneous equations than inversion techniques. The first six sections of this chapter will be concerned with elimination methods for linear equations in which pivot selection is unnecessary or undesirable.

Consider Gaussian elimination for the matrix equation **AX** = **B** (equation 1.39) where **A** and **B** are of order $n \times n$ and $n \times m$ respectively, and whose solution **X** is of order $n \times m$. Table 1.3 illustrates the hand solution procedure without pivot selection. It can be seen that the main part of the solution process (called the *reduction*) involves making a controlled series of modifications to the elements of **A** and **B** until **A** has been converted into upper triangular form. Let $a_{ij}^{(k)}$ be the coefficient in the i, j position at the k-th reduction step where $i \geqslant k$ and $j \geqslant k$, and let $b_{ij}^{(k)}$ represent the element b_{ij} at this stage where $i \geqslant k$. The k-th step in the elimination can be represented algebraically by

$$a_{ik}^{(k+1)} = 0 \qquad (k < i \leqslant n) \tag{4.1a}$$

$$a_{ij}^{(k+1)} = a_{ij}^{(k)} - \frac{a_{ik}^{(k)} a_{kj}^{(k)}}{a_{kk}^{(k)}} \qquad (k < i \leqslant n,\ k < j \leqslant n) \tag{4.1b}$$

$$b_{ij}^{(k+1)} = b_{ij}^{(k)} - \frac{a_{ik}^{(k)} b_{kj}^{(k)}}{a_{kk}^{(k)}} \qquad (k < i \leqslant n,\ 1 < j \leqslant m) \tag{4.1c}$$

On a computer the elements evaluated according to equations (4.1) may overwrite their predecessors at stage k, and hence the reduction can be carried out in the array storage which initially holds the matrices **A** and **B**. Since the eliminated elements play no part in the subsequent elimination there is no need to implement equation (4.1a). At stage k the arrays for **A** and **B** will be holding the elements shown in Figure 4.1, the contents of the stores marked ×, containing redundant information. The k-th step in the reduction process modifies the elements indicated in Figure 4.1. If the matrices **A** and **B** are stored in two-dimensional arrays A(N,N)

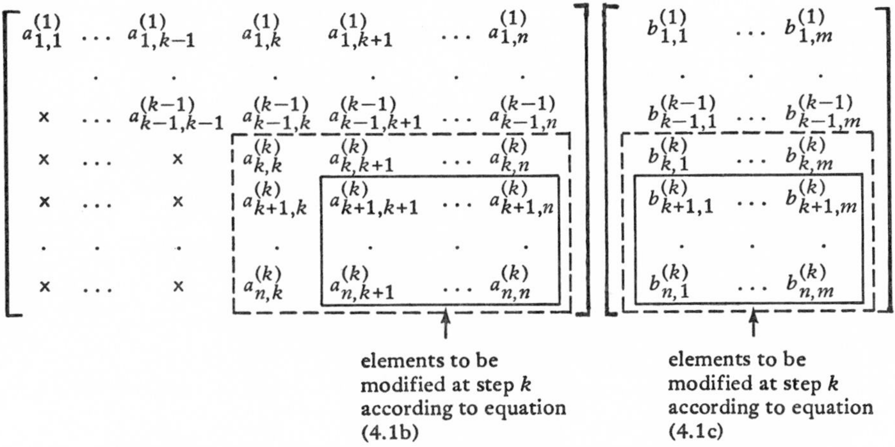

Figure 4.1 Contents of array stores for **A** and **B** at stage k of reduction

and B(N,M) then a program segment to perform the k-th step in the reduction process could be as follows:

REDUCTION STEP FOR GAUSSIAN ELIMINATION WITHOUT PIVOTING

ALGOL

```
for i:=k+1 step 1 until n do
begin x:=A[i,k];
      if x=0 then goto skip;
      x:=x/A[k,k];
      for j:=k+1 step 1 until n do
      A[i,j]:=A[i,j]-A[k,j]*x;
      for j:=1 step 1 until m do
      B[i,j]:=B[i,j]-B[k,j]*x;
skip:end step k of reduction;
```

FORTRAN

```
      KK=K+1
      DO 1 I=KK,N
      X=A(I,K)
      IF(X.EQ.0.0)GO TO 1
      X=X/A(K,K)
      DO 2 J=KK,N
    2 A(I,J)=A(I,J)-A(K,J)*X
      DO 1 J=1,M
      B(I,J)=B(I,J)-B(K,J)*X
    1 CONTINUE
```

Before entering on step k of the reduction it is advisable to test the magnitude of the pivot A(K,K) and exit with appropriate diagnostic printout if it is zero.

After the reduction phase has been completed the solution can be written into the array B beginning with the last row. This backsubstitution process is described algebraically by

$$x_{ij} = \frac{b_{ij}^{(i)} - \sum_{k=i+1}^{n} a_{ik}^{(i)} x_{kj}}{a_{ii}^{(i)}} \tag{4.2}$$

Figure 4.2 shows the contents of the array stores during backsubstitution at the stage at which row i of the solution is to be obtained. A program segment to

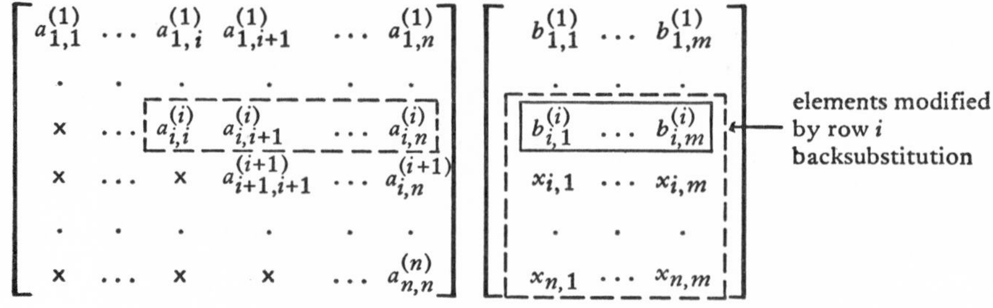

Figure 4.2 Contents of array stores before backsubstitution for row i

perform the backsubstitution for row i is:

ROW i BACKSUBSTITUTION

ALGOL

```
for j:=1 step 1 until m do
begin x:=B[i,j];
      for k:=i+1 step 1 until n do
      x:=x-A[i,k]*B[k,j];
      B[i,j]:=x/A[i,i]
end row i backsubstition;
```

FORTRAN

```
   II=I+1
   DO 1 J=1,M
   X=B(I,J)
   IF(I.EQ.N)GO TO 1
   DO 2 K=II,N
 2 X=X-A(I,K)*B(K,J)
 1 B(I,J)=X/A(I,I)
```

The total number of divisions required to obtain the solution can be reduced to n if the reciprocals of the pivot elements rather than the pivots themselves are stored on the diagonal of A. This will be of advantage if a division takes more computer time than a multiplication.

4.2 EQUIVALENCE OF GAUSSIAN ELIMINATION AND TRIANGULAR DECOMPOSITION

Let $\mathrm{A}^{(k)}$ be the coefficient matrix shown in Figure 4.1, but with the eliminated elements set to zero. Step k of the reduction forms $\mathrm{A}^{(k+1)}$ from $\mathrm{A}^{(k)}$ by subtracting multiples of the k-th (or pivotal) row from rows $k+1$ to n. This may be described by the matrix multiplication

$$\mathrm{A}^{(k+1)} = \mathrm{M}_k \mathrm{A}^{(k)} \tag{4.3}$$

where

$$\mathrm{M}_k = \begin{bmatrix} 1 & & & & \\ & \ddots & & & \\ & & 1 & & \\ & & -m_{k+1,k} & 1 & \\ & & \vdots & & \ddots & \\ & & -m_{n,k} & & & 1 \end{bmatrix} \tag{4.4}$$

and m_{ik} is a typical multiplier with value

$$m_{ik} = \frac{a_{ik}^{(k)}}{a_{kk}^{(k)}} \tag{4.5}$$

Since M_k is non-singular, it follows that

$$A^{(k)} = M_k^{-1} A^{(k+1)} \tag{4.6}$$

and by induction

$$A^{(1)} = M_1^{-1} M_2^{-1} \ldots M_{n-1}^{-1} A^{(n)} \tag{4.7}$$

It can be shown that the matrix product $M_1^{-1} M_2^{-1} \ldots M_{n-1}^{-1}$ is equal to a lower triangular matrix containing all of the multipliers, namely

$$L = \begin{bmatrix} 1 & & & & \\ m_{2,1} & 1 & & & \\ \cdot & \cdot & \cdot & & \\ m_{n-1,n} & m_{n-1,2} & \cdots & 1 & \\ m_{n,1} & m_{n,2} & \cdots & m_{n,n-1} & 1 \end{bmatrix} \tag{4.8}$$

Also, since $A^{(1)} = A$, equation (4.7) is equivalent to

$$A = LU \tag{4.9}$$

where U is equal to the upper triangular matrix $A^{(n)}$. The operations on the coefficient matrix during Gaussian elimination are therefore equivalent to a triangular decomposition of the matrix.

The solution of linear equations by triangular decomposition may be separated into three stages:

Stage 1 Factorization of the coefficient matrix into the triangular components specified by equation (4.9). The equations may then be written as

$$LUX = B \tag{4.10}$$

Stage 2 The solution for **Y** of the matrix equation

$$LY = B \tag{4.11}$$

This may be performed easily because of the triangular nature of **L**.

Stage 3 The solution for **X** of

$$UX = Y \tag{4.12}$$

which is easily performed because of the triangular nature of **U**.

The equivalence of Gaussian elimination and triangular decomposition is such that matrix **Y** is the same as the matrix to which **B** is reduced by the Gaussian reduction process. Hence the second stage in the triangular decomposition method corresponds to the right-hand reduction of Gaussian elimination and could be described as a forward substitution. The third stage in the triangular decomposition method is the same as the backsubstitution process of Gaussian elimination.

4.3 IMPLEMENTATION OF TRIANGULAR DECOMPOSITION

Storage of multipliers

The triangular decomposition represented by equations (4.9) has the expanded form

$$\begin{bmatrix} a_{11} & a_{12} & \cdots & a_{1n} \\ a_{21} & a_{22} & \cdots & a_{2n} \\ \cdot & \cdot & \cdot & \cdot \\ a_{n1} & a_{n2} & \cdots & a_{nn} \end{bmatrix} = \begin{bmatrix} 1 & & & \\ l_{21} & 1 & & \\ \cdot & \cdot & \cdot & \\ l_{n1} & l_{n2} & \cdots & 1 \end{bmatrix} \begin{bmatrix} u_{11} & u_{12} & \cdots & u_{1n} \\ & u_{22} & \cdots & u_{2n} \\ & & \cdot & \cdot \\ & & & u_{nn} \end{bmatrix} \qquad (4.13)$$

The variable elements of both **L** and **U** can be accommodated in the storage space originally occupied by **A**. After decomposition the utilization of storage space will correspond to a matrix $\mathbf{A}^F$ such that

$$\mathbf{A}^F = \begin{bmatrix} u_{11} & u_{12} & \cdots & u_{1n} \\ l_{21} & u_{22} & \cdots & u_{2n} \\ \cdot & \cdot & \cdot & \cdot \\ l_{n1} & l_{n2} & \cdots & u_{nn} \end{bmatrix} \quad (= \mathbf{L} + \mathbf{U} - \mathbf{I}) \qquad (4.14)$$

The significant difference with Gaussian elimination is that the storage space which falls into disuse in Gaussian elimination is used in triangular decomposition to store the multipliers. It is because these multipliers have been stored that the reduction of the right-hand matrix can be performed as a separate subsequent operation to the operations on the coefficient matrix.

Direct evaluation of the triangular factors

Equation (4.13) may be expanded to yield

$$\left.\begin{array}{llll} a_{11} = u_{11} & a_{12} = u_{12} & \cdots & a_{1n} = u_{1n} \\ a_{21} = l_{21}u_{11} & a_{22} = l_{21}u_{12} + u_{22} & \cdots & a_{2n} = l_{21}u_{1n} + u_{2n} \\ \cdots & \cdots & \cdots & \cdots \\ a_{n1} = l_{n1}u_{11} & a_{n2} = l_{n1}u_{12} + l_{n2}u_{22} & \cdots & a_{nn} = \sum_{k=1}^{n-1} l_{nk}u_{kn} + u_{nn} \end{array}\right\} \qquad (4.15)$$

By taking these element equations in turn row by row it is possible to evaluate directly the elements of $\mathbf{A}^F$ in row-wise order, or if the element equations are taken in turn column by column it is possible to directly evaluate the elements of $\mathbf{A}^F$ in column-wise order. The general formulae for the evaluation of elements of $\mathbf{A}^F$ are

$$l_{ij} = \frac{a_{ij} - \sum_{k=1}^{j-1} l_{ik} u_{kj}}{u_{jj}} \qquad (j<i) \tag{4.16}$$

$$u_{ij} = a_{ij} - \sum_{k=1}^{i-1} l_{ik} u_{kj} \qquad (j \geqslant i)$$

It will be noted that a computed element l_{ij} or u_{ij} may always overwrite its corresponding element a_{ij}.

Decomposition procedure

From equations (4.16) it may be seen that the only restriction on the order of evaluation of the elements of $\mathbf{A}^F$ is that to evaluate element (i, j) all other elements of $\mathbf{A}^F$ with both the row number less than or equal to i and the column number less than or equal to j must have been previously evaluated. The decomposition must therefore start and finish with the evaluation of u_{11} and u_{nn} respectively, but several algorithms are possible for ordering the evaluation of the intermediate elements, of which the row-wise and column-wise schemes mentioned above are two examples.

Variations of triangular decomposition

It is possible to factorize $\mathbf{U}$ into the product $\mathbf{D}\bar{\mathbf{U}}$ where $\mathbf{D}$ is a diagonal matrix such that $d_i = u_{ii}$. Consequently $\bar{u}_{ij} = u_{ij}/d_i$. The decomposition of $\mathbf{A}$ may then be written as

$$\mathbf{A} = \begin{bmatrix} 1 & & & \\ l_{21} & 1 & & \\ \cdot & \cdot & \cdot & \\ l_{n1} & l_{n2} & \cdots & 1 \end{bmatrix} \begin{bmatrix} d_1 & & & \\ & d_2 & & \\ & & \cdot & \\ & & & d_n \end{bmatrix} \begin{bmatrix} 1 & \bar{u}_{12} & \cdots & \bar{u}_{1n} \\ & 1 & \cdots & \bar{u}_{2n} \\ & & \cdot & \cdot \\ & & & 1 \end{bmatrix} = \mathbf{L}\mathbf{D}\bar{\mathbf{U}} \tag{4.17}$$

If $\bar{\mathbf{L}}$ is the product $\mathbf{LD}$ then

$$\mathbf{A} = \begin{bmatrix} \bar{l}_{11} & & & \\ \bar{l}_{21} & \bar{l}_{22} & & \\ \cdot & \cdot & \cdot & \\ \bar{l}_{n1} & \bar{l}_{n2} & \cdots & \bar{l}_{nn} \end{bmatrix} \begin{bmatrix} 1 & \bar{u}_{12} & \cdots & \bar{u}_{1n} \\ & 1 & \cdots & \bar{u}_{2n} \\ & & \cdot & \cdot \\ & & & 1 \end{bmatrix} = \bar{\mathbf{L}}\bar{\mathbf{U}} \tag{4.18}$$

where $\bar{l}_{ii} = d_i$ and $\bar{l}_{ij} = l_{ij} d_j$. Factorizations (4.17) and (4.18) can both be written into the store vacated by matrix A in a similar way to the procedure for factorization (4.9) and therefore provide alternative decompositions. Only slight differences in organization are entailed.

Compact elimination methods

In the hand solution of simultaneous equations, methods have been developed for writing down the reduced equations without evaluating all of the intermediate coefficients of the form $a_{ij}^{(k)}$. The resulting saving in transcribing times and tabulation layout gave rise to the general title *compact elimination methods*. Doolittle's method is equivalent to a row-wise standard triangular decomposition, while Crout's method is equivalent to a row-wise decomposition of the form of equation (4.18).

In the computer solution of equations there is no storage saved by directly evaluating the decomposed form, but there may be a slight saving in array access time over the conventional Gaussian elimination procedure. The different possible scanning procedures for performing the decomposition permit some flexibility in computer implementation.

4.4 SYMMETRIC DECOMPOSITION

If **A** is a symmetric matrix then the standard triangular decomposition of **A** to $\mathbf{A}^F$ according to equation (4.13) destroys the symmetry. For example,

$$\begin{bmatrix} 16 & 4 & 8 \\ 4 & 5 & -4 \\ 8 & -4 & 22 \end{bmatrix} = \begin{bmatrix} 1 & & \\ 0.25 & 1 & \\ 0.5 & -1.5 & 1 \end{bmatrix} \begin{bmatrix} 16 & 4 & 8 \\ & 4 & -6 \\ & & 9 \end{bmatrix} \tag{4.19}$$

However, it will be noted that the columns of **L** are proportional to the corresponding rows of **U**. In the general case the relationship between **L** and **U** is given by

$$l_{ij} = \frac{u_{ji}}{u_{jj}} \quad (j < i) \tag{4.20}$$

Alternatively, the decomposition of equation (4.17) may be adopted, which gives a symmetric factorization of the form

$$\mathbf{A} = \mathbf{L}\mathbf{D}\mathbf{L}^T \tag{4.21}$$

For the above example

$$\begin{bmatrix} 16 & 4 & 8 \\ 4 & 5 & -4 \\ 8 & -4 & 22 \end{bmatrix} = \begin{bmatrix} 1 & & \\ 0.25 & 1 & \\ 0.5 & -1.5 & 1 \end{bmatrix} \begin{bmatrix} 16 & & \\ & 4 & \\ & & 9 \end{bmatrix} \begin{bmatrix} 1 & 0.25 & 0.5 \\ & 1 & -1.5 \\ & & 1 \end{bmatrix} \tag{4.22}$$

If **A** is not only symmetric but also positive definite then it can be shown that all of the elements of $\mathbf{D}$ must be positive. To prove this theorem, let $\mathbf{y}$ be any vector and let $\mathbf{x}$ be related to it according to

$$\mathbf{L}^T\mathbf{x} = \mathbf{y} \tag{4.23}$$

(the existence of $\mathbf{x}$ is assured since $\mathbf{L}^T$ is non-singular). Then

$$\mathbf{y}^T\mathbf{D}\mathbf{y} = \mathbf{x}^T\mathbf{L}\mathbf{D}\mathbf{L}^T\mathbf{x} = \mathbf{x}^T\mathbf{A}\mathbf{x} > 0 \tag{4.24}$$

Hence $\mathbf{D}$ must also be positive definite. Since the elements of a diagonal matrix are its eigenvalues, the elements of $\mathbf{D}$ must all be positive, which proves the above theorem.

A diagonal matrix $\mathbf{D}^{1/2}$ may be defined such that the i-th element is equal to $\sqrt{d_i}$. Then writing

$$\tilde{\mathbf{L}} = \mathbf{L}\mathbf{D}^{1/2} \tag{4.25}$$

gives

$$\mathbf{A} = \mathbf{L}\mathbf{D}\mathbf{L}^T = \mathbf{L}\mathbf{D}^{1/2}\mathbf{D}^{1/2}\mathbf{L}^T = \tilde{\mathbf{L}}\tilde{\mathbf{L}}^T \tag{4.26}$$

The decomposition $\mathbf{A} = \tilde{\mathbf{L}}\tilde{\mathbf{L}}^T$ is attributed to Choleski. It will be noted that if $\mathbf{A}$ is positive definite $\mathbf{D}^{1/2}$ will be a real matrix and hence $\tilde{\mathbf{L}}$ will be real. Although Choleski decomposition of a matrix which is symmetric but not positive definite could be carried out, the generation of imaginary elements in $\tilde{\mathbf{L}}$ makes its implementation more complicated. The Choleski decomposition for the 3×3 example is

$$\begin{bmatrix} 16 & 4 & 8 \\ 4 & 5 & -4 \\ 8 & -4 & 22 \end{bmatrix} = \begin{bmatrix} 4 & & \\ 1 & 2 & \\ 2 & -3 & 3 \end{bmatrix} \begin{bmatrix} 4 & 1 & 2 \\ & 2 & -3 \\ & & 3 \end{bmatrix} \tag{4.27}$$

The general formulae for Choleski decomposition are

$$\left.\begin{aligned} l_{ii} &= \left(a_{ii} - \sum_{k=1}^{i-1} l_{ik}^2\right)^{1/2} \\ l_{ij} &= \frac{a_{ij} - \sum_{k=1}^{j-1} l_{ik}l_{jk}}{l_{jj}} \qquad (j < i) \end{aligned}\right\} \tag{4.28}$$

with the order of computation of the elements being governed by the same rules as for standard decomposition given in section 4.3.

Both Choleski decomposition and $\mathbf{LDL}^T$ decomposition may be implemented with about one-half of the computation required for the triangular decomposition of an unsymmetric matrix of the same order. It is also possible to perform both decompositions in a triangular store (section 3.16) which initially contains only one-half of the original matrix. Choleski's method has the disadvantage of requiring the evaluation of n square roots, which on a computer usually takes much longer than other arithmetical operations. However, it has the advantage of being simpler and may also be easier to implement with backing store transfers when the entire matrix cannot be held in the main store.

4.5 USE OF TRIANGULAR DECOMPOSITION

Triangular decomposition can be used not only for the solution of linear equations, as described in section 4.2, but also for evaluating matrix expressions involving inverses.

Consider the evaluation of a matrix **C**,

$$\mathbf{C} = \mathbf{A}^{-1}\mathbf{B}\mathbf{A}^{-T} \tag{4.29}$$

where **A** and **B** are square matrices of order n. If **A** has been decomposed into triangular matrices **L** and **U**, then

$$\mathbf{C} = \mathbf{U}^{-1}\mathbf{L}^{-1}\mathbf{B}\mathbf{L}^{-T}\mathbf{U}^{-T} \tag{4.30}$$

Letting $\mathbf{X} = \mathbf{L}^{-1}\mathbf{B}$, $\mathbf{Y} = \mathbf{X}\mathbf{L}^{-T}$ and $\mathbf{Z} = \mathbf{U}^{-1}\mathbf{Y}$, it follows that

$$\left.\begin{aligned} \mathbf{L}\mathbf{X} &= \mathbf{B} \\ \mathbf{L}\mathbf{Y}^T &= \mathbf{X}^T \\ \mathbf{U}\mathbf{Z} &= \mathbf{Y} \\ \text{and}\qquad & \\ \mathbf{U}\mathbf{C}^T &= \mathbf{Z}^T \end{aligned}\right\} \tag{4.31}$$

Each of these equations may be solved, yielding in turn **X**, **Y**, **Z** and then **C**. Since the required operations are either forward-substitutions or backsubstitutions with a square right-hand side, they each require approximately $n^3/2$ multiplications when n is large. Hence, including the initial decomposition, approximately $2\frac{1}{3}n^3$ multiplications will be required to determine **C**, as compared with approximately $3n^3$ multiplications if **C** is evaluated from the inverse of **A**.

If, in equation (4.29), **B** is a symmetric matrix, then **Y** and **C** will also be symmetric and it is possible to make further economies by making use of this symmetry. Table 4.1 describes how **C** may be formed from the inverse of **A** while Table 4.2 gives a more efficient method using triangular decomposition. These tables also include the stages of a computation in which **A** is the 3 × 3 matrix (1.63) whose triangular decomposition is

$$\begin{bmatrix} 10 & 1 & -5 \\ -20 & 3 & 20 \\ 5 & 3 & 5 \end{bmatrix} = \begin{bmatrix} 1 & & \\ -2 & 1 & \\ 0.5 & 0.5 & 1 \end{bmatrix} \begin{bmatrix} 10 & 1 & -5 \\ & 5 & 10 \\ & & 2.5 \end{bmatrix} \tag{4.32}$$

and **B** is the symmetric matrix (4.27).

Another example where triangular decomposition can be effectively employed is in the evaluation of

$$\mathbf{C} = \mathbf{B}^T\mathbf{A}^{-1}\mathbf{B} \tag{4.33}$$

Table 4.1 Computation of $\mathbf{C} = \mathbf{A}^{-1}\mathbf{B}\mathbf{A}^{-T}$ with symmetric $\mathbf{B}$ – inversion method

Operation	Notes	Approximate no. of multiplications	Result for 3 x 3 example
Invert $\mathbf{A} \rightarrow \mathbf{A}^{-1}$	–	n^3	as matrix (1.64)
Multiply $\mathbf{F} = \mathbf{A}^{-1}\mathbf{B}$	–	n^3	$\mathbf{F} = \begin{bmatrix} -4.16 & -3.36 & 3.92 \\ 21.6 & 12.6 & -7.2 \\ -7.2 & -5.0 & 4.8 \end{bmatrix}$
Transpose $\mathbf{A}^{-1} \rightarrow \mathbf{A}^{-T}$	–	–	–
Multiply $\mathbf{C} = \mathbf{F}\mathbf{A}^{-T}$	compute one triangle only	$\frac{n^3}{2}$	$\mathbf{C} = \begin{bmatrix} 3.1328 & \text{symmetric} & \\ -11.808 & 47.88 & \\ 4.736 & -18.36 & 7.24 \end{bmatrix}$
		Total = $2\frac{1}{2}n^3$	

Table 4.2 Computation of $\mathbf{C} = \mathbf{A}^{-1}\mathbf{B}\mathbf{A}^{-T}$ with symmetric $\mathbf{B}$ – a triangular decomposition method

Operation	Notes	Approximate no. of multiplications	Result for 3 x 3 example
Factorize $\mathbf{A} = \mathbf{L}\mathbf{U}$	–	$\frac{n^3}{3}$	as equation (4.32)
Solve $\mathbf{L}\mathbf{X} = \mathbf{B}$	forward-substitution	$\frac{n^3}{2}$	$\mathbf{X} = \begin{bmatrix} 16 & 4 & 8 \\ 36 & 13 & 12 \\ -18 & -12.5 & 12 \end{bmatrix}$
Transpose $\mathbf{X} \rightarrow \mathbf{X}^T$	–	–	–
Solve $\mathbf{L}\mathbf{Y} = \mathbf{X}^T$	forward-substitution, compute upper triangle only	$\frac{n^3}{6}$	$\mathbf{Y} = \begin{bmatrix} 16 & 36 & -18 \\ & 85 & -48.5 \\ \text{symmetric} & & 45.25 \end{bmatrix}$
Solve $\mathbf{U}\mathbf{Z} = \mathbf{Y}$	backsubstitution	$\frac{n^3}{2}$	$\mathbf{Z} = \begin{bmatrix} -4.16 & -11.68 & 11.84 \\ 21.6 & 55.8 & -45.9 \\ -7.2 & -19.4 & 18.1 \end{bmatrix}$
Transpose $\mathbf{Z} \rightarrow \mathbf{Z}^T$	–	–	–
Solve $\mathbf{U}\mathbf{C} = \mathbf{Z}^T$	backsubstitution, compute lower triangle only	$\frac{n^3}{6}$	$\mathbf{C} = \begin{bmatrix} 3.1328 & \text{symmetric} & \\ -11.808 & 47.88 & \\ 4.736 & -18.36 & 7.24 \end{bmatrix}$
		Total = $1\frac{2}{3}n^3$	

where $\mathbf{A}$ is a symmetric positive definite matrix of order n and $\mathbf{B}$ is of order $n \times m$. If a Choleski decomposition of $\mathbf{A}$ is performed such that $\mathbf{A} = \mathbf{L}\mathbf{L}^T$ then

$$\mathbf{C} = \mathbf{B}^T\mathbf{L}^{-T}\mathbf{L}^{-1}\mathbf{B} = \mathbf{X}^T\mathbf{X} \tag{4.34}$$

where $\mathbf{X}$ may be computed by forward-substitution according to the equation $\mathbf{L}\mathbf{X} = \mathbf{B}$. Forming $\mathbf{C}$ from $\mathbf{X}^T\mathbf{X}$ in this way requires a total of approximately

$(n/2)(n^2/3 + nm + m^2)$ multiplications compared with approximately $(n/2)(n^2 + 2nm + m^2)$ multiplications if it is formed from the inverse of **A**.

4.6 WHEN PIVOT SELECTION IS UNNECESSARY

The previous discussion of Gaussian elimination and related triangular decomposition methods has not included a pivot selection strategy. The use of elimination techniques without pivot selection may only be justified on a computer if it is known in advance that the leading diagonal pivots will be strong. (Here the word 'strong' is used without any strict mathematical definition.) Two classes of matrices are known to satisfy this criterion sufficiently well that pivot selection may be dispensed with, i.e. where the coefficient matrix is either symmetric positive definite or else diagonally dominant.

Symmetric positive definite matrices

In section 4.4 it was shown that the $\mathbf{LDL}^T$ decomposition of a symmetric positive definite matrix must give positive values for all of the diagonal elements of **D**. Since these elements are the full set of pivots (whether by this decomposition or by any equivalent decomposition or elimination), it is established that the reduction cannot break down through encountering a zero diagonal pivot element.

To examine the strength of the leading diagonal during elimination consider the partial decomposition of a matrix **A** of order 5 according to

$$\mathbf{A} = \mathbf{L}^{(3)}\mathbf{D}^{(3)}[\mathbf{L}^{(3)}]^T \tag{4.35}$$

where

$$\mathbf{L}^{(3)} = \begin{bmatrix} 1 & & & & \\ l_{21} & 1 & & & \\ l_{31} & l_{32} & 1 & & \\ l_{41} & l_{42} & & 1 & \\ l_{51} & l_{52} & & & 1 \end{bmatrix}$$

and

$$\mathbf{D}^{(3)} = \begin{bmatrix} d_1 & & & & \\ & d_2 & & & \\ & & a_{33}^{(3)} & a_{34}^{(3)} & a_{35}^{(3)} \\ & & a_{43}^{(3)} & a_{44}^{(3)} & a_{45}^{(3)} \\ & & a_{53}^{(3)} & a_{54}^{(3)} & a_{55}^{(3)} \end{bmatrix}$$

The elements of $\mathbf{D}^{(3)}$ with $i > 3$ and $j > 3$ have been shown as $a_{ij}^{(3)}$ because they are the same as the corresponding modified coefficients in the third stage of Gaussian elimination. By adopting the same reasoning as for the decomposition of equation

(4.21) it is possible to show that $D^{(3)}$ must be positive definite. Because leading principal minors of a symmetric positive definite matrix are also symmetric positive definite (section 1.21), it follows that

(a) the matrix of order 3 still to be reduced must also be positive definite,
(b) the leading diagonal elements must always remain positive throughout the reduction (by taking leading principal minors of order 1), and
(c) by taking the determinant of leading principal minors of order 2,

$$a_{ii}^{(k)}a_{jj}^{(k)} > (a_{ij}^{(k)})^2 \tag{4.36}$$

i.e. the product of any pair of *twin* off-diagonal elements at any stage in the elimination must always be less than the product of the corresponding diagonal elements. It is pertinent to examine the elements in pairs in this way since the choice of an off-diagonal element at position (i, j) as pivot would prevent pivots being chosen at both of the diagonal positions (i, i) and (j, j). Therefore there is sufficient justification to say that the leading diagonal is always strong throughout the reduction, even though the pivots are not necessarily the largest elements within their own row or column at the time when they are used. This may be illustrated by examining the positive definite matrix

$$\begin{bmatrix} 60 & -360 & 120 \\ -360 & 2162 & -740 \\ 120 & -740 & 640 \end{bmatrix} = \begin{bmatrix} 1 & & \\ -6 & 1 & \\ 2 & & 1 \end{bmatrix} \begin{bmatrix} 60 & -360 & 120 \\ & 2 & -20 \\ & -20 & 400 \end{bmatrix} \tag{4.37}$$

Although $a_{11} = 60$ is the element of smallest modulus in the matrix, the conditions $a_{11}a_{22} > a_{12}^2$ and $a_{11}a_{33} > a_{13}^2$ do hold. After elimination of a_{21} and a_{31}, a_{22} has been reduced from 2,162 to 2. Despite the fact that it is now much smaller than $|a_{23}^{(2)}|$, the strength of the remaining diagonal is confirmed by the fact that $a_{22}^{(2)}a_{33}^{(2)} > (a_{23}^{(2)})^2$.

In setting up equations for computer solution it is frequently known from physical properties or mathematical reasoning that the coefficient matrix must always be symmetric positive definite. In such cases the ability to proceed without pivot selection is particularly important, as this leads to a symmetric reduction, with possible savings in computing time and storage space.

Diagonally dominant matrices

A matrix is said to be diagonally dominant if

$$a_{ii} > \sum_{j \neq i} |a_{ij}| \tag{4.38}$$

for all rows of the matrix. In other words, the Gerschgorin row discs all lie within the positive region of the Argand diagram.

If a matrix is diagonally dominant it can be proved that, at every stage of the reduction,

$$a_{ii}^{(k)} - \sum_{j \neq i} |a_{ij}^{(k)}| \geqslant a_{ii} - \sum_{j \neq i} |a_{ij}| > 0 \tag{4.39}$$

Hence

$$a_{ii}^{(k)} a_{jj}^{(k)} > |a_{ij}^{(k)}| \, |a_{ji}^{(k)}| \tag{4.40}$$

i.e. throughout the reduction process, the product of any pair of complementary off-diagonal elements must always be less than the product of the corresponding diagonal elements.

Normally, even if all but one of the Gerschgorin discs touch the imaginary axis the diagonal dominance will still hold in the elimination to the extent that

$$a_{ii}^{(k)} \geqslant \sum_{j \neq i} |a_{ij}^{(k)}| \quad \text{and} \quad a_{ii}^{(k)} > 0 \tag{4.41}$$

For instance, the matrix

$$\begin{bmatrix} 0.2 & 0.2 & \\ 0.9 & 1.0 & -0.1 \\ & -9.0 & 10.0 \end{bmatrix} = \begin{bmatrix} 1 & & \\ 4.5 & 1 & \\ & & 1 \end{bmatrix} \begin{bmatrix} 0.2 & 0.2 & \\ & 0.1 & -0.1 \\ & -9.0 & 10.0 \end{bmatrix} \tag{4.42}$$

is only loosely diagonally dominant since two of its three Gerschgorin row discs touch the imaginary axis. Yet it exhibits the properties of equation (4.40) and (4.41) throughout the elimination, even though the pivots $a_{11}^{(1)}$ and $a_{22}^{(2)}$ are not the largest elements in their respective columns.

It is also possible to establish that a column-wise diagonal dominance of the form

$$a_{jj} > \sum_{i \neq j} |a_{ij}| \tag{4.43}$$

for all columns of the matrix, will be retained throughout an elimination and hence can be used to justify diagonal pivoting.

4.7 PIVOT SELECTION

Full pivoting

In section 1.8 the pivotal condensation version of the Gaussian elimination method was described and two methods of implementation were discussed. Either the equations and variables could be kept in the same sequence during elimination or they could both be subject to interchanges in order to place the selected pivots in the appropriate leading positions. On a computer the interchange method is the more common.

The Gaussian elimination procedure described in section 4.1 may be modified for row and column interchanges by including the following facilities:

(a) The declaration of a one-dimensional integer array of order n to store a *permutation* vector. The permutation vector is used to record the order of the variables. Initially it is set to $\{1 \quad 2 \quad \ldots \quad n\}$.

(b) Before the k-th reduction step the elements in the active part of the matrix must be scanned to determine the position of the element $a_{ij}^{(k)}$ of largest absolute magnitude. Suppose that it is situated on row p and column q.

(c) Rows k and p of arrays **A** and **B** should then be interchanged. Columns k and q of array **A** should also be interchanged and the revised position of the variables recorded by interchanging elements k and q of the permutation vector. After these interchanges have been carried out the k-th reduction step can be performed.

(d) After the reduction and backsubstitution have been completed the order of the variables will be indicated by the permutation vector. The permutation vector may therefore be used to arrange the solution vector so that it conforms with the original order of the variables. It is economical to combine the search operation for pivot number $k + 1$ with the k-th reduction step so that elements are only scanned once per reduction step. If this is done a dummy reduction step is required in order to determine the position of the first pivot.

Partial pivoting

Figure 4.3 shows the pattern of coefficients in a matrix after $k - 1$ reduction steps have been performed. If the original matrix of coefficients is non-singular then at least one of the elements $a_{k,k}^{(k)}, a_{k+1,k}^{(k)}, \ldots, a_{n,k}^{(k)}$ must be non-zero. Hence it is possible to use only column k as a search area for the next pivot. Furthermore, after the k-th row interchange, all of the multipliers m_{kj} will have a magnitude less than unity. This form of pivoting is easier to implement on a computer than full pivoting and is also easier to include in the triangular decomposition of a matrix.

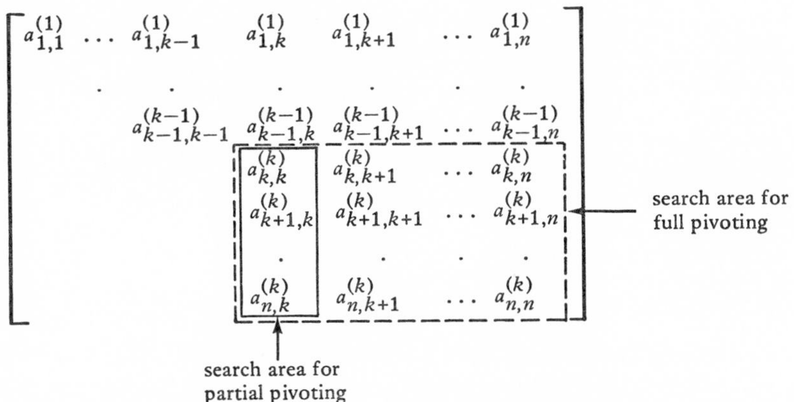

Figure 4.3 Search areas for full and partial pivoting strategies

Consider the triangular decomposition with row interchanges of a matrix **A**. Since any right-hand vector or matrix will be reduced in a subsequent operation, it is necessary to record the effect of row interchanges on the order of the equations by using a permutation vector. Assuming that the following storage space (specified as for FORTRAN) has been allocated:

N	order of matrix
A(N,N)	real array for matrix **A**
IPERM(N)	integer permutation vector
I,J	row and column indices
XMAX	magnitude of largest element in pivot search area
IPIV	row number of largest element in pivot search area
K,L	other integer working stores
X	real working store

and that the permutation vector contains integers $\{1 \quad 2 \quad \ldots \quad n\}$ initially, then a program segment to perform triangular decomposition with row interchanges is as follows:

PARTIAL PIVOTING GAUSSIAN ELIMINATION

ALGOL

```
for j:=1 step 1 until n do
begin l:=ipiv:=0;
      xmax:=0;
      for i:=1 step 1 until n do
      begin x:=A[i,j];
            for k:=1 step 1 until l do
            x:=x-A[i,k]*A[k,j];
            A[i,j]:=x;
            if i≥j then goto lower;
            l:=i;
            goto skip;
      lower:x:=abs(x);
            if x≤xmax then goto skip;
            xmax:=x;
            ipiv:=i;
skip: end multiplications and pivot
            selection for column j;
      if ipiv=j then goto leave;
      l:=iperm[j];
      iperm[j]:=iperm[ipiv];
      iperm[ipiv]:=l;
      for k:=1 step 1 until n do
      begin x:=A[j,k];
            A[j,k]:=A[ipiv,k];
            A[ipiv,k]:=x
      end row interchange;
leave: x:=A[j,j];
      for i:=j+1 step 1 until n do
      A[i,j]:=A[i,j]/x;
end decomposition with partial pivoting;
```

FORTRAN

```
  DO 1 J=1,N
  L=0
  IPIV=0
  XMAX=0.0
  DO 2 I=1,N
  X=A(I,J)
  IF(L.EQ.0)GO TO 3
  DO 4 K=1,L
4 X=X-A(I,K)*A(K,J)
  A(I,J)=X
3 IF(I.GE.J)GO TO 5
  L=I
  GO TO 2
5 X=ABS(X)
  IF(X.LE.XMAX)GO TO 2
  XMAX=X
  IPIV=I
2 CONTINUE
  IF(J.EQ.N)GO TO 1
  IF(IPIV.EQ.J)GO TO 6
  L=IPERM(J)
  IPERM(J)=IPERM(IPIV)
  IPERM(IPIV)=L
  DO 7 K=1,N
  X=A(J,K)
  A(J,K)=A(IPIV,K)
7 A(IPIV,K)=X
6 X=A(J,J)
  L=J+1
  DO 1 I=L,N
  A(I,J)=A(I,J)/X
1 CONTINUE
```

In this scheme the elements of **A** are modified in sequence by columns. The operations for column j involve firstly computing

$$\left.\begin{aligned} a'_{ij}(=u_{ij}) &= a_{ij} - \sum_{k=1}^{i-1} l_{ik}\,u_{kj} \qquad (i<j) \\ &\text{and} \\ a'_{ij} &= a_{ij} - \sum_{k=1}^{j-1} l_{ik}\,u_{kj} \qquad (i\geqslant j) \end{aligned}\right\} \tag{4.44}$$

(These operations have been performed with a single DO loop by setting the upper limit to L=I−1 when I<J and holding it at L=J−1 when I⩾J.) The pivot is selected from the set $a_{ij}^{(j)}$ $(i \geqslant j)$ and a row interchange performed so that the pivotal element becomes u_{jj}. The final operation for column j is to form the elements of **L** according to

$$l_{ij} = \frac{a_{ij}^{(j)}}{u_{jj}} \qquad (i>j) \tag{4.45}$$

4.8 ROW AND COLUMN SCALING

If each equation of a linear set **Ax** = **b** is scaled, this has the effect of scaling the rows of the coefficient matrix. If r_i is the scaling factor for equation i then the equations become

$$\begin{bmatrix} r_1a_{11} & r_1a_{12} & \cdots & r_1a_{1n} \\ r_2a_{21} & r_2a_{22} & \cdots & r_2a_{2n} \\ \cdot & \cdot & \cdot & \cdot \\ r_na_{n1} & r_na_{n2} & \cdots & r_na_{nn} \end{bmatrix} \begin{bmatrix} x_1 \\ x_2 \\ \cdot \\ x_n \end{bmatrix} = \begin{bmatrix} r_1b_1 \\ r_2b_2 \\ \cdot \\ r_nb_n \end{bmatrix} \tag{4.46}$$

On the other hand, if each variable x_i is replaced by x_i/c_i, the effect on the coefficient matrix is to scale the columns. With both row and column scaling the equations become

$$\begin{bmatrix} r_1a_{11}c_1 & r_1a_{12}c_2 & \cdots & r_1a_{1n}c_n \\ r_2a_{21}c_1 & r_2a_{22}c_2 & \cdots & r_2a_{2n}c_n \\ \cdot & \cdot & \cdot & \cdot \\ r_na_{n1}c_1 & r_na_{n2}c_2 & \cdots & r_na_{nn}c_n \end{bmatrix} \begin{bmatrix} x_1/c_1 \\ x_2/c_2 \\ \cdot \\ x_n/c_n \end{bmatrix} = \begin{bmatrix} r_1b_1 \\ r_2b_2 \\ \cdot \\ r_nb_n \end{bmatrix} \tag{4.47}$$

If **R** and **C** are diagonal matrices of row and column scaling factors respectively, the modified equations (4.47) may be expressed as

$$\bar{\mathbf{A}}\bar{\mathbf{x}} = \bar{\mathbf{b}} \tag{4.48}$$

where $\bar{\mathbf{A}} = \mathbf{RAC}$, $\bar{\mathbf{b}} = \mathbf{Rb}$ and $\mathbf{x} = \mathbf{C}\bar{\mathbf{x}}$.

A symmetric matrix may be scaled symmetrically by making $r_i = c_i$. If a symmetric positive definite matrix is scaled in such a way that $r_i = c_i = a_{ii}^{-1/2}$, then the resulting matrix will have a diagonal consisting entirely of unit elements. Furthermore, using the property of a positive definite matrix that $a_{ij}^2 < a_{ii}a_{jj}$, it follows that

$$|\bar{a}_{ij}| = |r_i a_{ij} c_j| = a_{ii}^{-1/2} a_{jj}^{-1/2} |a_{ij}| < 1$$

Thus the symmetric scaling of a symmetric positive definite matrix to give unit diagonal elements will ensure that all the off-diagonal elements have a modulus less than unity. Also, since $\bar{\mathbf{A}}$ has the form shown in equation (1.143) it can be proved that the scaling process cannot destroy the positive definite property.

Row and column scaling can have a marked effect on the choice of pivots where pivot selection is adopted. For instance, if the matrix

$$\mathbf{A} = \begin{bmatrix} -0.001 & 1 & 1 \\ 1 & 0.78125 & \\ 1 & & \end{bmatrix} \tag{4.49}$$

is scaled using the matrices $\mathbf{R} = \mathbf{C} = [2000 \quad 1 \quad 1]$, then

$$\bar{\mathbf{A}} = \begin{bmatrix} -4000 & 2000 & 2000 \\ 2000 & 0.78125 & \\ 2000 & & \end{bmatrix} \tag{4.50}$$

Hence what was the smallest non-zero element of $\mathbf{A}$ has been converted into the largest element of $\bar{\mathbf{A}}$. In fact it is possible to convert any non-zero element in a matrix into the element of largest magnitude by adopting suitable row and column scaling factors.

4.9 ON LOSS OF ACCURACY IN ELIMINATION

Why accuracy is lost if a weak pivot is chosen

The way in which accuracy can be lost through the choice of a weak pivot may be illustrated by considering the equations

$$\begin{bmatrix} -0.001 & 1 & 1 \\ 1 & 0.78125 & \\ 1 & & \end{bmatrix} \begin{bmatrix} x_1 \\ x_2 \\ x_3 \end{bmatrix} = \begin{bmatrix} 0.2 \\ 1.3816 \\ 1.9273 \end{bmatrix} \tag{4.51}$$

The triangular nature of the coefficient matrix may be easily recognized and used to compute the exact solution, namely $\{1.9273 \quad -0.698496 \quad 0.9004233\}$. If the solution is computed by taking pivots in the successive positions (1,3), (2,2) and (3,1) using five significant decimal places throughout the computation, the solution $\{1.9273 \quad -0.69850 \quad 0.90043\}$ is obtained, which is almost correct.

However, if the weak element in position (1,1) is chosen as the first pivot, then after the first reduction step the equations are transformed to

$$\begin{bmatrix} -0.001 & 1 & 1 \\ & 1000.78125 & 1000 \\ & 1000 & 1000 \end{bmatrix} \begin{bmatrix} x_1 \\ x_2 \\ x_3 \end{bmatrix} = \begin{bmatrix} 0.2 \\ 201.3816 \\ 201.9273 \end{bmatrix} \quad (4.52)$$

However, if the computation is carried out with five significant figures these equations will be rounded to

$$\begin{bmatrix} -0.001 & 1 & 1 \\ & 1000.8 & 1000 \\ & 1000 & 1000 \end{bmatrix} \begin{bmatrix} x_1 \\ x_2 \\ x_3 \end{bmatrix} = \begin{bmatrix} 0.2 \\ 201.38 \\ 201.93 \end{bmatrix} \quad (4.53)$$

This rounding operation has thrown away much important information and hence the equations must now yield an inaccurate solution. Furthermore, the coefficient matrix has been reduced to near singular form. (It would be singular if element $a_{22}^{(2)}$ were to be changed from 1000.8 to 1000.) If the computation is continued the erroneous solution {1.9300 −0.68557 0.88750} is obtained.

In the general case the choice of a small pivot at the k-th reduction step of a Gaussian elimination will cause digits in significant terms $a_{ij}^{(k)}$ and $b_{ij}^{(k)}$ to be lost when the much larger terms $a_{ik}^{(k)}a_{kj}^{(k)}/a_{kk}^{(k)}$ and $a_{ik}^{(k)}b_{kj}^{(k)}/a_{kk}^{(k)}$ are respectively subtracted. This is the reason why a pivoting strategy needs to be adopted when the coefficient matrix is neither symmetric positive definite nor diagonally dominant. (This result is just as valid for computation in floating-point binary as it is for decimal computation.)

Loss in accuracy when equations have been pre-scaled

The above discussion shows that the magnitude of the loss of accuracy depends on the magnitudes of factors of the form $|a_{ik}^{(k)}a_{kj}^{(k)}/a_{kk}^{(k)}a_{ij}^{(k)}|$ and $|a_{ik}^{(k)}b_{kj}^{(k)}/a_{kk}^{(k)}b_{ij}^{(k)}|$ throughout the reduction. The greater are the magnitudes of these factors the more significant will be the loss of accuracy. However, scaling the equations will not improve the situation since, using the notation of section 4.8,

$$\frac{\bar{a}_{ik}^{(k)}\bar{a}_{kj}^{(k)}}{\bar{a}_{kk}^{(k)}\bar{a}_{ij}^{(k)}} = \frac{r_i a_{ik}^{(k)} c_k r_k a_{kj}^{(k)} c_j}{r_k a_{kk}^{(k)} c_k r_i a_{ij}^{(k)} c_j} = \frac{a_{ik}^{(k)}a_{kj}^{(k)}}{a_{kk}^{(k)}a_{ij}^{(k)}} \quad (4.54)$$

It can also be shown that scaling does not substantially affect any possible loss of accuracy in the backsubstitution. Hence the loss of accuracy will be of the same order of magnitude whether or not the equations are pre-scaled. (This principle was established in a more rigorous way by Bauer in 1963 and is applicable provided that scaling does not affect the order in which the pivots are selected.)

An illustration of the principle may be obtained by scaling equations (4.51) so that the coefficient matrix corresponds to matrix (4.50), in which case the equations are transformed to

$$\begin{bmatrix} -4000 & 2000 & 2000 \\ 2000 & 0.78125 & \\ 2000 & & \end{bmatrix} \begin{bmatrix} \bar{x}_1 \\ \bar{x}_2 \\ \bar{x}_3 \end{bmatrix} = \begin{bmatrix} 400 \\ 1.3816 \\ 1.9273 \end{bmatrix} \tag{4.55}$$

Despite the fact that coefficient $\bar{a}_{11}$ is now the largest coefficient in the matrix, it is still not advisable to choose it for the first pivot.

4.10 ON PIVOT SELECTION

Pivot strengths for full and partial pivoting

It may be conjectured that if the relative magnitudes of the coefficients in a set of equations approximately reflect their relative importance, then the pivots selected by either a full or partial pivoting strategy will be strong. However, if this condition does not hold, weak pivots may be chosen.

An example where partial pivoting leads to the choice of a weak pivot is in the solution of the equations

$$\begin{bmatrix} 1000000 & 1000 & \\ 1000 & 2.041 & 2.041 \\ & 2.041 & 1000 \end{bmatrix} \begin{bmatrix} x_1 \\ x_2 \\ x_3 \end{bmatrix} = \begin{bmatrix} 27.11 \\ 1.367 \\ 329.9 \end{bmatrix} \tag{4.56}$$

In this case the first reduction step yields $a_{22}^{(2)} = 1.041$ and so a_{32} is chosen as the second pivot. If computation is carried out to four significant decimal figures the erroneous solution $\{-0.0005608 \quad 0.5879 \quad 0.3287\}$ is obtained. However, the coefficient matrix is symmetric positive definite and hence from section 4.6 pivot selection is unnecessary. If the equations are solved with the same arithmetical precision, but without interchanges, the almost correct solution $\{-0.0006159 \quad 0.6430 \quad 0.3286\}$ is obtained. In this example the diagonal coefficients have most influence on the solution. Since the diagonal coefficients are of very different orders of magnitudes this example violates the principle for strong pivoting proposed above.

If full pivoting were to be adopted for the solution of equations (4.56) diagonal pivots would be chosen and hence the solution would be satisfactory. Although the adoption of full pivoting rather than partial pivoting is generally less likely to lead to weak pivots being chosen, it is still possible for this to happen. For instance, if either partial or full pivoting is used for the solution of equations (4.55), element $a_{11} = -4000$ will be chosen for the first pivot. This element can reasonably be called a weak element since, for all values of i and j not equal to 1, $a_{11}a_{ij} \ll a_{i1}a_{1j}$, and it has been previously established that the choice of this element as a pivot leads to considerable loss of accuracy.

The use of pre-scaling to improve pivot selection

In cases where a pivot selection strategy is liable to result in the selection of weak pivots it may be possible to influence beneficially the selection process by pre-scaling the equations. One simple technique commonly advocated is that, before a partial pivoting strategy is employed, the equations should be scaled by rows in such a way that the largest coefficients in every equation have the same magnitude (a matrix so scaled is described as *row equilibrated*). Although such a technique will normally help to provide a satisfactory choice of pivots it cannot be guaranteed to do so. For example, consider the following row-equilibrated set of equations obtained by row scaling equations (4.56):

$$\begin{bmatrix} 1000 & 1 & \\ 1000 & 2.041 & 2.041 \\ & 2.041 & 1000 \end{bmatrix} \begin{bmatrix} x_1 \\ x_2 \\ x_3 \end{bmatrix} = \begin{bmatrix} 0.02711 \\ 1.367 \\ 329.9 \end{bmatrix} \tag{4.57}$$

If elimination with partial pivoting is carried out with these equations the weak element in position (3,2) will still be chosen as pivot on the second reduction step. No simple automatic pre-scaling technique is available which completely ensures that strong pivots will always be chosen during elimination when using either partial or full pivoting.

Conclusions about pivoting strategy

(a) If the coefficient matrix of a set of equations is either symmetric and positive definite or diagonally dominant, then the use of partial pivoting is not only unnecessary but may also be detrimental to the accuracy of the solution.

(b) If it is necessary to adopt a pivoting strategy then it may be advisable to try to ensure that the relative magnitudes of the coefficients approximately reflect their relative importance. This condition is most likely to be violated when the coefficients do not all have the same physical characteristics (as in equations 2.3).

(c) If an elimination involving pivoting yields insufficient accuracy some improvement in accuracy may be gained by changing the strategy (e.g. from partial to full pivoting) or by pre-scaling the equations.

4.11 ILL-CONDITIONING

Even with the most suitable choice of pivots it may not be possible to obtain a very accurate solution to a set of linear equations. If a significant loss of accuracy is inevitable then the equations are described as *ill-conditioned*.

A simple example of a set of ill-conditioned equations is

$$\begin{bmatrix} 1.012671 & 1.446949 \\ 1.446949 & 2.068528 \end{bmatrix} \begin{bmatrix} x_1 \\ x_2 \end{bmatrix} = \begin{bmatrix} 0.006324242 \\ 0.002755853 \end{bmatrix} \quad (4.58)$$

The coefficient matrix of these equations is symmetric positive definite and the solution, correct to five significant figures, is {8.4448 −5.9059}. However, if only five significant figures are available throughout the computation, the equations must firstly be rounded to

$$\begin{bmatrix} 1.0127 & 1.4469 \\ 1.4469 & 2.0685 \end{bmatrix} \begin{bmatrix} x_1 \\ x_2 \end{bmatrix} = \begin{bmatrix} 0.0063242 \\ 0.0027559 \end{bmatrix} \quad (4.59)$$

and the reduced equations will be

$$\begin{bmatrix} 1.0127 & 1.4469 \\ & 0.00090000 \end{bmatrix} \begin{bmatrix} x_1 \\ x_2 \end{bmatrix} = \begin{bmatrix} 0.0063242 \\ -0.0062807 \end{bmatrix} \quad (4.60)$$

giving the erroneous solution {9.9773 −6.9786}.

In the solution of ill-conditioned equations accuracy is lost due to cancellation in the elimination. In the above example the excessive loss of accuracy is due to cancellation incurred in the evaluation of the single coefficient $a_{22}^{(2)}$. Clearly the computed value of 0.00090000 (equation 4.60) would be much different if the whole computation were to be carried out to a larger number of significant figures. In the solution of larger systems of ill-conditioned equations a high degree of cancellation may occur in off-diagonal as well as in diagonal coefficients. Whereas the effects of such cancellations may have a cumulative effect on the loss of accuracy of the solution, it is usually cancellation in the pivotal elements which has the greater adverse effect.

A set of linear equations having a singular coefficient matrix does not have a unique solution (section 1.7). Hence it is not difficult to accept the principle that a set of ill-conditioned equations is one having a nearly singular coefficient matrix. Certainly equations (4.58) conform to this criterion. However, this principle has sometimes been interpreted as implying that the magnitude of the determinant of the coefficient matrix is a direct measure of the conditioning of the equations. This interpretation can easily be shown to be invalid for any general set of equations by scaling the whole set of equations, since the determinant is affected considerably whereas the loss of accuracy is essentially unaltered. Furthermore, even with the matrix scaled in such a way that the maximum element in each row and each column is unity, the determinant does not give a good measure of the condition of the equations. For example, the determinants of the two matrices

$$\begin{bmatrix} 1 & 0.9995 & & \\ 0.9995 & 1 & & \\ & & 1 & 0.9995 \\ & & 0.9995 & 1 \end{bmatrix}$$

and

$$\begin{bmatrix} 1 & & & \\ & 1 & 0.9999995 & \\ & 0.9999995 & 1 & \\ & & & 1 \end{bmatrix}$$

are both approximately equal to 0.000001, yet, if they appear as coefficient matrices in two sets of linear equations, losses of accuracy in the solution of three and six figures, respectively, are likely to be obtained. The former matrix can therefore be described as better conditioned than the latter matrix.

An important property of an ill-conditioned set of equations is that the solution is highly sensitive to small changes in the coefficients. For example, in the solution of equations (4.58, small changes to any of the coefficients makes a very significant change in $a_{22}^{(2)}$. In fact, the mere process of rounding the initial equations to the form shown in equation (4.59) has introduced large errors into the solution. (The accurate solution of the rounded equations specified to five significant figures is $\{9.4634 \quad -6.6187\}$ as compared with $\{8.4448 \quad -5.9059\}$ for the unrounded equations.)

Various measures of the condition of a matrix called *condition numbers* have been specified for matrices. The condition numbers provide bounds for the sensitivity of the solution of a set of equations to changes in the coefficient matrix. Unfortunately, the evaluation of any of the condition numbers of a matrix **A** is not a trivial task since it is necessary first to obtain either its inverse in explicit form or else the largest and smallest eigenvalues of **A** or $\mathbf{AA}^T$ (the latter being used if **A** is unsymmetric).

A computer program for solving a set of equations by elimination or triangular decomposition does not normally include, alongside the standard computation, any parallel computation to provide an assessment of the accuracy. Hence the only indication that a user will normally obtain from the execution of the elimination that significant errors are present is if a pivot is so affected that it is identically zero or, in the case of a Choleski decomposition, if a pivot is not positive. In either of these cases the elimination procedure will fail. In all other cases the computation will proceed to yield an erroneous solution.

4.12 ILL-CONDITIONING IN PRACTICE

For many problems the initial data is of limited accuracy (due, for instance to the difficulty of obtaining precise physical measurements) and hence the meaningful accuracy of the solution is correspondingly limited. Suppose that the data for a given problem has an accuracy of no more than four decimal places. If this problem is solved on a computer with a relative precision of 10^{-11}, then the error magnification in the solution would have to be greater than seven decimal places in order to make any significant difference to the solution. There are large classes of problems for which the degree of ill-conditioning required to produce this order of error magnification can only be caused by very unusual and easily recognizable circumstances. For instance, in the analysis of electrical resistance networks by the node conductance method, very ill-conditioned equations are only likely to be encountered if the conductances of the branches have widely varying magnitudes. Thus if in the network of Figure 2.4 the conductance of the branch connecting nodes 1 and 2 is increased from 3.2 mhos to, say, 32,000 mhos while the other branches remain unaltered, then ill-conditioned equations are obtained for which about four figures of accuracy are lost in the solution. Even this loss of accuracy is not likely to be of any concern unless the computer has a relative precision of 10^{-8} or less.

However, there are other types of problem for which ill-conditioning can cause serious difficulties, e.g. in the case of curve fitting. In section 2.7 a simple curve fitting problem was investigated which led to the normal equations (2.39). On elimination the coefficient in position (3,3) reduced from 3.64 to 0.0597, indicating that the loss of accuracy due to cancellation is of the order of two decimal places. Curve fitting operations using higher order polynomials may give rise to normal equations which are much more severely ill-conditioned. A classic problem is the fitting of a polynomial curve of the form

$$y = \sum_{k=1}^{n} c_k x^k$$

to a series of m points whose x coordinates span, at equal intervals, the range from 0 to 1. If m is large the coefficient matrix of the normal equations, scaled by a factor $1/m$, can be shown to approximate to the Hilbert matrix

$$\begin{bmatrix} 1 & \frac{1}{2} & \frac{1}{3} & \cdots & \frac{1}{n} \\ \frac{1}{2} & \frac{1}{3} & \frac{1}{4} & \cdots & \frac{1}{n+1} \\ \frac{1}{3} & \frac{1}{4} & \frac{1}{5} & \cdots & \frac{1}{n+2} \\ \cdot & \cdot & \cdot & \cdots & \cdot \\ \frac{1}{n} & \frac{1}{n+1} & \frac{1}{n+2} & \cdots & \frac{1}{2n-1} \end{bmatrix} \tag{4.61}$$

This matrix is notoriously ill-conditioned. The reason for the severe ill-conditioning can be attributed to the fact that, when expressed in the form $y = \Sigma c_i f_i$ the functions f_i have similar shapes. This is illustrated in Figure 4.4. The similarity in these functions over the range of x causes the columns of the rectangular coefficient matrix A of the error equations (2.20) to have a strong degree of linear dependence, which in turn makes the coefficient matrix of the normal equations near singular.

It can be misleading to say that such problems are naturally ill-conditioned. It is only the choice of variables which make them ill conditioned. In the curve fitting

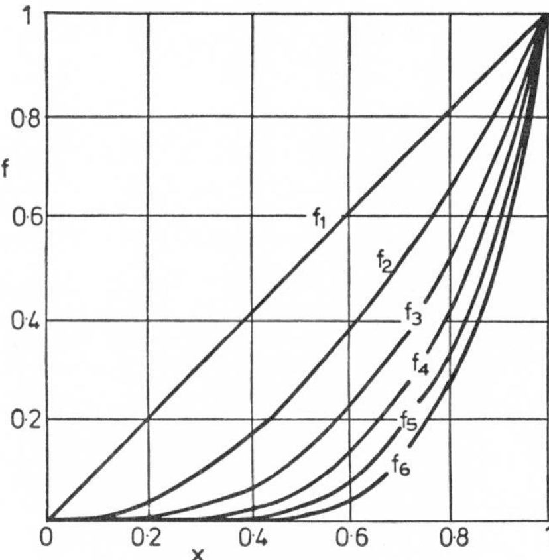

Figure 4.4 Constituent functions of a simple polynomial $y = \sum_{i=1}^{6} c_i f_i$ where $f_i = x^i$

example of section 2.7 well-conditioned equations can be obtained by the simple expedient of subtracting 0.7 from all of the x coordinates. If this is done the normal equations become

$$\begin{bmatrix} 6 & & 0.7 \\ & 0.7 & \\ 0.7 & & 0.1414 \end{bmatrix} \begin{bmatrix} c_0' \\ c_1' \\ c_2' \end{bmatrix} = \begin{bmatrix} 19 \\ 2.1 \\ 1.71 \end{bmatrix} \tag{4.62}$$

from which the coefficients $\{c_0' \quad c_1' \quad c_2'\}$ of the revised polynomial can easily be determined without any significant loss of accuracy. For curve fitting with higher order polynomials it will be necessary to use more effective means to prevent serious ill-conditioning problems than shifting the origin. This may be done by choosing functions f_i which are mutually orthogonal over the range of x (e.g. Chebyshev polynomials). An alternative procedure is to convert the functions into a mutually orthogonal set by numerical operations on their vectors, as described in section 4.19.

4.13 RESIDUALS AND ITERATIVE IMPROVEMENT

For problems yielding linear equations which are not known in advance to be well conditioned, it is necessary to determine whether a computed solution is sufficiently accurate and also to improve the solution if the accuracy is insufficient. The straightforward way of determining whether a set of linear equations $\mathbf{Ax} = \mathbf{b}$ have been solved with sufficient accuracy is to determine the *residual vector*,

$$\mathbf{r}^{(1)} = \mathbf{b} - \mathbf{Ax}^{(1)} \tag{4.63}$$

where $\mathbf{x}^{(1)}$ is the computed solution. If the elements of $\mathbf{r}^{(1)}$ are very small compared with those of b it can normally be assumed that the solution is accurate. Furthermore, the physical significance of the equations may help in interpreting the magnitudes of the residuals, e.g. the residuals obtained from an electrical network node conductance analysis would be the values by which the currents do not balance at the nodes.

If the solution to a particular problem does not appear to be sufficiently accurate, one possible remedy is to repeat the whole computation (not just the solution of the equations) using double precision arithmetic throughout. However, a more efficient remedy may be to adopt an iterative improvement scheme.

It follows from equation (4.63) that

$$\mathbf{A}(\mathbf{x} - \mathbf{x}^{(1)}) = \mathbf{r}^{(1)} \tag{4.64}$$

Thus, if the residual vector is taken as a revised right-hand side to the original equations, a correction to the variables is obtained. But just as the original solution was inaccurate, so also will be the computed correction vector. Let $\mathbf{y}^{(1)}$ be the correction vector computed from

$$\mathbf{A}\mathbf{y}^{(1)} = \mathbf{r}^{(1)} \tag{4.65}$$

Then the next approximate solution will be

$$\mathbf{x}^{(2)} = \mathbf{x}^{(1)} + \mathbf{y}^{(1)} \tag{4.66}$$

Figure 4.5 shows a flow diagram for the complete iteration sequence in which the initial solution of the equations has been included as the first step in the iteration cycle.

Notes on the iterative improvement scheme

(a) The decomposition of the coefficient matrix only needs to be performed once and hence has been placed outside the iteration loop.

(b) Because, in an ill-conditioned set of equations, the rounding of the coefficients of A may cause large errors in the solution, it is essential that the coefficient matrix is constructed with double precision arithmetic and also stored in double precision. It is also necessary to compute the residuals with double precision arithmetic in order that significant errors are not introduced during this computation.

(c) It is possible that the method will not converge at all if the error magnification in the decomposition is excessively large. In the flow diagram the relative error

$$e^{(k)} = \frac{\|\mathbf{r}^{(k)}\|}{\|\mathbf{b}\|} \tag{4.67}$$

has been monitored and a failure exit provided in the case where $e^{(k+1)} > e^{(k)}$.

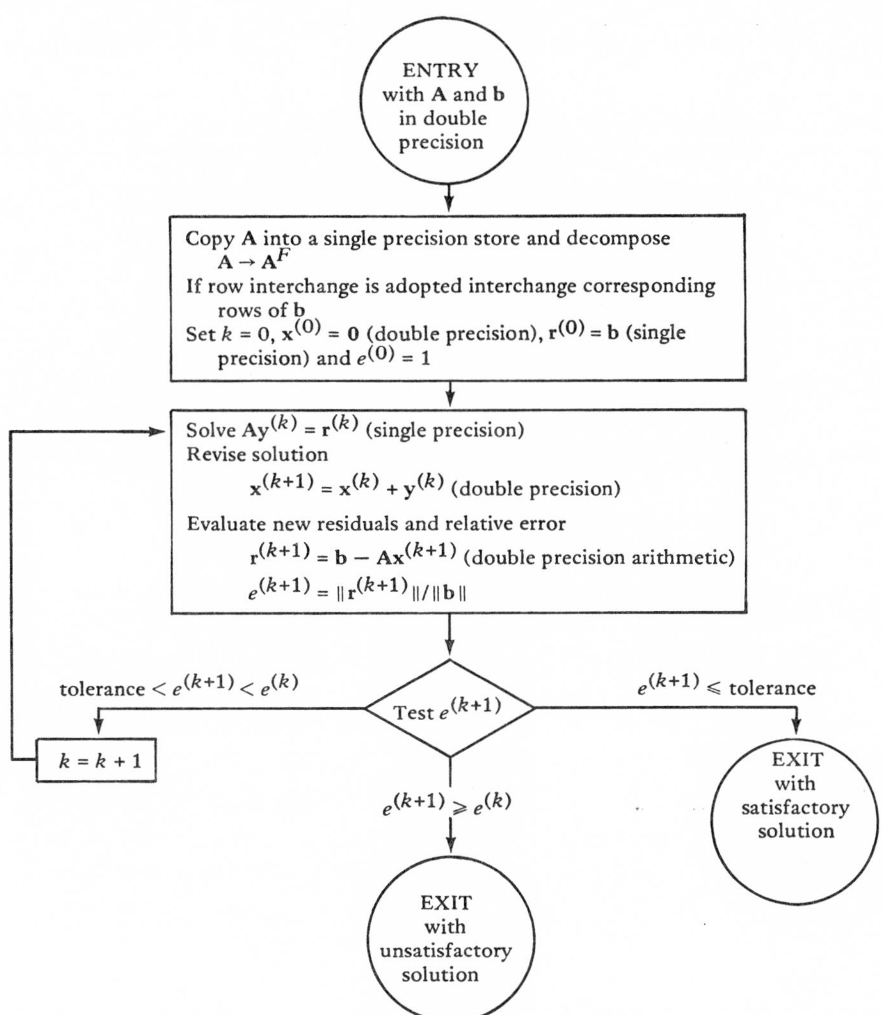

Figure 4.5 Flow diagram for iterative improvement

Examples of iterative improvement

Tables 4.3 and 4.4 show the progress of the iterative improvement scheme when applied to two examples discussed previously in which inaccurate solutions were obtained by elimination. In both cases the decomposition and equation-solving segments were performed using only five significant figures, while each of the residual vectors were computed to higher precision and then rounded to five figures before re-solution. The relative error specified in the last column in both tables is based on largest element norms.

Table 4.3 Iterative improvement for equations (4.51) with weak pivot selection

k	$\mathbf{x}^{(k)}$	$\mathbf{r}^{(k)}$	$e^{(k)}$
0	{0 0 0}	{0.2 1.3816 1.9273}	1
1	{1.9300 −0.68557 0.88750}	{0 −0.012798 −0.0027}	0.0067
2	{1.9270 −0.69818 0.90011}	{0 0.000055469 0.0003}	0.00016
3	{1.9273 −0.69849 0.90042}	{0 −0.0000059141 0.0000}	0.0000031
	etc.		
Correct	{1.9273 −0.69850 0.90042}		

Table 4.4 Iterative improvement for equations (4.58) which are ill-conditioned

k	$\mathbf{x}^{(k)}$	$\mathbf{r}^{(k)}$	$e^{(k)}$
0	{0 0}	{0.0063242 0.0027559}	1
1	{9.9773 −6.9786}	{0.00028017 0.0015411}	0.24
2	{8.1665 −5.7110}	{−0.00012774 −0.00038975}	0.062
3	{8.4954 −5.9413}	{−0.000024094 0.000072077}	0.011
4	{8.4356 −5.8994}	{−0.0000035163 −0.000011861}	0.0019
	etc.		
Correct	{8.4448 −5.9059}		

4.14 TWIN PIVOTING FOR SYMMETRIC MATRICES

In section 4.6 it was established that pivot selection was unnecessary for symmetric positive definite matrices. However, for matrices which are symmetric but not positive definite it is desirable and sometimes necessary to adopt a strategy which allows off-diagonal pivots to be selected. The set of equations

$$\begin{bmatrix} 0 & 10 \\ 10 & 0 \end{bmatrix} \begin{bmatrix} x_1 \\ x_2 \end{bmatrix} = \begin{bmatrix} 5 \\ 6 \end{bmatrix} \tag{4.68}$$

is sufficient to show that off-diagonal pivot selection is sometimes necessary to accomplish a solution by elimination. If pivots are selected individually then the row and/or column interchanges necessary to shift an off-diagonal pivot into the leading position will almost invariably destroy the symmetry of the matrix. In this section a modification to the symmetric $\mathbf{LDL}^T$ decomposition (equation 4.21) will be given which can be applied to general symmetric matrices. The method has some similarities with the one given by Bunch and Parlett (1971).

The reduction procedure

Suppose that element a_{21} of a symmetric matrix $\mathbf{A}$ is strong and that

$$a_{21}^2 > a_{11}a_{22} \tag{4.69}$$

If the second row of the matrix is replaced by (row 2) $-f$(row 1) and in addition the second column is replaced by (column 2) $-f$(column 1), the modified element in position (2,2) will be

$$a_{22}^{(2)} = a_{22} - 2fa_{21} + f^2 a_{11} \tag{4.70}$$

Hence this element will be eliminated if f is chosen to be either of the real values

$$f = \frac{a_{21} \pm (a_{21}^2 - a_{11}a_{22})^{1/2}}{a_{11}} \tag{4.71}$$

Less modification of the matrix will be caused by taking the value having the least absolute magnitude, which is best computed as

$$f = \frac{a_{22}}{a_{21} + (\text{sign}\, a_{21})(a_{21}^2 - a_{11}a_{22})^{1/2}} \tag{4.72}$$

The element in position (2,1) will be modified to

$$p = a_{21}^{(2)} = a_{21} - fa_{11} = (\text{sign}\, a_{21})(a_{21}^2 - a_{11}a_{22})^{1/2} \tag{4.73}$$

On account of conditions (4.69) this element and its companion $a_{12}^{(2)}$ will be non-zero and both can act as pivots. The two reduction steps for these pivots are illustrated as a three-stage operation in Figure 4.6. The modifications to the lower triangular elements can be summarized as follows:

$$\text{Stage (a)} \left\{ \begin{aligned} & a_{21}^{(2)} = p \\ & a_{22}^{(2)} = 0 \\ & a'_{i2} = a_{i2} - fa_{i1}, \quad \text{for } i = 3 \to n \\ & l_{21} = f \end{aligned} \right. \tag{4.74a}$$

$$\text{Stage (b)} \left\{ \begin{aligned} & a_{i1}^{(2)} = a_{i1} - \frac{a'_{i2}a_{11}}{p}, \quad \text{for } i = 3 \to n \\ & a_{i2}^{(2)} = 0 \\ & a_{ij}^{(2)} = a_{ij} - \frac{a'_{i2}a_{j1} + a'_{j2}a_{i1}^{(2)}}{p}, \quad \text{for } i = 3 \to n,\ j = 3 \to i \\ & l_{i1} = \frac{a'_{i2}}{p}, \quad \text{for } i = 3 \to n \end{aligned} \right. \tag{4.74b}$$

$$\text{Stage (c)} \left\{ \begin{aligned} & a_{i1}^{(3)} = 0 \\ & l_{i2} = \frac{a_{i1}^{(2)}}{p} \end{aligned} \right\} \quad \text{for } i = 3 \to n \tag{4.74c}$$

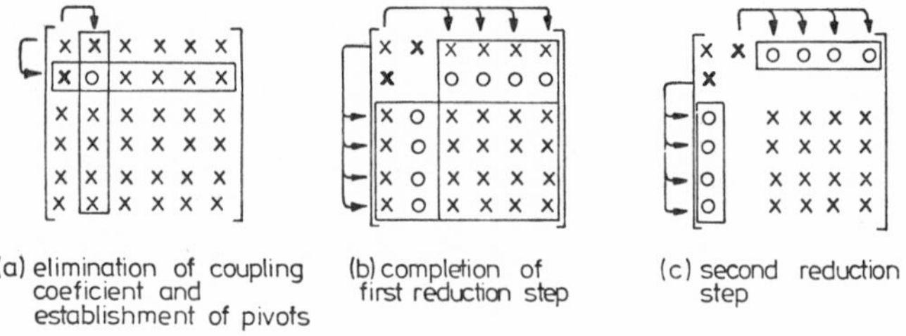

(a) elimination of coupling coeficient and establishment of pivots

(b) completion of first reduction step

(c) second reduction step

Figure 4.6 Two reduction steps corresponding to twin off-diagonal pivots

Unspecified coefficients can be assumed to remain unaltered, for instance

$$a_{11}^{(3)} = a_{11}^{(2)} = a_{11}$$

Pivot selection

If an off-diagonal element a_{ij} and its companion a_{ji} are selected for shifting into the leading off-diagonal positions, the corresponding pivots have the magnitude $(a_{ij}^2 - a_{ii}a_{jj})^{1/2}$. Alternatively, a diagonal pivot could be chosen and corresponding row and column interchanges carried out so that the pivot is placed in the leading position of the leading diagonal. In the latter case a normal $\mathbf{LDL}^T$ reduction step could follow. Hence, if the values of all the expressions $a_{ij}^2 - a_{ii}a_{jj}$ are computed for $i > j$ and the maximum of these values is compared with the square of the maximum value of a_{ii}, it is possible to determine how to proceed in order to obtain the maximum pivot at the next reduction step.

General procedure

After the pivot(s) has been shifted to the appropriate leading position(s) and the eliminations corresponding to the pivot selection have been performed, the active part of the matrix has then been reduced to one of order one or two less. The whole process can be repeated on the reduced matrix in a recursive way until the decomposition is complete. The result of the decomposition is to factorize the permuted matrix $\bar{\mathbf{A}}$ into components satisfying

$$\bar{\mathbf{A}} = \mathbf{L}\bar{\mathbf{D}}\mathbf{L}^T \tag{4.75}$$

For example, if in the case of a 6×6 matrix two diagonal pivots were selected in the first two reduction steps and this was followed by the selection of two sets of off-diagonal pivots, $\bar{\mathbf{D}}$ would have the form

$$\bar{\mathbf{D}} = \begin{bmatrix} d_1 & & & & & \\ & d_2 & & & & \\ & & d_3 & \bar{d}_4 & & \\ & & \bar{d}_4 & & & \\ & & & & d_5 & \bar{d}_6 \\ & & & & \bar{d}_6 & \end{bmatrix} \tag{4.76}$$

$$
\mathbf{A} = \begin{array}{c} \\ 1 \\ 2 \\ 3 \\ 4 \end{array}
\begin{array}{cccc} 1 & 2 & 3 & 4 \\ \hline -292 & & & \\ -280 & 47 & \text{symmetric} & \\ \boxed{660} & -200 & 700 & \\ 2 & -175 & 390 & 152 \end{array}
\qquad
\mathbf{A}^{(1)} = \begin{array}{c} \\ 3 \\ 1 \\ 2 \\ 4 \end{array}
\begin{array}{cccc} 3 & 1 & 2 & 4 \\ \hline 700 & & & \\ \boxed{660} & -292 & \text{symmetric} & \\ -200 & -280 & 47 & \\ 390 & 2 & -175 & 152 \end{array}
$$

Note. Pivots are placed in leading off-diagonal position with $a_{11} \geqslant a_{22}$.

$$
\mathbf{L}^{(2)} = \begin{array}{c} \\ 3 \\ 1 \\ 2 \\ 4 \end{array}
\begin{array}{cccc} 3 & 1 & 2 & 4 \\ \hline 1 & & & \\ -0.2 & 1 & & \\ -0.4 & & 1 & \\ 0.1 & & & 1 \end{array}
\qquad
\mathbf{A}^{(2)} = \begin{array}{c} \\ 3 \\ 1 \\ 2 \\ 4 \end{array}
\begin{array}{cccc} 3 & 1 & 2 & 4 \\ \hline 700 & 800 & 80 & 320 \\ 800 & & & \\ 80 & & -1 & -27 \\ 320 & & -27 & 81 \end{array}
$$

Note. $l_{21} = f$.

$$
\mathbf{L}^{(3)} = \begin{array}{c} \\ 3 \\ 1 \\ 2 \\ 4 \end{array}
\begin{array}{cccc} 3 & 1 & 2 & 4 \\ \hline 1 & & & \\ -0.2 & 1 & & \\ -0.4 & 0.1 & 1 & \\ 0.1 & 0.4 & & 1 \end{array}
\qquad
\mathbf{A}^{(3)} = \begin{array}{c} \\ 3 \\ 1 \\ 2 \\ 4 \end{array}
\begin{array}{cccc} 3 & 1 & 2 & 4 \\ \hline 700 & 800 & & \\ 800 & & & \\ & & -1 & -27 \\ & & -27 & \boxed{81} \end{array}
$$

Note. Last two rows need to be interchanged to place diagonal pivot in leading position.

$$
\mathbf{L} = \begin{array}{c} \\ 3 \\ 1 \\ 4 \\ 2 \end{array}
\begin{array}{cccc} 3 & 1 & 4 & 2 \\ \hline 1 & & & \\ -0.2 & 1 & & \\ 0.1 & 0.4 & 1 & \\ -0.4 & 0.1 & -0.3333 & 1 \end{array}
\qquad
\bar{\mathbf{D}} = \begin{array}{c} \\ 3 \\ 1 \\ 4 \\ 2 \end{array}
\begin{array}{cccc} 3 & 1 & 4 & 2 \\ \hline 700 & 800 & & \\ 800 & & & \\ & & 81 & \\ & & & -10 \end{array}
$$

Figure 4.7 An example of symmetric decomposition with twin pivoting (the row and column order is indicated alongside each matrix)

Once the decomposition is complete the solution of any set of linear equations of the form $\mathbf{Ax} = \mathbf{b}$ can easily be carried out by three substitution processes, remembering that the right-hand vector and the variables must be permuted to be consistent with the row and column permutations of the decomposed form of the matrix.

Figure 4.7 shows the stages in the symmetric decomposition of a 4 x 4 matrix.

Notes

(a) The pivot selection procedure cannot break down unless the matrix is singular.

(b) If the method is applied to a positive definite matrix, diagonal pivots will always be selected (since $a_{ij}^2 < a_{ii}a_{jj}$).

(c) If off-diagonal pivots are selected the interchange can be carried out in two possible ways. The method which makes $|a_{11}| \geqslant |a_{22}|$ would seem to be the more expedient alternative since this ensures that $|f| < 1$.

(d) The decomposition may be carried out in a triangular store.

(e) The decomposition itself requires less computation than decompositions based on the normal pivoting strategies, and may give a more accurate result. However, the amount of computation saved by the symmetric reduction is offset by the computation penalty involved in the pivot selection procedure if the full search procedure outlined above is implemented. (On the other hand, it is not reliable to select each pivot just from a single column.)

4.15 EQUATIONS WITH PRESCRIBED VARIABLES

In the solution of linear problems it is frequently possible to write the equations in the basic form $\mathbf{Ax} = \mathbf{b}$ when some of the elements of $\mathbf{x}$ have known values and are not therefore true variables. Suppose that a basic set of equations of order 4 has a prescribed value of v for x_3. Only three of the original equations will be required to determine the three unknowns. Assuming that the third equation is the one to be omitted, the remaining equations can be specified in the form

$$\begin{bmatrix} a_{11} & a_{12} & & a_{14} \\ a_{21} & a_{22} & & a_{24} \\ & & & \\ a_{41} & a_{42} & & a_{44} \end{bmatrix} \begin{bmatrix} x_1 \\ x_2 \\ \\ x_4 \end{bmatrix} = \begin{bmatrix} b_1 - a_{13}v \\ b_2 - a_{23}v \\ \\ b_4 - a_{43}v \end{bmatrix} \tag{4.77}$$

Gaps have been left in the equations so that the original structure of the coefficients is maintained. In order to solve these equations with a standard program it would be necessary to compact the storage so that no gaps were present.

An alternative procedure is to include $x_3 = v$ as the third equation, in which case the third column of $\mathbf{A}$ can be returned to the left-hand side, i.e.

$$\begin{bmatrix} a_{11} & a_{12} & a_{13} & a_{14} \\ a_{21} & a_{22} & a_{23} & a_{24} \\ & & 1 & \\ a_{41} & a_{42} & a_{43} & a_{44} \end{bmatrix} \begin{bmatrix} x_1 \\ x_2 \\ x_3 \\ x_4 \end{bmatrix} = \begin{bmatrix} b_1 \\ b_2 \\ v \\ b_4 \end{bmatrix} \tag{4.78}$$

This method has the disadvantage that, if $\mathbf{A}$ was originally a symmetric matrix, the symmetry would be destroyed. A novel modification which can be adopted with floating-point storage is to leave the third row of $\mathbf{A}$ in place, but to override any influence that these coefficients may have by applying a scaling factoı to the equation $x_3 = v$. The scaling factor should be very much larger than the coefficients of $\mathbf{A}$. Normally 10^{20} would be a suitable value. The equations would therefore become

$$\begin{bmatrix} a_{11} & a_{12} & a_{13} & a_{14} \\ a_{21} & a_{22} & a_{23} & a_{24} \\ a_{31} & a_{32} & 10^{20} & a_{34} \\ a_{41} & a_{42} & a_{43} & a_{44} \end{bmatrix} \begin{bmatrix} x_1 \\ x_2 \\ x_3 \\ x_4 \end{bmatrix} = \begin{bmatrix} b_1 \\ b_2 \\ 10^{20}v \\ b_4 \end{bmatrix} \tag{4.79}$$

the solution of which will be the same as the solution of equations (4.77).

An example of the possible use of this technique is in the analysis of electrical networks by the node conductance method. The node conductance equations may first be automatically constructed for the whole network with a voltage variable allocated to every node whether its voltage is known or not. Then for every node (such as a datum node) which has a known voltage the corresponding equation could be modified in the way described above. Clearly this technique will require the solution of a set of equations of larger order than is strictly necessary. However, the technique can often result in simpler programs, particularly where the matrix **A** has a systematic structure which is worth preserving.

It may be useful to appreciate the physical interpretation of the additional terms. If equation (4.79) represents the node conductance equations of an electrical network, then the term 10^{20} in the coefficient matrix would arise if node 3 has an earth connection with a very high conductance of 10^{20}. The term $10^{20}v$ in the right-hand side would be caused by a very large current input at node 3 of $10^{20}v$. Hence the modifications to the equations correspond to the modification to the network shown in Figure 4.8.

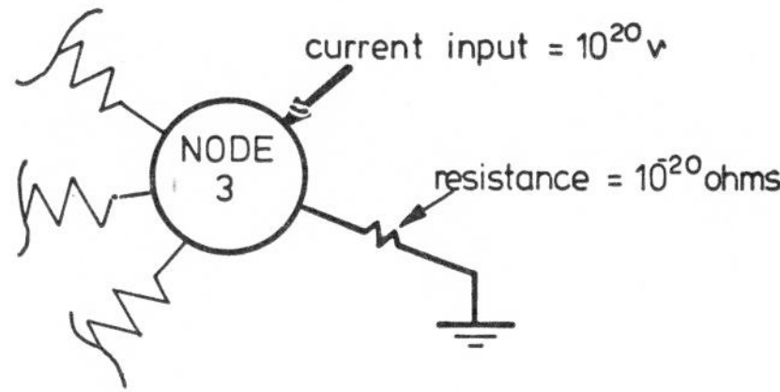

Figure 4.8 Modification to an electrical resistance network to constrain node 3 to a voltage v

4.16 EQUATIONS WITH A SINGULAR COEFFICIENT MATRIX

It is possible to express the solution of a set of linear equations in terms of the eigenvalues and eigenvectors of the coefficient matrix. Suppose that the coefficient matrix **A** is symmetric with eigenvalues and eigenvectors satisfying equation (1.112). Then the right-hand vector may be expressed as a linear combination of the eigenvectors, i.e.

$$\mathbf{b} = c_1\mathbf{q}_1 + c_2\mathbf{q}_2 + \cdots + c_n\mathbf{q}_n \tag{4.80}$$

Alternatively, if $\mathbf{c} = \{c_1 c_2 \ldots c_n\}$ and $\mathbf{Q} = [\mathbf{q}_1 \mathbf{q}_2 \ldots \mathbf{q}_n]$,

$$\mathbf{b} = \mathbf{Q}\mathbf{c} \tag{4.81}$$

then the linear equations may be written in the form

$$\mathbf{A}\mathbf{x} = \mathbf{Q}\mathbf{c} \tag{4.82}$$

Since $\mathbf{Q}$ is an orthogonal matrix (section 1.20) it follows that

$$\mathbf{A}\mathbf{Q}\mathbf{Q}^T\mathbf{x} = \mathbf{Q}\mathbf{c} \tag{4.83}$$

Hence, from equation (1.112),

$$\mathbf{Q}\boldsymbol{\Lambda}\mathbf{Q}^T\mathbf{x} = \mathbf{Q}\mathbf{c}. \tag{4.84}$$

Premultiplying by $\mathbf{Q}\boldsymbol{\Lambda}^{-1}\mathbf{Q}^T$ gives

$$\mathbf{x} = \mathbf{Q}\boldsymbol{\Lambda}^{-1}\mathbf{c} = \frac{c_1}{\lambda_1}\mathbf{q}_1 + \frac{c_2}{\lambda_2}\mathbf{q}_2 + \cdots + \frac{c_n}{\lambda_n}\mathbf{q}_n \tag{4.85}$$

If A is singular then one or more of its eigenvalues will be zero and the theoretical solution will normally be infinite. However, there are special cases which have finite solutions because the right-hand side is such that for every zero eigenvalue the corresponding coefficient c_i is also zero.

An example of such a set of equations is the node conductance equations for the Julie bridge (Figure 2.1) specified without a datum node. The full set of equations can be derived in terms of the conductances of the branches and the voltage input as on the facing page.

If an elimination is performed on these equations the last reduction step should theoretically yield

$$0 \times e_E = 0 \tag{4.87}$$

In practice, rounding errors will normally affect the computation on both the left- and right-hand sides of the equations, so giving an arbitrary value for e_E. Back-substitution should then give a valid solution for the voltages. The fact that the valid computed solution is not unique will not be of much concern since the relative voltages and the branch voltages will still be unique and computable.

It can easily be ascertained that the node conductance matrix shown above has a zero eigenvalue with corresponding eigenvector $\mathbf{q}_1 = \{1 \quad 1 \quad 1 \quad 1 \quad 1\}$, and is consequently singular. Using the eigenvector orthogonality property for symmetric matrices, it is possible to show that equation (4.80) gives

$$\mathbf{q}_1^T\mathbf{b} = c_1 \tag{4.88}$$

Hence it is not difficult to show that the right-hand vector of equations (4.86) is such that $c_1 = 0$, proving that these equations fall into the special category discussed.

This section should not be considered as an exhortation to deliberately risk solving sets of equations with singular coefficient matrices, but rather as an explanation of why such equations could yield a valid solution if they are accidentally constructed. The conclusions are also true for equations having unsymmetric coefficient matrices, although the eigenvector orthogonality properties will not be the same.

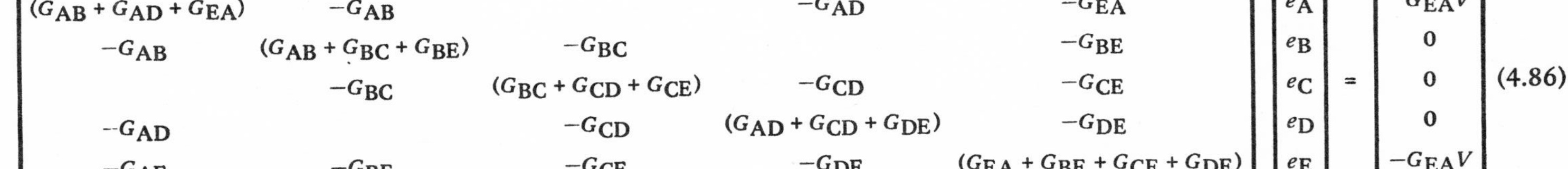

$$\begin{bmatrix} (G_{AB}+G_{AD}+G_{EA}) & -G_{AB} & & -G_{AD} & -G_{EA} \\ -G_{AB} & (G_{AB}+G_{BC}+G_{BE}) & -G_{BC} & & -G_{BE} \\ & -G_{BC} & (G_{BC}+G_{CD}+G_{CE}) & -G_{CD} & -G_{CE} \\ -G_{AD} & & -G_{CD} & (G_{AD}+G_{CD}+G_{DE}) & -G_{DE} \\ -G_{AE} & -G_{BE} & -G_{CE} & -G_{DE} & (G_{EA}+G_{BE}+G_{CE}+G_{DE}) \end{bmatrix} \begin{bmatrix} e_A \\ e_B \\ e_C \\ e_D \\ e_E \end{bmatrix} = \begin{bmatrix} G_{EA}V \\ 0 \\ 0 \\ 0 \\ -G_{EA}V \end{bmatrix} \qquad (4.86)$$

4.17 THE SOLUTION OF MODIFIED EQUATIONS

This section is concerned with the solution of a set of equations for which a similar, but not identical, coefficient matrix has already been decomposed. The equations to be solved may be written in the form

$$(\mathbf{A} + \mathbf{C})\mathbf{x} = \mathbf{b} \tag{4.89}$$

where **A** is the matrix whose decomposition is available. Two situations arise in which it can be of advantage to make use of the decomposition of **A** rather than to solve the new equations by the standard elimination procedure. Namely, when all of the elements of **C** are small compared with those of **A** and when **C** contains only a limited number of non-zero elements.

Use of the iterative improvement method

If the elements of **C** are small compared with those of **A** and in addition **A** is well conditioned, then the decomposition of **A** can be regarded as an inaccurate decomposition of **A** + **C** and the solution to equation (4.89) obtained by the iterative improvement method (section 4.13). The necessary modifications to the flow diagram, Figure 4.5, will be as follows:

(a) The decomposition $\mathbf{A} \to \mathbf{A}^F$ will already be available.

(b) The residuals must be computed from $\mathbf{r}^{(k+1)} = \mathbf{b} - (\mathbf{A} + \mathbf{C})\mathbf{x}^{(k+1)}$.

(c) The accuracy of the result will normally be satisfactory if the computation is carried out with single precision arithmetic throughout.

It can be shown that the iterative improvement method can be written in the alternative form:

$$\mathbf{A}\mathbf{x}^{(k+1)} = \mathbf{b} - \mathbf{C}\mathbf{x}^{(k)} \tag{4.90}$$

As a simple example consider the solution of the equations

$$\left\{\begin{bmatrix} 16 & 4 & 8 \\ 4 & 5 & -4 \\ 8 & -4 & 22 \end{bmatrix} + \begin{bmatrix} 1 & 1 & 0 \\ 1 & 1 & 0 \\ 0 & 0 & 2 \end{bmatrix}\right\}\begin{bmatrix} x_1 \\ x_2 \\ x_3 \end{bmatrix} = \begin{bmatrix} 20 \\ -5 \\ 36 \end{bmatrix} \tag{4.91}$$

using the iterative improvement method and making use of the decomposition specified by equation (4.27). Table 4.5 gives the convergence of the solution, which in this case can be described as steady. Iterative improvement is likely to be most useful for large-order equations in which a large amount of computation will be required for a direct solution of the equations. If **A** and **C** are fully populated $n \times n$ matrices then it is of advantage to use iterative improvement rather than re-solution if the number of iterations is less than $n/6$ (unsymmetric **A** and **C**) or $n/12$ (symmetric **A** and **C**).

Table 4.5 Iterative improvement for modified equations (4.91)

k	$\mathbf{x}^{(k)}$	$\mathbf{r}^{(k)}$	$e^{(k)}$
0	{0 0 0}	{20 −5 36}	1
1	{0.8056 −0.6667 1.2222}	{−0.1389 −0.1389 −2.4444}	0.068
2	{1.0495 −1.1146 0.9410}	{0.2040 0.2040 0.5625}	0.016
3	{0.9856 −0.9613 1.0176}	{−0.0894 −0.0894 −0.1533}	0.0043
4	{1.0045 −1.0130 0.9944}	{0.0327 0.0327 0.0465}	0.0013
5	{0.9985 −0.9957 1.0018}	{−0.0113 −0.0113 −0.0149}	0.00041
6	{1.0005 −1.0014 0.9994}	{0.0038 0.0038 0.0049}	0.00014
	etc.		
Correct	{1 −1 1}		

A supplementary equation method

Both the supplementary equation method and the revised elimination method (to be described later) are non-iterative in character and are applicable when the elements of **C** are large, provided that **C** contains a limited number of non-zero elements. Both **A** and **A** + **C** must be non-singular.

If equation (4.89) is premultiplied by $\mathbf{A}^{-1}$ then the equation

$$(\mathbf{I} - \mathbf{A}^{-1}\mathbf{C})\mathbf{x} = \mathbf{A}^{-1}\mathbf{b} \tag{4.92}$$

is obtained. When some of the columns of **C** are null, $\mathbf{A}^{-1}\mathbf{C}$ is found to have corresponding columns null and hence equation (4.92) has a form in which the left-hand side is already partly reduced. Consider the case where the decomposition

$$\mathbf{A} = \begin{bmatrix} 4 & 2 & 0 & 2 & 0 \\ 2 & 5 & 2 & -1 & 0 \\ 0 & 2 & 5 & -3 & 2 \\ 2 & -1 & -3 & 7 & 1 \\ 0 & 0 & 2 & 1 & 6 \end{bmatrix} = \begin{bmatrix} 1 & & & & \\ \tfrac{1}{2} & 1 & & & \\ 0 & \tfrac{1}{2} & 1 & & \\ \tfrac{1}{2} & -\tfrac{1}{2} & -\tfrac{1}{2} & 1 & \\ 0 & 0 & \tfrac{1}{2} & \tfrac{1}{2} & 1 \end{bmatrix} \begin{bmatrix} 4 & 2 & 0 & 2 & 0 \\ & 4 & 2 & -2 & 0 \\ & & 4 & -2 & 2 \\ & & & 4 & 2 \\ & & & & 4 \end{bmatrix} \tag{4.93}$$

is already available and that a solution of the equations

$$\left\{ \begin{bmatrix} & & \\ & \mathbf{A} & \\ & & \end{bmatrix} + \begin{bmatrix} 0 & 0 & 0 & 0 & 0 \\ 0 & 0 & 0 & 0 & 0 \\ 0 & 0 & 8 & 0 & 4 \\ 0 & 0 & 4 & 0 & -10 \\ 0 & 0 & 8 & 0 & -18 \end{bmatrix} \right\} \begin{bmatrix} x_1 \\ x_2 \\ x_3 \\ x_4 \\ x_5 \end{bmatrix} = \begin{bmatrix} 0 \\ 14 \\ -15 \\ 10 \\ 4 \end{bmatrix} \tag{4.94}$$

is required. By using the decomposition of A it is found that

$$\mathbf{A}^{-1}\mathbf{C} = \begin{bmatrix} 0 & 0 & -0.75 & 0 & 0.75 \\ 0 & 0 & -0.5 & 0 & -1.5 \\ 0 & 0 & 3 & 0 & 3 \\ 0 & 0 & 2 & 0 & 0 \\ 0 & 0 & 0 & 0 & -4 \end{bmatrix} \quad \text{and} \quad \mathbf{A}^{-1}\mathbf{b} = \begin{bmatrix} -3.5 \\ 7 \\ -7 \\ 0 \\ 3 \end{bmatrix} \tag{4.95}$$

Substitution in equation (4.92) gives

$$\begin{bmatrix} 1 & & -0.75 & & 0.75 \\ & 1 & -0.5 & & -1.5 \\ & & 4 & & 3 \\ & & 2 & 1 & 0 \\ & & 0 & & -3 \end{bmatrix} \begin{bmatrix} x_1 \\ x_2 \\ x_3 \\ x_4 \\ x_5 \end{bmatrix} = \begin{bmatrix} -3.5 \\ 7 \\ -7 \\ 0 \\ 3 \end{bmatrix} \tag{4.96}$$

Extracting the third and fifth element equations give supplementary equations

$$\begin{bmatrix} 4 & 3 \\ 0 & -3 \end{bmatrix} \begin{bmatrix} x_3 \\ x_5 \end{bmatrix} = \begin{bmatrix} -7 \\ 3 \end{bmatrix} \tag{4.97}$$

whose solution is $\{-1 \quad -1\}$. Substituting these values into equation (4.96) yields the values of the other variables, giving the complete solution $\{-3.5 \quad 5 \quad -1 \quad 2 \quad -1\}$.

It will be noticed that the non-zero columns of $\mathbf{A}^{-1}\mathbf{C}$ can be obtained by processing the non-zero columns of **C** as if they were supplementary right-hand vectors to the original equations. The order of the supplementary equations will be equal to the number of non-zero columns of **C**. It is also possible to process multiple right-hand sides to the equations in parallel by expanding **b** and **x** from column vectors to $n \times m$ matrices **B** and **X** respectively. If **A** is a previously decomposed fully populated unsymmetric matrix of order $n \times n$ and **C** contains p non-zero columns, then the solution of equation (4.92) with m right-hand vectors require approximately $n^2(p + m) + npm + p^3/3$ multiplications. Hence it is likely to be more economical to use the above method rather than to solve the equations directly by elimination if

$$p < \frac{n^2}{3(n + m)} \tag{4.98}$$

Revising the decomposition

If a decomposition is carried out on the coefficient matrix **A** + **C** of equations (4.94) the first two rows and columns of the factors will be identical to those obtained in the decomposition of A. Hence the solution of these equations by elimination can be accelerated by entering the decomposition procedure at an appropriate point so that only the revised part of the decomposition needs to be performed. For equation (4.94) it would be most convenient if the decomposition of A were carried out by the standard Gaussian elimination method in which the multipliers were retained, rather than by a direct procedure based on equations (4.15) or (4.28). After the second reduction step the decomposition could then be written as

$$\mathbf{A} = \begin{bmatrix} 1 & & & & \\ 0.5 & 1 & & & \\ 0 & 0.5 & 1 & & \\ 0.5 & -0.5 & & 1 & \\ 0 & 0 & & & 1 \end{bmatrix} \begin{bmatrix} 4 & 2 & 0 & 2 & 0 \\ & 4 & 2 & -2 & 0 \\ & & 4 & -2 & 2 \\ & & -2 & 5 & 1 \\ & & 2 & 1 & 6 \end{bmatrix} \tag{4.99}$$

The active part of the matrix can then be modified and the decomposition revised to give

$$\mathbf{A} + \mathbf{C} = \begin{bmatrix} 1 & & & & \\ 0.5 & 1 & & & \\ 0 & 0.5 & 1 & & \\ 0.5 & -0.5 & & 1 & \\ 0 & 0 & & & 1 \end{bmatrix} \begin{bmatrix} 4 & 2 & 0 & 2 & 0 \\ & 4 & 2 & -2 & 0 \\ & & 12 & -2 & 6 \\ & & 2 & 5 & -9 \\ & & 10 & 1 & -12 \end{bmatrix}$$

$$= \begin{bmatrix} 1 & & & & \\ 0.5 & 1 & & & \\ 0 & 0.5 & 1 & & \\ 0.5 & -0.5 & 0.16667 & 1 & \\ 0 & 0 & 0.83333 & 0.5 & 1 \end{bmatrix} \begin{bmatrix} 4 & 2 & 0 & 2 & 0 \\ & 4 & 2 & -2 & 0 \\ & & 12 & -2 & 6 \\ & & & 5.3333 & -10 \\ & & & & -12 \end{bmatrix} \tag{4.100}$$

The solution to equation (4.94) can then be determined by forward-substitution and backsubstitution in the normal way.

The revision of the decomposition is rapid if the only non-zero elements of C occur in the last rows and columns. If it is known, before a set of equations are first solved, which coefficients are to be modified, the rows and columns can be ordered in such a way that the revised solutions may be obtained rapidly. It may also be possible to store the partially eliminated form of the equations at an appropriate stage in the reduction if this will facilitate restarting the elimination. If it is not known in advance where modifications to the coefficients will occur until after the first decomposition has taken place, not only may most of the elimination have to be revised but also results of the intermediate reduction steps in the Gaussian elimination will not be available. However, it will be noted that $a_{ij}^{(k)}$ at the k-th reduction step can be recovered from the decomposed form. Where the standard

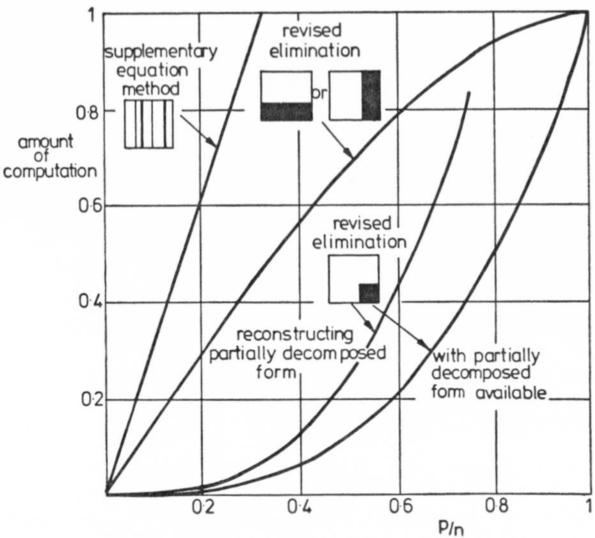

Figure 4.9 Solution of modified equations: the amount of computation as a proportion of computation for direct elimination assuming that A is of order $n \times n$, fully populated and unsymmetric and that C has p non-zero rows or columns

LU decomposition has been carried out then

$$\left.\begin{aligned} a_{ij}^{(k)} &= \sum_{r=k}^{j} l_{ir} u_{rj}, \qquad &&\text{for } j < i \\ &\text{and} \\ a_{ij}^{(k)} &= u_{ij} + \sum_{r=k}^{i-1} l_{ir} u_{rj}, \qquad &&\text{for } j \geqslant i \end{aligned}\right\} \tag{4.101}$$

In Figure 4.9 a comparison is made of the amounts of computation required to solve modified equations where only p rows or columns are to be modified, and there is a single right-hand side. Revised elimination is more efficient than the supplementary equation method but has the restriction that the modified rows or columns must be placed last. If the last p rows or columns are to be modified then triangular decomposition can easily be revised by direct evaluation of the revised part of $\mathbf{A}^F$, the number of multiplications required being approximately $(p/2)(n^2 - p^2/3)$. If only the trailing $p \times p$ submatrix is to be modified then approximately $p^3/3$ multiplications will be required to revise the solution of the appropriate partially reduced form. However, if this is not available, a total of approximately $2p^3/3$ multiplications will be required to recover the active coefficients (by means of equation (4.101) with $k = n - p + 1$) and then to revise the solution. An advantage of the revised elimination is that further modifications can be carried out easily, whereas with the supplementary equation method this is more difficult.

Methods for modifying the triangular decomposition of a matrix are also given by Gill, Murray and Saunders (1975). Although these methods are more complicated than those given in this section, they are efficient for gross modifications of a small number of elements situated anywhere in the matrix.

4.18 ORTHOGONAL DECOMPOSITION

It is possible to decompose an $m \times n$ matrix $\mathbf{A}$ ($m \geqslant n$) into the product of an $m \times n$ matrix $\mathbf{Q}$ satisfying $\mathbf{Q}^T\mathbf{Q} = \mathbf{I}$ and an $n \times n$ upper triangular matrix $\mathbf{U}$, i.e.

$$\mathbf{A} = \mathbf{QU} \tag{4.102}$$

The matrix $\mathbf{U}$ will be non-singular provided that the columns of $\mathbf{A}$ are not linearly dependent. If $\mathbf{A}$ is the coefficient matrix of a set of linear equations $\mathbf{Ax} = \mathbf{b}$ then

$$\mathbf{QUx} = \mathbf{b} \tag{4.103}$$

Premultiplying by $\mathbf{Q}^T$ and using the orthogonality condition gives

$$\mathbf{Ux} = \mathbf{Q}^T\mathbf{b} \tag{4.104}$$

Since $\mathbf{U}$ is non-singular, equation (4.104) may always be solved to determine $\mathbf{x}$. Hence, if there is a solution to the equations $\mathbf{Ax} = \mathbf{b}$, it may be obtained in the following way:

$$\left.\begin{array}{ll} \text{(a)} & \text{Decompose } \mathbf{A} \text{ according to } \mathbf{A} = \mathbf{QU}. \\ \text{(b)} & \text{Premultiply the right-hand vector by } \mathbf{Q}^T \text{ to give } \bar{\mathbf{b}} = \mathbf{Q}^T\mathbf{b}. \\ \text{(c)} & \text{Backsubstitute in equation } \mathbf{Ux} = \bar{\mathbf{b}}. \end{array}\right\} \tag{4.105}$$

If $m = n$ the original linear equations have a solution, which this orthogonal decomposition procedure obtains. Although it is not, strictly speaking, an elimination method, there are some similarities with the method of triangular decomposition (sections 4.2 and 4.3).

The *Gram–Schmidt procedure* may be used to carry out the orthogonal decomposition of $\mathbf{A}$ where $m \geqslant n$. If $\mathbf{a}_i$ and $\mathbf{q}_i$ are the i-th columns of $\mathbf{A}$ and $\mathbf{Q}$ respectively, the decomposition may be expressed as

$$\left[\begin{array}{c|c|c|c} & & & \\ \mathbf{a}_1 & \mathbf{a}_2 & \cdots & \mathbf{a}_n \\ & & & \end{array}\right] = \left[\begin{array}{c|c|c|c} & & & \\ \mathbf{q}_1 & \mathbf{q}_2 & \cdots & \mathbf{q}_n \\ & & & \end{array}\right] \begin{bmatrix} u_{11} & u_{12} & \cdots & u_{1n} \\ & u_{22} & \cdots & u_{2n} \\ & & \ddots & \vdots \\ & & & u_{nn} \end{bmatrix} \tag{4.106}$$

the first vector equation of which is

$$\mathbf{a}_1 = u_{11}\mathbf{q}_1 \tag{4.107}$$

Since $\mathbf{q}_1^T\mathbf{q}_1 = 1$ it follows that u_{11} is the Euclidean norm of $\mathbf{a}_1$. Thus $\mathbf{q}_1$ may be obtained from $\mathbf{a}_1$ by Euclidean normalization. The second vector equation of (4.106) is

$$\mathbf{a}_2 = u_{12}\mathbf{q}_1 + u_{22}\mathbf{q}_2 \tag{4.108}$$

Premultiplying this equation by $\mathbf{q}_1^T$ and making use of the relationships $\mathbf{q}_1^T\mathbf{q}_1 = 1$ and $\mathbf{q}_1^T\mathbf{q}_2 = 0$ gives

$$u_{12} = \mathbf{q}_1^T\mathbf{a}_2 \tag{4.109}$$

Hence a modified vector

$$\mathbf{a}_2^{(2)} = \mathbf{a}_2 - u_{12}\mathbf{q}_1 \tag{4.110}$$

may be obtained which, from equation (4.108), is proportional to $\mathbf{q}_2$. The procedure for obtaining u_{22} and $\mathbf{q}_2$ from $\mathbf{a}_2^{(2)}$ is identical to that of obtaining u_{11} and $\mathbf{q}_1$ from $\mathbf{a}_1$. In general, $\mathbf{q}_j$ may be obtained from $\mathbf{a}_j$ by a series of $j - 1$ steps to make it orthogonal to each of the vectors $\mathbf{q}_1, \mathbf{q}_2, \ldots, \mathbf{q}_{j-1}$ in turn. The last step is followed by a Euclidean normalization. A flow diagram for the full orthogonal decomposition is given in Figure 4.10, in which $\mathbf{a}_j^{(i)}$ represents the vector obtained from $\mathbf{a}_j$ by orthogonalizing it with respect to $\mathbf{q}_1, \mathbf{q}_2, \ldots$ and $\mathbf{q}_{i-1}$. When

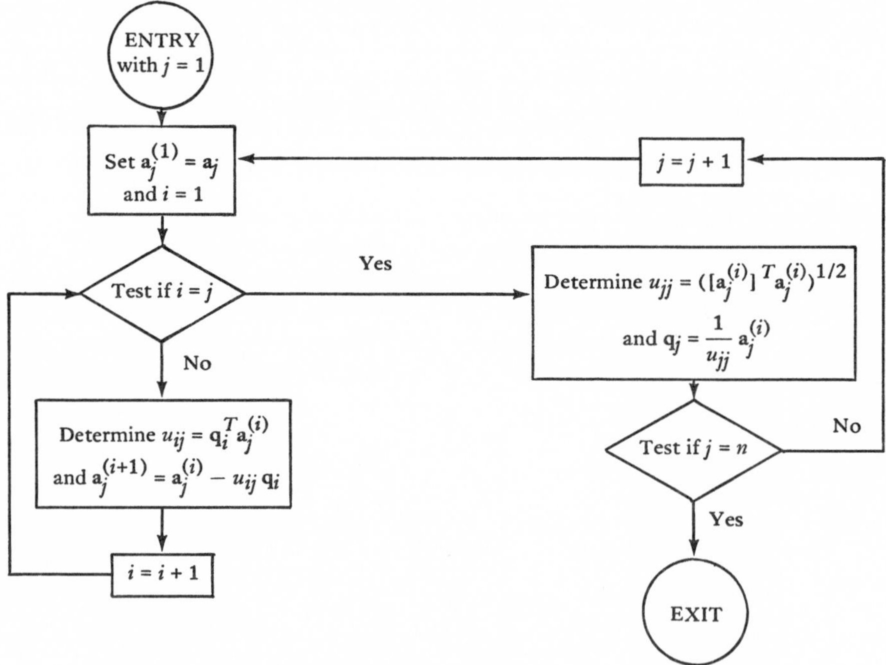

Figure 4.10 Flow diagram for Gram–Schmidt orthogonal decompostion A = QU

implemented on a computer the store used initially for $\mathbf{a}_j$ may be used in turn to store $\mathbf{a}_j^{(2)}, \ldots, \mathbf{a}_j^{(j)}$ and then $\mathbf{q}_j$. However, extra storage space will be needed for the matrix U.

If this orthogonal decomposition procedure is used to solve linear simultaneous equations (with $m = n$), it is found to be much less efficient than elimination, and consequently has not been used extensively. However, as compared with elimination procedures, it does avoid the need for pivot selection.

4.19 ORTHOGONAL DECOMPOSITION FOR LEAST SQUARES EQUATIONS (Golub, 1965)

Application of orthogonal decomposition

It has already been noted in section 4.18 that, if a set of overdetermined equations $\mathbf{Ax} = \mathbf{b}$ (where A is of order $m \times n$) has a solution, then the orthogonal decomposition procedure can be used to obtain this solution. Since a solution may always be obtained to equation (4.104), it is pertinent to consider what this solution represents when the original equations have no solution. If equation (4.104) is premultiplied by $\mathbf{U}^T$ then

$$\mathbf{U}^T\mathbf{U}\mathbf{x} = \mathbf{U}^T\mathbf{Q}^T\mathbf{b} \tag{4.111}$$

and hence

$$\mathbf{U}^T\mathbf{Q}^T\mathbf{Q}\mathbf{U}\mathbf{x} = \mathbf{U}^T\mathbf{Q}^T\mathbf{b} \tag{4.112}$$

Substituting $\mathbf{A} = \mathbf{QU}$ gives

$$\mathbf{A}^T\mathbf{A}\mathbf{x} = \mathbf{A}^T\mathbf{b} \tag{4.113}$$

Hence, if the orthogonal decomposition procedure is applied to a set of over-determined equations, the least squares solution with unit weighting factors will be obtained (compare with equation 2.25).

If A is fully populated with non-zero elements and $m \geqslant n \geqslant 1$, the orthogonal decomposition procedure requires approximately mn^2 multiplications to obtain a solution. Alternatively, if the least squares solution for the equations is obtained by forming $\mathbf{A}^T\mathbf{A}$ and $\mathbf{A}^T\mathbf{b}$ explicitly and then solving equation (4.113) by elimination, approximately $\frac{1}{2}mn^2 + \frac{1}{6}n^3$ multiplications will be required (making use of the symmetry of $\mathbf{A}^T\mathbf{A}$). Consequently the solution by orthogonal decomposition will be less efficient than the solution by elimination. (This will also be true when A is sparse and when the vectors x and b are replaced by matrices to accommodate multiple right-hand sides). However, the particular advantage of the orthogonal decomposition method is that it provides a solution of acceptable accuracy in many cases where an elimination solution does not.

Example

For the first curve fitting problem in section 2.7 the Gram–Schmidt decomposition of A is given by

$$\begin{bmatrix} 1 & 0.2 & 0.04 \\ 1 & 0.4 & 0.16 \\ 1 & 0.6 & 0.36 \\ 1 & 0.8 & 0.64 \\ 1 & 1.0 & 1.00 \\ 1 & 1.2 & 1.44 \end{bmatrix} = \begin{bmatrix} 0.4082 & -0.5976 & 0.5455 \\ 0.4082 & -0.3586 & -0.1091 \\ 0.4082 & -0.1195 & -0.4364 \\ 0.4082 & 0.1195 & -0.4364 \\ 0.4082 & 0.3586 & -0.1091 \\ 0.4082 & 0.5976 & 0.5455 \end{bmatrix} \begin{bmatrix} 2.4495 & 1.7146 & 1.4860 \\ & 0.8367 & 1.1713 \\ & & 0.2444 \end{bmatrix} \tag{4.114}$$

Thus the equations $\mathbf{Ux} = \mathbf{Qb}$ are

$$\begin{bmatrix} 2.4495 & 1.7146 & 1.4860 \\ & 0.8367 & 1.1713 \\ & & 0.2444 \end{bmatrix} \begin{bmatrix} c_0 \\ c_1 \\ c_2 \end{bmatrix} = \begin{bmatrix} 7.7567 \\ 2.5100 \\ -2.0730 \end{bmatrix} \tag{4.115}$$

which yield the correct solution for $\{c_0 \quad c_1 \quad c_2\}$. It can easily be verified that **U** is the same as the upper triangular matrix obtained by Choleski decomposition of $\mathbf{A}^T\mathbf{A}$. (This may be proved theoretically for the general case, but in practice the build-up of rounding errors in the arithmetical processes may affect numerical comparisons when the columns of **A** are almost linearly dependent.)

Ill-conditioning

Consider a set of least squares equations in which

$$\mathbf{A} = \begin{bmatrix} 0.5 & 0.5(1+\alpha) \\ 0.5 & 0.5(1+\alpha) \\ 0.5 & 0.5(1-\alpha) \\ 0.5 & 0.5(1-\alpha) \end{bmatrix} \quad \text{and} \quad \mathbf{b} = \begin{bmatrix} 2 \\ 1 \\ -1 \\ 0 \end{bmatrix} \tag{4.116}$$

The solution is $\{x_1 \quad x_2\} = \{(1 - 2/\alpha) \quad 2/\alpha\}$. Since the columns of **A** approach linear dependence as the parameter α decreases in magnitude, the effect of ill-conditioning can be investigated by considering the numerical solution of the equations when α is very small. For instance, if $\alpha = 0.000062426$ and the precision of the computer is eight decimal places, the last significant figure of α will be lost immediately the matrix **A** is formed. If the equations are then solved numerically by orthogonal decomposition the solution will be accurate to four decimal places, which is as accurate as can be expected considering the given magnitude of α. However, if the coefficient matrix

$$A^T A = \begin{bmatrix} 1 & 1 \\ 1 & 1+\alpha^2 \end{bmatrix} \tag{4.117}$$

is computed, α^2 will be so small compared with unity that its contribution to element (2,2) will not be represented in the computer. Hence the matrix, as stored, will be singular. If an elimination solution is attempted, then either the elimination procedure will break down or the results will be meaningless.

As a rough guide, it may be expected that the loss of accuracy will be approximately twice as many figures in an elimination solution as it will be in a solution by orthogonal decomposition. If orthogonal decomposition can provide a solution of acceptable accuracy to a least squares problem, and the same accuracy can only be achieved using elimination by adopting double-length arithmetic, then the orthogonal decomposition procedure is preferable.

Extensions of the basic procedure

Consider the equations

$$\mathbf{A}^T\mathbf{W}\mathbf{A}\mathbf{x} = \mathbf{A}^T\mathbf{W}\mathbf{b} \tag{4.118}$$

where $\mathbf{W}$ is symmetric and positive definite. If the Choleski decomposition of $\mathbf{W}$ is given by

$$\mathbf{W} = \mathbf{L}\mathbf{L}^T \tag{4.119}$$

then equations (4.118) can be expressed in the form

$$\bar{\mathbf{A}}^T\bar{\mathbf{A}}\mathbf{x} = \bar{\mathbf{A}}^T\bar{\mathbf{b}} \tag{4.120}$$

where $\bar{\mathbf{A}} = \mathbf{L}^T\mathbf{A}$ and $\bar{\mathbf{b}} = \mathbf{L}^T\mathbf{b}$. Hence, by obtaining $\bar{\mathbf{A}}$ and $\bar{\mathbf{b}}$, it is possible to convert equations (4.118) into the form of equations (4.113) which may be solved by means of orthogonal decomposition. Least squares problems in which the weighting factors are not equal yield equations of the same form as (4.118), but with a diagonal matrix for $\mathbf{W}$. Consequently $\mathbf{L}$ will be a diagonal matrix with typical element, l_{ii}, being the square root of the corresponding weighting factor w_i.

Alternatively the orthogonal decomposition $\mathbf{A} = \mathbf{Q}\mathbf{U}$ may be performed with $\mathbf{Q}$ being obtained as a product of factors as in either Givens or Householder transformations for the QR method (section 8.13). It is also possible to take advantage of sparseness in the matrix $\mathbf{A}$ (Duff and Reid, 1976; Gentleman, 1973).

BIBLIOGRAPHY

Bauer, F. L. (1963). 'Optimally scaled matrices'. *Numer. Math.*, **5**, 73–87. (Discusses row and column scaling).

Beale, E. M. L. (1971). 'Sparseness in linear programming'. In J. K. Reid (Ed.), *Large Sparse Sets of Linear Equations*, Academic Press, London. (Discusses re-solution of equations in which a column of the coefficient matrix has been modified).

Bunch, J. R., and Parlett, B. N. (1971). 'Direct methods for solving symmetric indefinite systems of linear equations'. *SIAM J. Numer. Anal.*, 8, 639–655.

Duff, I. S., and Reid, J. K. (1976). 'A comparison of some methods for the solution of sparse overdetermined systems of linear equations'. *J. Inst. Maths. Applics.*, 17, 267–280.

Fadeev, D. F., and Fadeeva, V. N. (1960). *Computational Methods of Linear Algebra*, State Publishing House for Physico-Mathematical Literature, Moscow (English translation by R. C. Williams, W. H. Freeman and Co., San Francisco, 1963).

Forsythe, G. E., and Moler, C. B. (1967). *Computer Solution of Linear Algebraic Systems*, Prentice-Hall, Englewood Cliffs, New Jersey.

Fox, L. (1964). *Introduction to Numerical Linear Algebra*, Clarendon Press, Oxford.

Gentleman, W. M. (1973). 'Least squares computations by Givens transformations without square roots'. *J. Inst. Maths. Applics.*, 12, 329–336.

Gill, P. E., Murray, W., and Saunders, M. A. (1975). 'Methods for computing and modifying the *LDV* factors of a matrix'. *Mathematics of Computation*, 29, 1051–1077.

Golub, G. H. (1965). 'Numerical methods for solving linear least squares problems'. *Numer. Math.*, 7, 206–216.

Kavlie, D., and Powell, G. H. (1971). 'Efficient reanalysis of modified structures'. *Proc. ASCE (J. of the Struct. Div.)* 97, 377–392.

Lawson, C. L., and Hanson, R. J. (1974). *Solving Least Squares Problems*, Prentice-Hall, Englewood Cliffs, New Jersey.

Noble, B. (1969). *Applied Linear Algebra*, Prentice-Hall, Englewood Cliffs, New Jersey.

Ortega, J. M. (1972). *Numerical Analysis, A Second Course*, Academic Press, New York. (Chapter 9 discusses equilibrated matrices.)

Stewart, G. W. (1973). *Introduction to Matrix Computations*, Academic Press, New York.

Westlake, J. R. (1968). *A Handbook of Numerical Matrix Inversion and Solution of Linear Equations*, Wiley, New York.

Wilkinson, J. H. (1965). *The Algebraic Eigenvalue Problem*, Clarendon Press, Oxford. (Chapter 4 is on the solution of linear algebraic equations.)

Chapter 5
Sparse Matrix Elimination

5.1 CHANGES IN SPARSITY PATTERN DURING ELIMINATION

Elimination methods for solving simultaneous equations can be executed rapidly if the order of the equations is not large. However, the n^3 law governing the amount of computation can make the execution of large-order problems very time consuming. If large-order sets of equations have sparse coefficient matrices it is highly desirable that advantage should be taken of the sparseness in order to reduce both the computation time and the storage requirements. Most of this chapter (sections 5.1 to 5.9) will be concerned with equations having symmetric positive definite coefficient matrices. Because in this case pivot selection is unnecessary, simpler elimination procedures may be used than in the general case. However, before large-order sets of equations are considered, it is useful to investigate the characteristics of sparse elimination by examining a simple example.

Consider a set of eight simultaneous equations in the variables $x_A, x_B, \ldots, x_H$ whose coefficient matrix is symmetric and positive definite, having non-zero entries in the positions shown in Figure 5.1(a). Such a set of equations could, for instance, be obtained from the node conductance analysis of the electrical resistance network shown in Figure 5.2. (In setting up the equation the order of the variables has been deliberately scrambled.) The first Gaussian elimination step will produce extra non-zero coefficients as shown in Figure 5.1(b) and the final reduced form

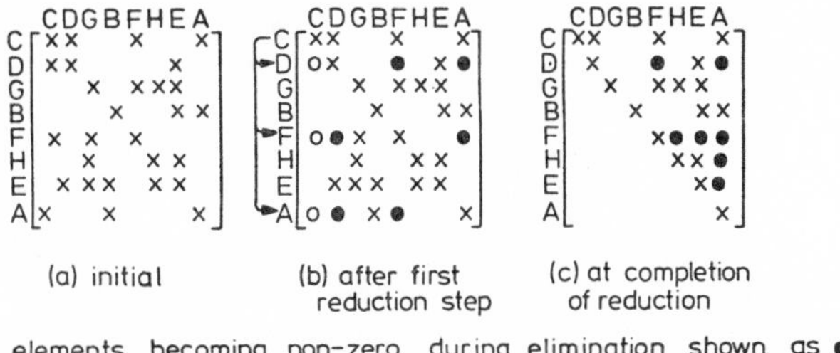

Figure 5.1 An example of sparse matrix reduction showing the pattern of non-zero elements

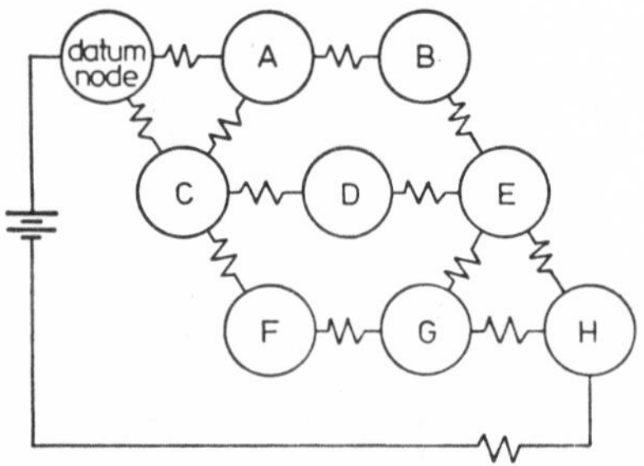

Figure 5.2 Electrical resistance network which gives a node conductance matrix of the form shown in Figure 5.1

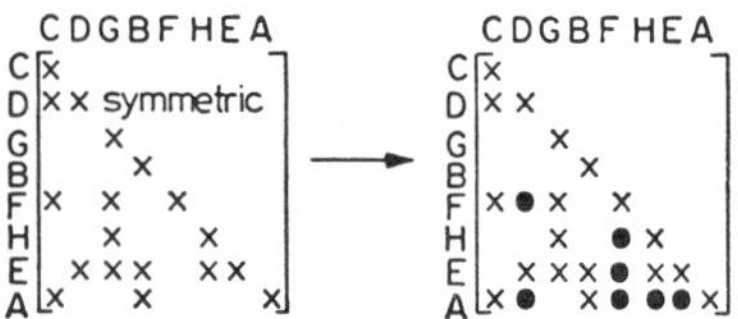

Figure 5.3 Non-zero elements in the Choleski decomposition of the matrix, Figure 5.1(a)

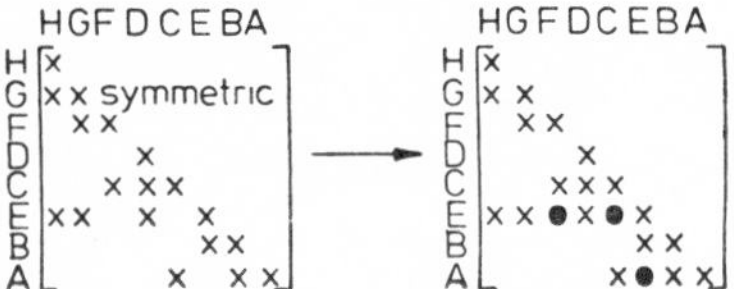

Figure 5.4 Choleski decomposition, as Figure 5.3, but with variables reordered

will contain non-zero coefficients as shown in Figure 5.1(c), seven of which will have been introduced during the elimination. If the coefficient matrix is decomposed into one of the forms **LU**, **LDL**^{T} or **LL**^{T}, then in each case **L** will have a pattern of non-zero elements corresponding to the transpose of Figure 5.1(c), as shown in Figure 5.3. If, on the other hand, the variables are arranged in the order $x_H, x_G, x_F, x_D, x_C, x_E, x_B, x_A$ and the equations are rearranged so that symmetry is preserved, the Choleski decomposition will yield the pattern shown in Figure 5.4, in which only three zero elements have become non-zero during the elimination.

If, at the start of the k-th reduction step, there are b_k non-zero elements in column k of the lower triangle, then $\frac{1}{2}(b_k^2 + b_k - 2)$ multiplications and divisions are required to eliminate the $b_k - 1$ elements below the diagonal (there will also be one square root to evaluate if the Choleski decomposition is used). Hence the total number of multiplications and divisions for the whole decomposition is $\frac{1}{2}\sum_{k=1}^{n}(b_k^2 + b_k - 2)$. The order in which the variables are arranged is likely to affect the total amount of computation more than it will affect the total number of non-zero elements $\sum_{k=1}^{n} b_k$. The comparison between the first and second ordering schemes for the example of Figures 5.3 and 5.4 is shown in Table 5.1

Table 5.1 A comparison of the effect of the variable ordering of Figures 5.3 and 5.4 on sparse decomposition

Ordering scheme	No. of non-zero elements	No. of multiplications and divisions
(a) as Fig. 5.3	25	48
(b) as Fig. 5.4	21	32
Ratio (b)/(a)	84%	66.7%

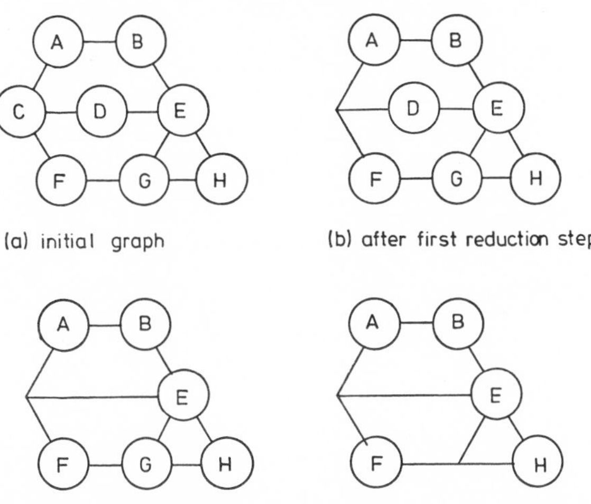

Figure 5.5 Graphical interpretation of the elimination shown in Figure 5.1

5.2 GRAPHICAL INTERPRETATION OF SPARSE ELIMINATION

The 8 x 8 matrix of coefficients considered in the above section may be represented graphically by Figure 5.5(a). Here each twin pair of off-diagonal elements gives rise to one link of the graph. (The graph is, in fact, the equivalent network with all links to datum nodes missing.) It may be noted from Figure 5.1(b) that if x_C is specified first, then the first reduction step creates new connections between nodes

A, D and F. The active part of the matrix can therefore be represented by the modified graph shown in Figure 5.5(b) where nodes A, D and F are seen to be interconnected. Similarly, the active part of the matrix after the second and third reduction steps can be represented by the graphs shown in Figure 5.5(c) and 5.5(d) respectively. At this stage node E is connected to the other nodes, although node B, the next node to be eliminated, has only its original two connections. For a simple example, an examination of the graph indicates which ordering schemes will give rise to the least number of additional nodal connections during the elimination. An optimum or near optimum ordering scheme may be obtained by eliminating, at each reduction step, one of the nodes which has the least number of connections. In the example under discussion it is advisable to start with A, B, D, F or H, each of which has two connections. If H is chosen no new connections are generated by the

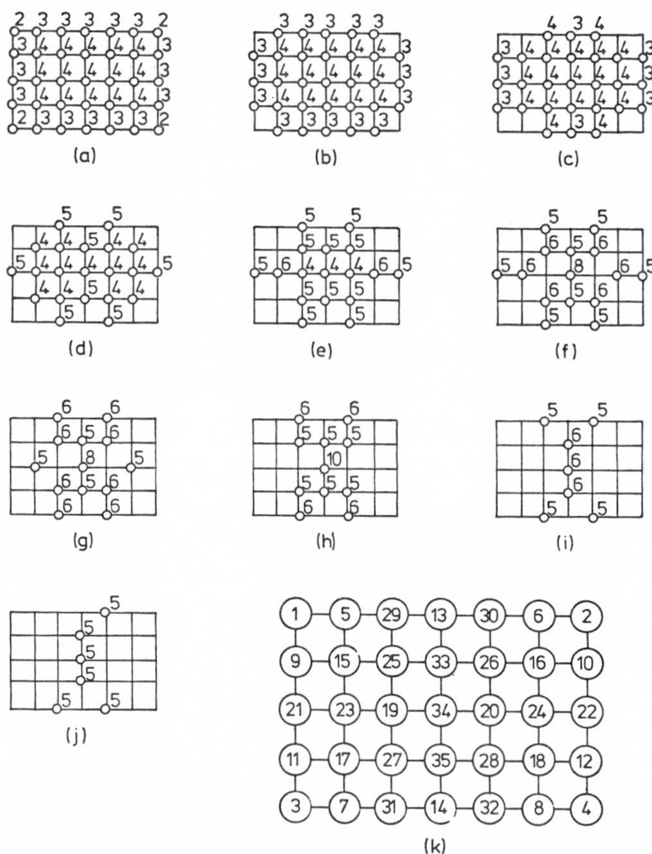

Figure 5.6 Graphical method of ordering the nodes for a 5 x 7 grid using the minimum connection rule: (a) shows initial graph with numbers of connections alongside each node; (b) to (j) show graphs of stages during the reduction; and (k) shows the resulting node numbering scheme

first reduction step. If G and F are then eliminated in turn, connections EF and EC are formed. Node D may then be eliminated without introducing any new connections, leaving a ring of four nodes which can be eliminated in any order with equal efficiency. If the last nodes are eliminated in the order C, E, B, A, the ordering scheme shown in Figure 5.4 is obtained, which is an optimum for the system. There are several other ordering schemes which also yield the optimum of three extra nodal connections during the reduction, for instance, A, B, D, F, C, E, G, H or F, H, A, G, B, C, E, D.

Consider the application of this ordering principle to the set of equations whose coefficient matrix conforms with the graph shown in Figure 5.6(a). Beside each node is recorded the number of connections which the node has. The corner nodes will be eliminated first, modifying the graph to that shown in Figure 5.6(b). Since all of the edge nodes have then the same number of connections, there are various possible choices for the next set of nodes to be eliminated. Figure 5.6(c) to (j) shows a possible continuation scheme which results in the nodal numbering shown in Figure 5.6(k) and an L matrix of the form shown in Figure 5.7.

Figure 5.8(a) shows the graph of an alternative ordering scheme for the variables. If the elimination is interpreted graphically it will be seen that at any elimination step between the fourth and the thirtieth there are a set of five consecutive nodes

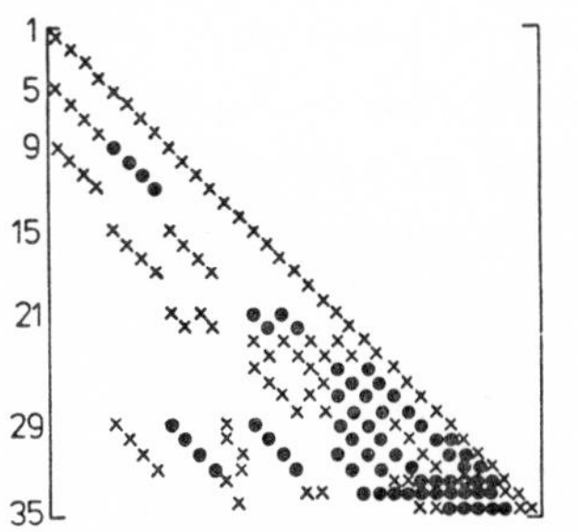

Figure 5.7 Non-zero element pattern of the L matrix for the 5 x 7 grid with the order of the nodes as shown in Figure 5.6(k)

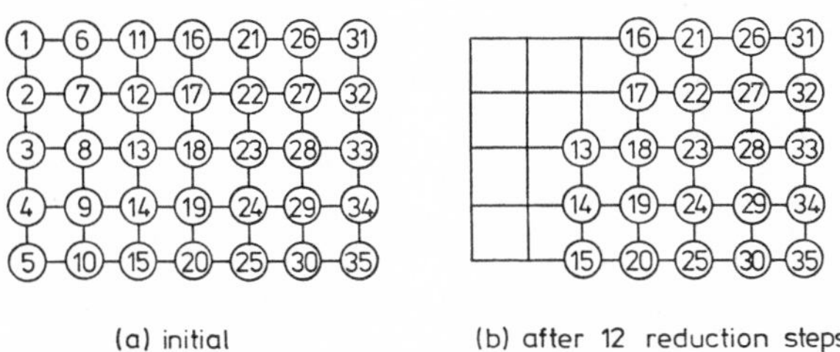

Figure 5.8 Graphical interpretation of a simple ordering for the nodes of the 5 x 7 grid

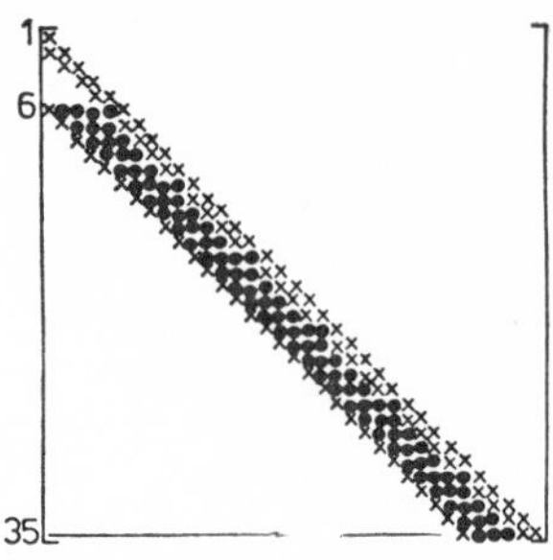

Figure 5.9 Non-zero element pattern of the L matrix for the 5 x 7 grid with the simple ordering scheme for the nodes shown in Figure 5.8(a)

which are interconnected (e.g. Figure 5.8(b)). The **L** matrix associated with this ordering scheme has the regular band pattern shown in Figure 5.9. The contrast in the patterns of non-zero elements in Figures 5.7 and 5.9 is interesting. For the former pattern, a packing scheme would give the most effective storage compaction (a diagonal band scheme not being useful). However, the latter pattern could fit easily into a diagonal band store (with a packing scheme being unnecessarily complicated and inefficient).

A number of alternative storage schemes will now be investigated.

5.3 DIAGONAL BAND ELIMINATION

It is a property of elimination without pivoting that elements before the first non-zero element on any row of a matrix must remain zero throughout the elimination. Hence, if the non-zero elements of a symmetric positive definite matrix lie entirely within a diagonal band, then the decomposition of the matrix can be performed entirely within the band.

Figure 5.10 illustrates the procedure for Gaussian-type elimination of a 6 x 6 symmetric diagonal band matrix of half-bandwidth $b = 3$ in which the Choleski lower triangular matrix is formed. The formulae for the second reduction step are as follows:

$$\left.\begin{aligned} &l_{22} = (a_{22}^{(2)})^{1/2} \\ &l_{32} = a_{32}^{(2)}/l_{22}, \quad a_{33}^{(3)} = a_{33}^{(2)} - l_{32}^2 \\ &l_{42} = a_{42}/l_{22}, \quad a_{43}^{(3)} = a_{43} - l_{42}l_{32}, \quad a_{44}^{(3)} = a_{44} - l_{42}^2 \end{aligned}\right\} \tag{5.1}$$

Alternatively, a compact elimination may be performed, one step of which is illustrated in Figure 5.11. The relevant formulae are

$$\left.\begin{aligned} &l_{42} = a_{42}/l_{22} \\ &l_{43} = (a_{43} - l_{42}l_{32})/l_{33} \\ &l_{44} = (a_{44} - l_{42}^2 - l_{43}^2)^{1/2} \end{aligned}\right\} \tag{5.2}$$

In both cases the significant part of the matrix is restricted to a triangle of elements. The full decomposition involves operating with a succession of such triangles which

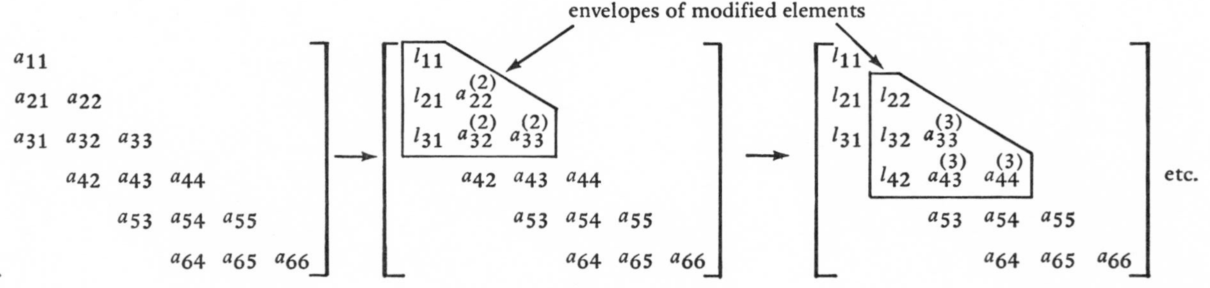

Figure 5.10 Symmetric band Choleski decomposition using Gaussian elimination

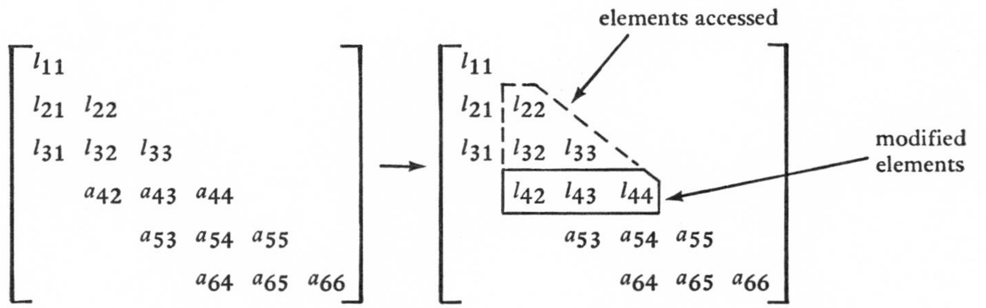

Figure 5.11 One step of a symmetric band Choleski decomposition using compact elimination

are situated progressively further down the band. Over the main part of the decomposition these triangles will all have the dimensions $b \times b$. Hence the programming of symmetric diagonal band decomposition for the general case of a matrix of order n and half-bandwidth b is easily carried out using any of the three types of diagonal band storage described in section 3.16. The total number of multiplications and divisions required for the decomposition is

$$\frac{n}{2}(b^2 + b - 2) - \frac{b}{3}(b^2 - 1) \simeq \frac{nb^2}{2} \tag{5.3}$$

Alternatively, the decomposition may be expressed in the form $\mathbf{LDL}^T$ with little modification to the procedure. This would avoid the need to evaluate any square roots.

Since the storage requirement is proportional to b and the amount of computation for decomposition is approximately proportional to b^2, it is particularly important that the bandwidth should be kept as small as possible. If the i-th and j-th nodes of the graph are interconnected then, in order that the corresponding lower triangular coefficient lies within the diagonal band of the matrix,

$$b \geqslant i - j + 1 \tag{5.4}$$

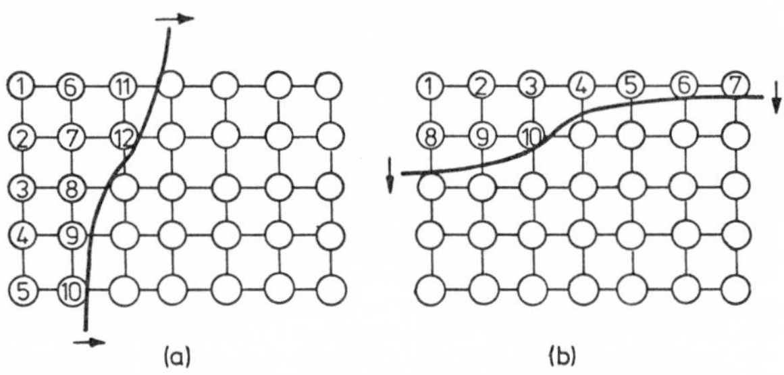

Figure 5.12 Two frontal sweep methods of ordering nodes

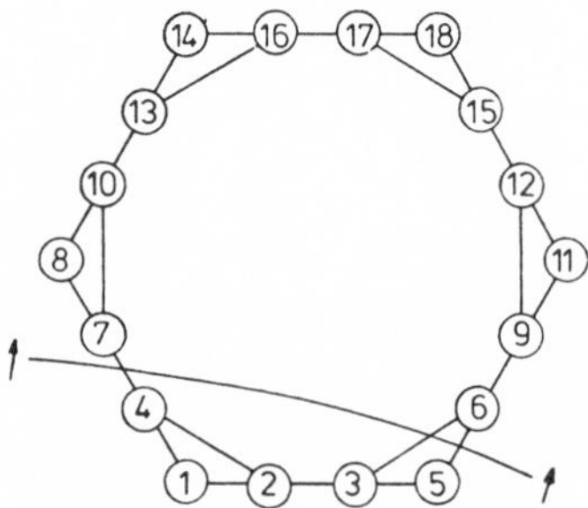

Figure 5.13 A possible frontal ordering for a type of ring network

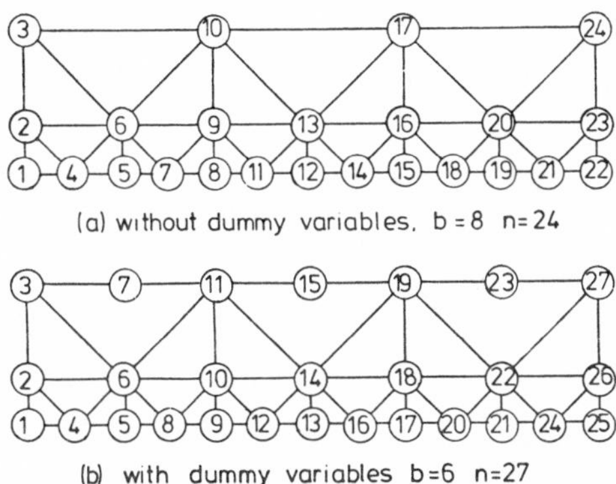

Figure 5.14 Use of dummy variables to reduce the bandwidth of a triangulated network

Hence the bandwidth must be no less than the maximum value of $i - j + 1$ appertaining to any of the interconnections. Ordering schemes for the variables which minimize the bandwidth will generally be of a *frontal* form. The numbering scheme shown in Figure 5.8(a) which gives a half-bandwidth $b = 6$ for the 5×7 grid may be obtained from the movement of a front as it sweeps across the graph, as shown in Figure 5.12(a). The front shown in Figure 5.12(b) is not so suitable because it produces a half-bandwidth $b = 8$. Figure 5.13 shows a frontal ordering scheme for a ring network such that $b = 4$. It is sometimes only possible to achieve the minimum bandwidth by introducing *dummy variables*. Figure 5.14(a) shows a network which has a frontal ordering scheme such that $b = 8$. By using three dummy nodes, as shown in Figure 5.14(b), the half-bandwidth b is reduced to 6 and hence a more efficient form of diagonal band solution is obtained.

In general it may be stated that the front should move forward regularly across the graph in such a way that it does not dwell long on any particular branch.

5.4 A VARIABLE BANDWIDTH ELIMINATION ALGORITHM

Although diagonal band elimination makes use of the property that zero elements situated before the first non-zero element on any row always remain zero, it does not make as effective use of this property as does variable bandwidth elimination. The matrix represented by the graph of Figure 5.14(a) is particularly well suited to variable bandwidth elimination. Figure 5.15 shows the variable bandwidth store required for this. It will be noted that for this matrix only fifteen stored elements are originally zero, all of which become non-zero during the elimination. The maximum number of stored elements in any column is four, indicating a significant saving over the diagonal band representations, both with and without the use of dummy variables.

Variable bandwidth stores can be classified into those with re-entrant rows (illustrated by Figure 5.15) and those without re-entrant rows (illustrated by

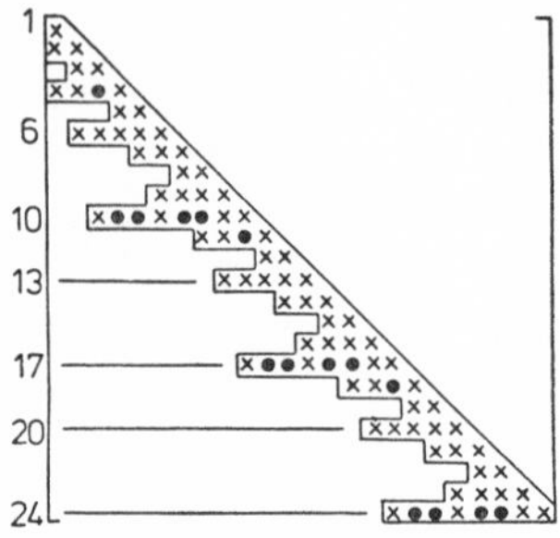

Figure 5.15 Variable bandwidth store for decomposition of the matrix whose graph is given by Figure 5.14(a). In this example the store has re-entrant rows

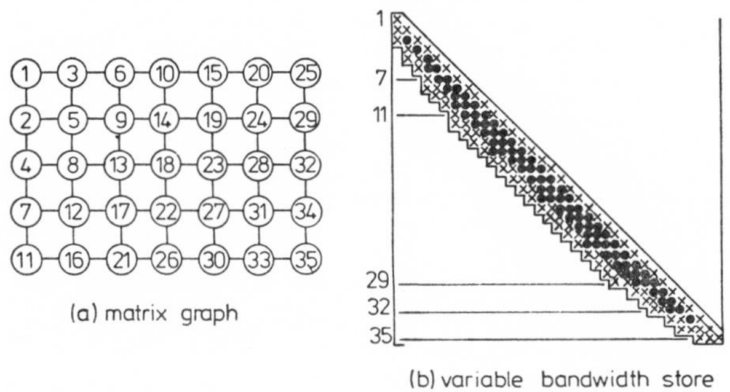

Figure 5.16 Use of variable bandwidth store for the 5 x 7 grid with corner to corner node ordering. In this example the store does not have re-entrant rows

Figure 5.16(b)). Row i is re-entrant when the column number of the first element in the row is smaller than the corresponding column number for row $i-1$. If there are no re-entrant rows in the matrix a Gaussian elimination can easily be carried out with the active area of the matrix at reduction step k being a $b_k \times b_k$ triangle, where b_k is the number of stored elements in column k (illustrated by Figure 5.17(a)). However, if re-entrant rows are present, the active area of the matrix at any reduction step is much more awkward to specify in relation to the variable bandwidth store (Figure 5.17(b)). If, on the other hand, a row-wise compact elimination is used, the direct determination of row i of $\mathbf{L}$ entails accessing stored elements within a triangle of dimension equal to the length of row i, as shown in Figure 5.17(c). All of the elements in row i will be altered, and hence may be described as active, while all of the other elements within the triangle are not modified and may be described as passive.

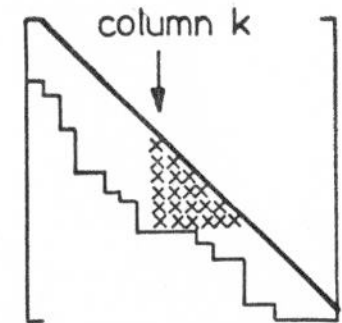

(a) Gaussian type elimination without re-entrant rows showing active triange for kth reduction step

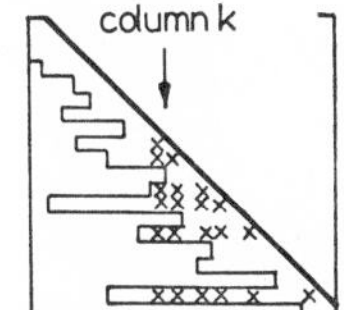

(b) Gaussian type elimination with re-entrant rows showing active elements for kth reduction step

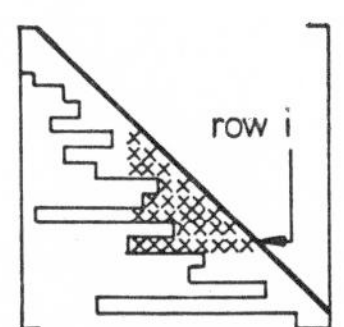

(c) Compact elimination with re-entrant rows showing active and passive elements for computation of row i

Figure 5.17 Alternative schemes for elimination in a variable bandwidth store.

The following decomposition algorithm uses a row-wise compact elimination procedure to obtain the Choleski triangular factor of a symmetric positive definite variable bandwidth matrix. The decomposition is carried out within the variable bandwidth store and will operate satisfactorily when re-entrant rows are present. The formulae used for the decomposition are as equations (4.28) except that the summations start at $k = \bar{l}$, where $\bar{l}$ is the first column number for which there are stored elements in positions l_{ik} and l_{jk}. With storage space allocated as follows:

N	order of matrix
A(NN)	real array containing the variable bandwidth matrix
KDIAG(N)	integer array of the addresses within A of the diagonal elements
I,J	row and column indices
L	column number of first stored element in row i
LBAR	column number to start the summation
KI,KJ	fictitious addresses of elements a_{i0} and a_{j0}
K	column number for summations
X	real working store

the program segment to perform Choleski decomposition is:

VARIABLE BANDWIDTH CHOLESKI DECOMPOSITION

ALGOL

```
A[1]:=sqrt(A[1]);
for i:=2 step 1 until n do
begin ki:=kdiag[i]-i;
      l:=kdiag[i-1]-ki+1;
      for j:=l step 1 until i do
      begin x:=A[ki+j];
            kj:=kdiag[j]-j;
            if j=1 then goto col1;
            lbar:=kdiag[j-1]-kj+1;
            if l>lbar then lbar=l;
            for k:=lbar step 1 until j-1 do
            x:=x-A[ki+k]*A[kj+k];
      col1: A[ki+j]:=x/A[kj+j];
      end row but with wrong pivot;
      A[ki+i]:=sqrt(x)
end variable band decomposition;
```

FORTRAN

```
  A(1)=SQRT(A(1))
  DO 1 I=2,N
  KI=KDIAG(I)-I
  L=KDIAG(I-1)-KI+1
  DO 2 J=L,I
  X=A(KI+J)
  KJ=KDIAG(J)-J
  IF(J.EQ.1)GO TO 2
  LBAR=KDIAG(J-1)-KJ+1
  LBAR=MAX0(L,LBAR)
  IF(LBAR.EQ.J)GO TO 2
  DO 3 K=LBAR,J-1
3 X=X-A(KI+K)*A(KJ+K)
2 A(KI+J)=X/A(KJ+J)
1 A(KI+I)=SQRT(X)
```

If B is a real array of dimension n containing a right-hand vector **b**, then a program segment for forward-substitution in which the vector **b** is overwritten by the solution **y** of the equation $\mathbf{L}\mathbf{y} = \mathbf{b}$ is as follows:

VARIABLE BANDWIDTH FORWARD-SUBSTITUTION

ALGOL

```
b[1]:=b[1]/A[1];
for i:=2 step 1 until n do
begin ki:=kdiag[i]-i;
      l:=kdiag[i-1]-ki+1;
      x:=b[i];
      for j:=l step 1 until i-1 do
      x:=x-A[ki+j]*b[j];
      b[i]:=x/A[ki+i]
end variable band forward-substitution;
```

FORTRAN

```
  B(1)=B(1)/A(1)
  DO 4 I=2,N
  KI=KDIAG(I)-I
  L=KDIAG(I-1)-KI+1
  X=B(I)
  IF(L.EQ.I)GO TO 4
  DO 5 J=L,I-1
5 X=X-A(KI+J)*B(J)
4 B(I)=X/A(KI+I)
```

A program segment for backsubstitution in which the vector **y** in array B is overwritten by the solution **x** of the equation $\mathbf{L}^T\mathbf{x} = \mathbf{y}$ is as follows:

VARIABLE BANDWIDTH BACKSUBSTITUTION

ALGOL

```
for i:=n step -1 until 2 do
begin ki:=kdiag[i]-i;
      b[i]:=x:=b[i]/A[ki+i];
      l:=kdiag[i-1]-ki+1;
      for k:=l step 1 until i-1 do
      b[k]:=b[k]-x*A[ki+k]
end;
b[1]:=b[1]/A[1];
comment variable band backsub complete;
```

FORTRAN

```
  DO 6 IT=2,N
  I=N+2-IT
  KI=KDIAG(I)-I
  X=B(I)/A(KI+I)
  B(I)=X
  L=KDIAG(I-1)-KI+1
  IF(L.EQ.I)GO TO 6
  DO 7 K=L,I-1
7 B(K)=B(K)-X*A(KI+K)
6 CONTINUE
  B(1)=B(1)/A(1)
```

Notes on the three algorithms

(a) The most heavily used part of the procedures is the inner loop of the decomposition process. This part is indicated by being enclosed in a box.

(b) The FORTRAN version contains various non-standard DO loop parameters and array subscripts.

(c) In the decomposition it is advisable to include a test and suitable exit instructions for the case where the pivot element is not positive.

(d) The backsubstitution is not carried out in the same way as in section 4.1, in order that the elements of **L** may be accessed row by row.

5.5 ON THE USE OF THE VARIABLE BANDWIDTH ALGORITHM

An efficient decomposition is obtained by the variable bandwidth scheme for the ring network of Figure 5.13 if the simple alternative node numbering scheme shown in Figure 5.18 is used. This node numbering scheme can be considered as being

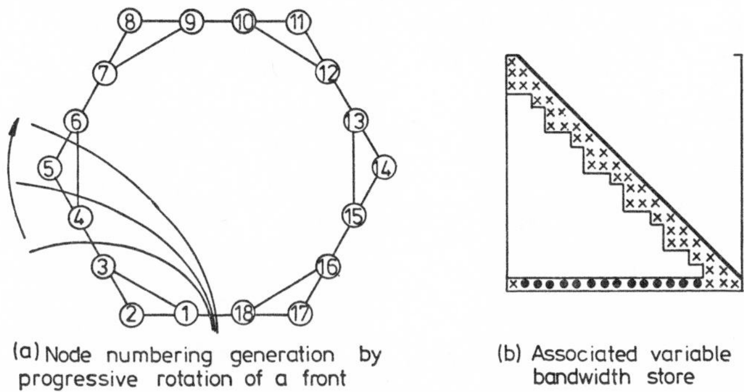

Figure 5.18 A simple alternative scheme for the ring network of Figure 5.13

generated by a front which rotates rather than sweeps across the network as shown in Figure 5.18(a). In general it may be stated that, for variable bandwidth storage, a frontal ordering scheme for the variables will prove to be the most efficient. However, in contrast with diagonal band storage ordering, it is not necessary to move the front forward in a regular way.

The advantages of using variable bandwidth storage as opposed to diagonal band storage for sparse matrix decomposition are as follows:

(a) Greater flexibility is permissible in the choice of ordering scheme for the variables.

(b) Optimum variable bandwidth ordering schemes will often provide more efficient decompositions than optimum diagonal band schemes.

(c) Dummy variables are never required to obtain efficient decompositions using variable bandwidth storage.

(d) Whatever ordering scheme is used for the variables some advantage will be gained by using variable bandwidth storage instead of full matrix or triangular storage (diagonal band storage may not be more efficient than full matrix or triangular storage).

The main disadvantages are that the address sequence not only takes up a small amount of additional storage space, but also needs to be specified before the variable bandwidth store itself can be used. In most problems which give rise to sparse coefficient matrices the geometry or topology of the system being analysed will be defined by input data (as, for instance, with the network analysis data of Table 2.1). In this case the specific shape of the variable bandwidth store will not be known at the programming stage. However, it is possible to develop a program segment which automatically constructs the appropriate address sequence by inspection of the input data, and therefore acts as a suitable preliminary to the main part of the analysis program. Using the form of input data described in section 2.3, two program segments, one to determine the address sequence for the node conductance matrix associated with an electrical resistance network and the other to construct the node conductance matrix itself in the variable bandwidth store defined by the address sequence, are as follows:

FORM ADDRESS SEQUENCE FOR NETWORK

ALGOL

```
      for i:=1 step 1 until n do kdiag[i]:=0;
      for k:=1 step 1 until m do
      begin i:=nodeA[k];
            j:=nodeB[k];
            if i=0 or j=0 then goto skip;
            if j−i>kdiag[j] then kdiag[j]:=j−i;
            if i−j>kdiag[i] then kdiag[i]:=i−j;
skip: end branch inspections;
      kdiag[1]:=1;
      for i:=2 step 1 until n do
      kdiag[i]:=kdiag[i−1]+kdiag[i]+1;
```

FORTRAN

```
  DO 1 I=1,N
1 KDIAG(I)=0
  DO 2 K=1,M
  I=NODEA(K)
  J=NODEB(K)
  IF(I.EQ.0.OR.J.EQ.0)GO TO 2
  KDIAG(J)=MAX0(KDIAG(J),J−I)
  KDIAG(I)=MAX0(KDIAG(I),I−J)
2 CONTINUE
  KDIAG(1)=1
  DO 3 I=2,N
3 KDIAG(I)=KDIAG(I−1)+KDIAG(I)+1
```

CONSTRUCT NODE CONDUCTANCE MATRIX IN A VARIABLE BANDWIDTH STORE

ALGOL

```
    kn:=kdiag[n];
    for i:=1 step 1 until kn do A[i]:=0;
    for k:=1 step 1 until m do
    begin i:=nodeA[k];
          j:=nodeB[k];
          x:=conduc[k];
          if j≠0 then kj=kdiag[j];
          if i=0 then goto A0;
          ki:=kdiag[i];
          A[ki]:=A[ki]+x;
          if j=0 then goto B0;
          if j>i then A[kj−j+i]:=−x
          else A[ki−i+j]:=−x;
      A0: A[kj]:=A[kj]+x;
B0: end forming node conductance matrix;
```

FORTRAN

```
  KN=KDIAG(N)
  DO 4 I=1,KN
4 A(I)=0.0
  DO 5 K=1,M
  I=NODEA(K)
  J=NODEB(K)
  X=CONDUC(K)
  IF(J.NE.0)KJ=KDIAG(J)
  IF(I.EQ.0)GO TO 6
  KI=KDIAG(I)
  A(KI)=A(KI)+X
  IF(J.EQ.0)GO TO 5
  IF(J.GT.I)A(KJ−J+I)=−X
  IF(J.LT.I)A(KI−I+J)=−X
6 A(KJ)=A(KJ)+X
5 CONTINUE
```

Some notes on the above algorithms

(a) The integer array KDIAG is initially used to record the number of off-diagonal elements to be stored in each row and then, on execution of the last statement of the first program segment, is converted into the address sequence.

(b) On entry to the second program segment it is assumed that the one-dimensional real array A, in which the node conductance matrix is to be formed, is of dimension greater than or equal to KDIAG(N).

5.6 AUTOMATIC FRONTAL ORDERING SCHEMES

In the above schemes the order of the variables is taken to be that specified in the input data. Where the number of variables is sufficiently large for storage space and computing time to be important, it is advisable for the user to arrange the order of the variables so as to give an economical solution. An alternative procedure is to allow the variables to be specified in an arbitrary order within the input data, and to include an initial segment in the program which automatically rearranges the variables in a way that should give an efficient solution.

The Cuthill–McKee algorithm (1969)

Cuthill and McKee's algorithm provides a simple scheme for renumbering the variables. It may be used in conjunction with diagonal band storage, although it is most effective if used in conjunction with variable bandwidth storage. The renumbering scheme may be described with reference to the graphical interpretation of the coefficient matrix as follows:

(a) Choose a node to be relabelled 1. This should be located at an extremity of the graph and should have, if possible, few connections to other nodes.
(b) The nodes connected to the new node are relabelled 2, 3, etc., in the order of their increasing degree (the degree of a node is the number of nodes to which it is connected).
(c) The sequence is then extended by relabelling the nodes which are directly connected to the new node 2 and which have not previously been relabelled. The nodes are again listed in the order of their increasing degree.
(d) The last operation is repeated for the new nodes 3, 4, etc., until the renumbering is complete.

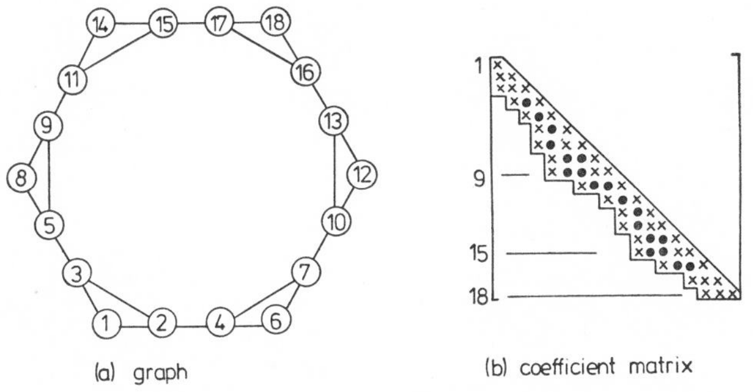

Figure 5.19 Cuthill–McKee renumbering for the ring network

If the algorithm is applied to the 5 x 7 grid problem starting with a corner node, the renumbering will yield the graph shown in Figure 5.16. This not only gives the optimum bandwidth for the coefficient matrix but also gives an efficient numbering scheme for use with variable bandwidth store. Cuthill–McKee renumbering of the ring network and the triangulated network are shown in Figures 5.19 and 5.20 respectively. Also shown are the associated variable bandwidth stores required for the coefficient matrices. Table 5.2 compares the variable bandwidth store using the

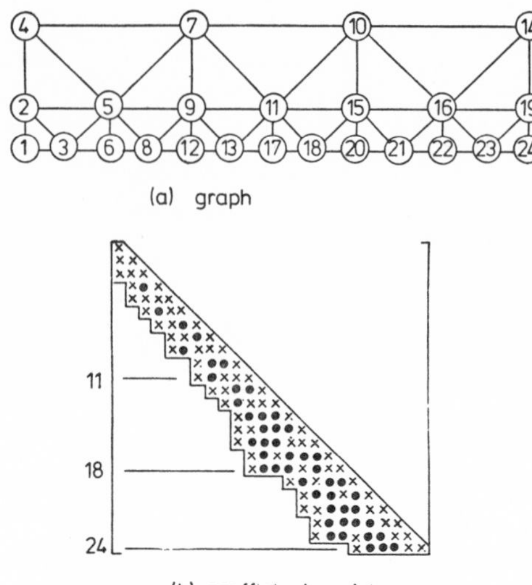

Figure 5.20 Cuthill–McKee renumbering for the triangulated network

Table 5.2 Comparison of some alternative variable bandwidth schemes

		Hand numbering (Figs. 5.18a and 5.14a)	Cuthill–McKee	Reverse Cuthill–McKee
Ring network	Storage requirement	56	61	55
	No. of multiplications for decomposition	102	124	98
Triangulated network	Storage requirement	49	122	101
	No. of multiplications for decomposition	192	385	256

Cuthill–McKee algorithm with variable bandwidth stores for these problems which have been described previously.

Discussion of the Cuthill–McKee algorithm

Figure 5.21 shows a possible initial numbering scheme for the triangulated network and Table 5.3 shows a node connection list which may be automatically constructed from the appropriate branch data. The Cuthill–McKee algorithm may be more easily implemented by referring to such a node connection list than by referring directly to the branch data.

Consider the renumbering of the nodes using the information presented in

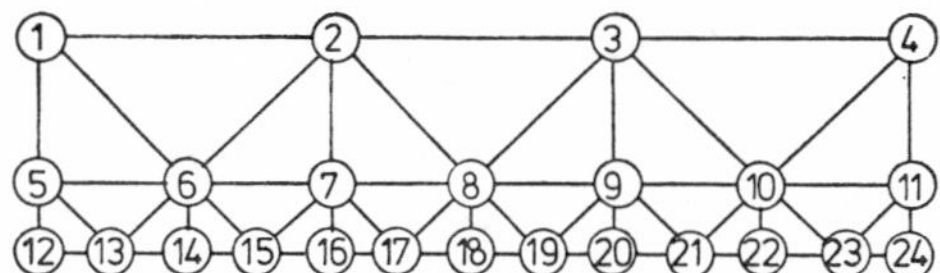

Figure 5.21 An initial numbering scheme for the triangulated network

Table 5.3 A node connection list for the triangulated network with initial numbering as Figure 5.21

Node	No. of connections	Connection list
1	3	2,5,6
2	5	1,3,6,7,8
3	5	2,4,8,9,10
4	3	3,10,11
5	4	1,6,12,13
6	7	1,2,5,7,13,14,15
7	6	2,6,8,15,16,17
8	7	2,3,7,9,17,18,19
9	6	3,8,10,19,20,21
10	7	3,4,9,11,21,22,23
11	4	4,10,23,24
12	2	5,13
13	4	5,6,12,14
14	3	6,13,15
—	—	—
etc.		

Table 5.3 and starting with node 12. By referring to row 12 of the node connection list it is apparent that nodes 5 and 13 need to be renumbered 2 and 3 (node 5 may be renumbered before node 13 since both have the same number of node connections and therefore the choice is arbitrary). An examination of row 5 of the list then reveals that nodes 1 and 6 should be renumbered 4 and 5. According to the number of connections node 1 should be renumbered first, and so on. It is also possible to specify the address sequence for the variable bandwidth store as the nodes are being renumbered.

The variable bandwidth store formed from the Cuthill–McKee algorithm cannot have re-entrant rows and may be considered to be a set of overlapping triangles as shown in Figure 5.22(a). It can be shown that, if the sides of these triangles span $p_1, p_2, \ldots, p_r$ rows and the sides of the triangular overlaps (Figure 5.22(b)) span $q_1, q_2, \ldots, q_{r-1}$ rows, the number of storage locations required for the primary array is

$$s = \sum_{i=1}^{r} \frac{p_i(p_i + 1)}{2} - \sum_{i=1}^{r-1} \frac{q_i(q_i + 1)}{2} \tag{5.5}$$

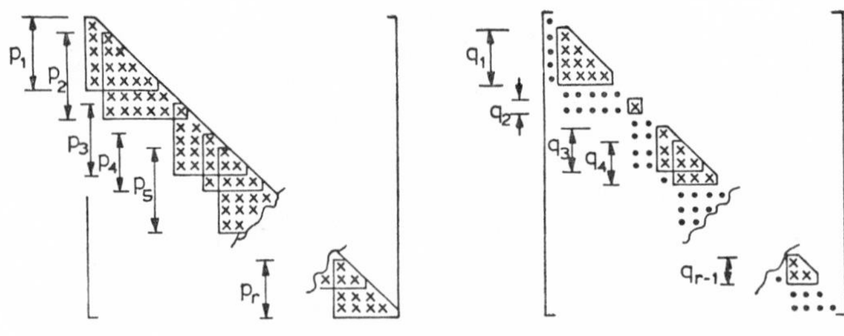

(a) dimensions of positive triangles (b) dimensions of negative triangles

Figure 5.22 A typical variable bandwidth store without re-entrant rows

The total number of multiplications and divisions for triangular decomposition is

$$m = \sum_{i=1}^{r} \frac{p_i(p_i - 1)(p_i + 4)}{6} - \sum_{i=1}^{r-1} \frac{q_i(q_i - 1)(q_i + 4)}{6} \tag{5.6}$$

The reverse Cuthill–McKee algorithm

It has been recognized by George (Cuthill, 1972) that if a Cuthill–McKee numbering is reversed a more efficient scheme is often obtained. Reversal of the Cuthill–McKee numbering for the ring network and the triangulated network both give more efficient variable bandwidth stores, as shown in Table 5.2.

Figure 5.23 shows the reverse Cuthill–McKee numbering for the triangulated network. It will be noted from Figure 5.23(b) that, if the variable bandwidth store is

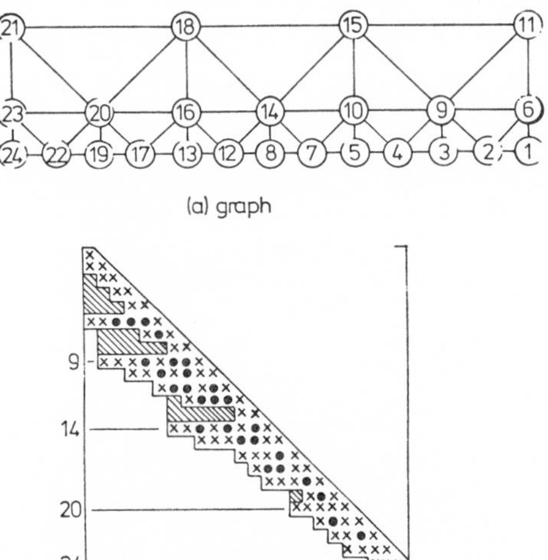

(b) coefficient matrix – shaded areas show storage space saved in variable bandwidth store by the reversal

Figure 5.23 Reverse Cuthill–McKee renumbering for the triangulated network

not allowed to contain re-entrant rows (i.e. the storage including the shaded areas), then the storage pattern is the reverse of the variable bandwidth store obtained from the direct Cuthill–McKee algorithm (Figure 5.20). Here 'reverse' is used in the sense that element (i, j) moves to $(n - j + 1, n - i + 1)$. In general it may be stated that the non-zero elements in the coefficient matrix for the reverse numbering can always be contained within the reverse storage pattern. Since reversing the storage pattern simply involves reversing the sequences of triangles $p_1, p_2, \ldots, p_r$ and $q_1, q_2, \ldots, q_{r-1}$, the storage space requirement (equation 5.5) and the computational requirement for triangular decomposition (equation 5.6) will remain unaltered. However, when the reverse algorithm is used it is often possible to improve on the above storage scheme by using a re-entrant row type of storage. Thus in Figure 5.23 the shaded areas need not be stored. Hence the reverse Cuthill–McKee algorithm cannot give a less efficient variable bandwidth scheme than the direct algorithm and will often give a more efficient variable bandwidth scheme. An alternative proof of this theorem is given by Liu and Sherman (1976). It is likely to be advantageous to relabel nodes according to decreasing (instead of increasing) degree in (b) and (c) of the main algorithm if the reverse node ordering is to be adopted.

Some other automatic renumbering schemes have been discussed by Cuthill (1972).

5.7 ELIMINATION IN A PACKED STORE

If the coefficient matrix of a set of equations is stored by means of one of the sparse packing schemes (sections 3.10 to 3.14) there is no restriction on the pattern of non-zero elements in the matrix. Thus for the 5 × 7 grid problem it is possible to adopt the ordering scheme for the variables represented graphically by Figure 5.6(k), which was developed with the intention of keeping the number of non-zero elements involved in the elimination as small as possible. With reference to the grid problem, Table 5.4 shows the total number of non-zero elements to be stored by the packing scheme and also the total number of multiplications and divisions required for triangular decomposition. Both totals are less than the corresponding totals for the optimum variable bandwidth scheme in which the nodes are labelled from corner to corner (Figure 5.16).

Table 5.4 Comparison of packed storage and variable bandwidth storage for the 5 × 7 grid

Type of storage	Total no. of elements to be stored	Total no. of multiplications for decomposition
Packed (as Figure 5.7)	157	424
Variable bandwidth (as Figure 5.16b)	175	520

In a procedure given by Brameller, Allan and Hamam (1976) the non-zero off-diagonal elements of the coefficient matrix are randomly packed using a primary array and two secondary arrays (one for row identifiers and the other for column links). All of the diagonal elements are stored consecutively in a one-dimensional array having n locations, with three corresponding secondary arrays containing:

(a) pointers to define the addresses of the first off-diagonal non-zero elements on each row,
(b) the total numbers of non-zero elements on each row, and
(c) integers specifying the order of elimination.

This last array is compiled automatically by simulating the elimination process without actually performing the arithmetical operations. During the simulation extra storage addresses are prepared for elements which become non-zero during the elimination. The use of a random packing scheme enables these addresses to be included at the end of the list. As the storage pattern changes during the simulated elimination the totals for the numbers of non-zero elements on each row are revised. These running totals provide the criterion for the automatic selection of the order of elimination, since at any stage the row with the least current number of non-zero elements may be found. Once the simulation is complete the decomposition is accomplished without modifying the storage pattern.

As compared with diagonal band and variable bandwidth techniques, elimination in a packed store has the following disadvantages:

(a) Every element in a packed store needs extra storage space for identifiers. If two integer identifiers use an equivalent amount of storage space to one non-zero element, then Brameller, Allan and Hamam's scheme requires $2T_1 + \frac{1}{2}n$ storage locations for the decomposition, where T_1 is the total number of non-zero elements in **L**. In contrast the variable bandwidth method only requires $T_2 + \frac{1}{2}n$ storage locations, where T_2 is the total number of elements stored in the variable bandwidth array. Using values of T_1 and T_2 obtained from Table 5.4 it follows that a decomposition for the 5×7 grid requires 331½ locations using a packing scheme, compared with only 192½ locations using a variable bandwidth scheme.

(b) The book-keeping operations necessary to perform a decomposition in a packed store will extend the computing time beyond that necessary to perform the non-zero arithmetical operations. (Examples of book-keeping operations are index inspections and searches to identify which particular arithmetical operations need to be performed.)

(c) If an automatic procedure for ordering the equations is adopted it adds complexity to the program and extends the computing time. (Note, however, that George's (1973) nested dissection method gives a suitable numbering procedure which can be specified manually.)

In view of these factors, the apparent benefit of using a packed store for elimination

Table 5.5 Comparison of storage requirements using sparse packing and variable bandwidth schemes for five-point finite difference grids ($c \simeq 5$)

		No. of elements in **L**		Storage locations required	
Grid size	No. of equations	Packed	Variable band	Packed	Variable band
5 x 6	30	131*	145	277	160
10 x 11	110	715*	915	1,485	970
15 x 16	240	1,919*	2,810	3,958	2,930
30 x 31	930	10,937*	20,245	22,339	20,710

*Figures given by Brameller, Allan and Hamam (1976). See also Reid (1974) for a similar comparison which includes computer times.

of the 5 x 7 grid equations, that might be inferred from Table 5.4, cannot be realized in practice. Table 5.5 shows that, for rectangular grids up to size 30 x 31, a packing scheme of the Brameller, Allan and Hamam type will be less efficient in the use of storage space than the variable bandwidth scheme with corner-to-corner frontal ordering of the nodes. It can be inferred from the trend in the figures that the packing scheme may be more efficient in the use of storage for large grids of this type (i.e. with the least dimension of the grid larger than about 35). However, in considering the solution of equations of order 1,000 or more it should be noted that the use of random packing schemes with address links is likely to require a large number of backing store transfers if it is not possible to hold the entire coefficient matrix within the main store. Therefore there is no strong case for using the more complicated packing schemes for elimination of five-point finite difference equations or equations arising from simple plane rectangular grid networks. This conclusion must be modified to some extent if the ratio of the storage space required for the primary and secondary arrays differs from that assumed above. (For instance, if the elements need to be stored double length the comparison will be more favourable to the packing scheme.)

Packing schemes are most efficient when c, the average number of non-zero elements per row of the coefficient matrix, is small. This is usually the case with distribution networks for electricity, gas or water supply. However, in such cases, optimum variable bandwidth storage is also particularly efficient, provided that typical fronts required for node ordering have only to pass through a few branches. Thus for the two networks shown in Figure 5.24 the first would require less storage space using a packed scheme and the second using a variable bandwidth scheme, although the advantage in either case would not be very large. Thus, even for networks which have a low value of c, the number of nodes needs to be larger than 100 before a packing scheme is likely to be more economical in storage space than a well-ordered variable bandwidth scheme.

Equations arising from finite element analyses almost invariably have c values much larger than 5. Hence a packing scheme is unlikely to prove the most efficient

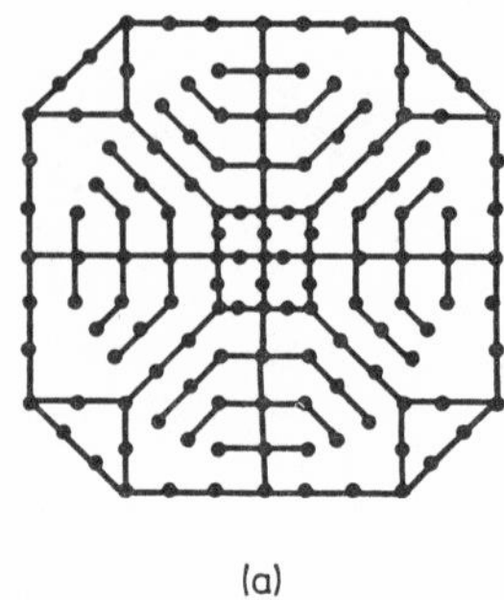

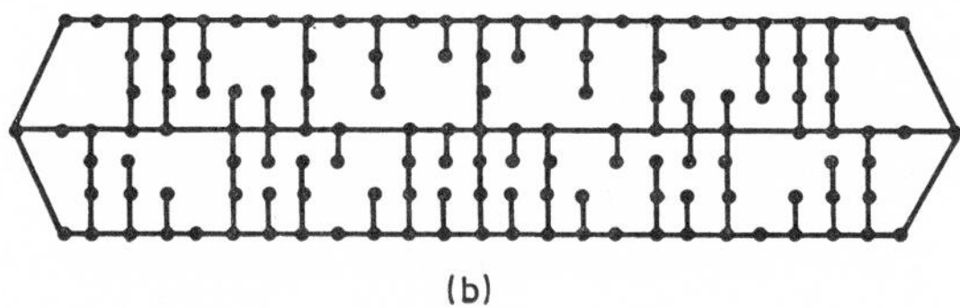

Figure 5.24 Two networks yielding coefficient matrices of order 137 with $c = 3.22$

method of obtaining their solution by elimination, except, possibly, where the problems are three-dimensional rather than plane.

5.8 ELIMINATION USING SUBMATRICES

An advantage of using submatrix (or partitioning) techniques is that the submatrices are convenient data packages for transfer to and from backing store. In this section two alternative submatrix solutions will be developed for equations in which the coefficient matrix has non-null submatrices arranged in a tridiagonal pattern.

Consider the set of simultaneous equations

$$\begin{bmatrix} \mathbf{A}_{11} & \mathbf{A}_{21}^T & & & \\ \mathbf{A}_{21} & \mathbf{A}_{22} & \mathbf{A}_{32}^T & & \\ & \mathbf{A}_{32} & \mathbf{A}_{33} & \mathbf{A}_{43}^T & \\ & & \cdot & \cdot & \cdot \\ & & & \cdots & \mathbf{A}_{nn} \end{bmatrix} \begin{bmatrix} \mathbf{x}_1 \\ \mathbf{x}_2 \\ \mathbf{x}_3 \\ \cdot \\ \mathbf{x}_n \end{bmatrix} = \begin{bmatrix} \mathbf{b}_1 \\ \mathbf{b}_2 \\ \mathbf{b}_3 \\ \cdot \\ \mathbf{b}_n \end{bmatrix} \tag{5.7}$$

where $\mathbf{A}_{11}$, $\mathbf{A}_{21}$, etc., are $b \times b$ submatrices and $\mathbf{x}_1$, $\mathbf{b}_1$, $\mathbf{x}_2$, $\mathbf{b}_2$, etc., are $b \times 1$ subvectors. It will be assumed that the coefficient matrix is symmetric and positive definite. Consequently each of the submatrices $\mathbf{A}_{11}$, $\mathbf{A}_{22}$, etc., in leading diagonal positions, will themselves be symmetric and positive definite. Multiplying the first submatrix equation by $\mathbf{A}_{11}^{-1}$ gives

$$\left.\begin{aligned} &\mathbf{x}_1 + \mathbf{C}_{21}^T\mathbf{x}_2 = \mathbf{d}_1 \\ &\text{where} \\ &\mathbf{C}_{21}^T = \mathbf{A}_{11}^{-1}\mathbf{A}_{21}^T \quad \text{and} \quad \mathbf{d}_1 = \mathbf{A}_{11}^{-1}\mathbf{b}_1 \end{aligned}\right\} \tag{5.8}$$

Using this equation to eliminate $\mathbf{x}_1$ from the second submatrix equation gives

$$\mathbf{A}_{22}^*\mathbf{x}_2 + \mathbf{A}_{32}^T\mathbf{x}_3 = \mathbf{b}_2^* \tag{5.9a}$$

where

$$\mathbf{A}_{22}^* = \mathbf{A}_{22} - \mathbf{A}_{21}\mathbf{C}_{21}^T \quad \text{and} \quad \mathbf{b}_2^* = \mathbf{b}_2 - \mathbf{A}_{21}\mathbf{d}_1 \tag{5.9b}$$

The matrices $\mathbf{C}_{21}^T$, $\mathbf{A}_{22}^*$ and the vectors $\mathbf{d}_1$ and $\mathbf{b}_2^*$ are all computable, with the result that matrix equation (5.7) can be modified to

$$\begin{bmatrix} \mathbf{I} & \mathbf{C}_{21}^T & & & \\ & \mathbf{A}_{22}^* & \mathbf{A}_{32}^T & & \\ & \mathbf{A}_{32} & \mathbf{A}_{33} & \mathbf{A}_{43}^T & \\ & & \cdot & \cdot & \cdot \\ & & & \cdots & \mathbf{A}_{nn} \end{bmatrix} \begin{bmatrix} \mathbf{x}_1 \\ \mathbf{x}_2 \\ \mathbf{x}_3 \\ \cdot \\ \mathbf{x}_n \end{bmatrix} = \begin{bmatrix} \mathbf{d}_1 \\ \mathbf{b}_2^* \\ \mathbf{b}_3 \\ \cdot \\ \mathbf{b}_n \end{bmatrix} \tag{5.10}$$

This process can be applied recursively so that after n steps the equations become

$$\begin{bmatrix} \mathbf{I} & \mathbf{C}_{21}^T & & & \\ & \mathbf{I} & \mathbf{C}_{32}^T & & \\ & & \mathbf{I} & \mathbf{C}_{43}^T & \\ & & & \cdot & \cdot \\ & & & & \mathbf{I} \end{bmatrix} \begin{bmatrix} \mathbf{x}_1 \\ \mathbf{x}_2 \\ \mathbf{x}_3 \\ \cdot \\ \mathbf{x}_n \end{bmatrix} = \begin{bmatrix} \mathbf{d}_1 \\ \mathbf{d}_2 \\ \mathbf{d}_3 \\ \cdot \\ \mathbf{d}_n \end{bmatrix} \tag{5.11}$$

At this stage $\mathbf{x}_n = \mathbf{d}_n$ and the other subvectors comprising the solution vector may be obtained in reverse order from

$$\mathbf{x}_i = \mathbf{d}_i - \mathbf{C}_{i+1,i}^T\mathbf{x}_{i+1} \tag{5.12}$$

In cases where n and b are large the total number of multiplications required to obtain the solution is approximately $(3n - 2)b^3$ if the submatrix operations are performed by standard matrix inversion, multiplication and subtraction procedures. If, on the other hand, the inversion is replaced by an equation-solving procedure for determining $\mathbf{C}_{i+1,i}^T$ from $\mathbf{A}_{ii}\mathbf{C}_{i+1,i}^T = \mathbf{A}_{i+1,i}^T$ and if, in addition, the determination of $\mathbf{d}_i$ from $\mathbf{A}_{ii}\mathbf{d}_i = \mathbf{b}_i$ is treated as a supplementary right-hand side to the same set of equations, the total number of multiplications is reduced to approximately $\frac{5}{3}(n - 1)b^3$ allowing for symmetry where relevant.

An alternative procedure is to determine the lower triangular matrix $\mathbf{L}_{11}$ obtained by the Choleski decomposition of $\mathbf{A}_{11}$. Then premultiplying the first submatrix equation by $\mathbf{L}_{11}^{-1}$ gives

$$\mathbf{L}_{11}^T\mathbf{x}_1 + \mathbf{L}_{21}^T\mathbf{x}_2 = \mathbf{h}_1 \tag{5.13}$$

where $\mathbf{L}_{21}^T$ and $\mathbf{h}_1$ may be obtained by forward-substitution within the equations

$$\left.\begin{array}{l}\mathbf{L}_{11}\mathbf{L}_{21}^T = \mathbf{A}_{21}^T \\ \text{and} \\ \mathbf{L}_{11}\mathbf{h}_1 = \mathbf{b}_1\end{array}\right\} \tag{5.14}$$

Since $\mathbf{A}_{21} = \mathbf{L}_{21}\mathbf{L}_{11}^T$, premultiplication of equation (5.13) by $\mathbf{L}_{21}$ and subtraction from the second submatrix equation yields equation (5.9a) with

$$\left.\begin{array}{l}\mathbf{A}_{22}^* = \mathbf{A}_{22} - \mathbf{L}_{21}\mathbf{L}_{21}^T \\ \text{and} \\ \mathbf{b}_2^* = \mathbf{b}_2 - \mathbf{L}_{21}\mathbf{h}_1\end{array}\right\} \tag{5.15}$$

The reduction process can be continued to give

$$\begin{bmatrix} \mathbf{L}_{11}^T & \mathbf{L}_{21}^T & & & \\ & \mathbf{L}_{22}^T & \mathbf{L}_{32}^T & & \\ & & \mathbf{L}_{33}^T & \mathbf{L}_{43}^T & \\ & & & \cdot & \cdot \\ & & & & \mathbf{L}_{nn}^T \end{bmatrix} \begin{bmatrix} \mathbf{x}_1 \\ \mathbf{x}_2 \\ \mathbf{x}_3 \\ \cdot \\ \mathbf{x}_n \end{bmatrix} = \begin{bmatrix} \mathbf{h}_1 \\ \mathbf{h}_2 \\ \mathbf{h}_3 \\ \cdot \\ \mathbf{h}_n \end{bmatrix} \tag{5.16}$$

in which the coefficient matrix is simply a submatrix representation of the Choleski triangular decomposition of the coefficient matrix of the original set of equations. At this stage $\mathbf{x}_n$ may be obtained by backsubstitution within

$$\mathbf{L}_{nn}^T\mathbf{x}_n = \mathbf{h}_n \tag{5.17}$$

and the other subsets of the variables may be obtained in reverse order by backsubstitution within

$$\mathbf{L}_{ii}^T\mathbf{x}_i = \mathbf{h}_i - \mathbf{L}_{i+1,i}^T\mathbf{x}_{i+1} \tag{5.18}$$

In cases where n and b are large the total number of multiplications required is now approximately $\frac{7}{6}(n-1)b^3$, which is more efficient than the previous scheme. This method involves the same amount of computation as that required to solve the full set of equations directly by the Choleski decomposition, provided that only non-zero operations are taken into account. The method is still valid even if the submatrices have different dimensions, provided that the submatrices in the leading diagonal positions are square. However, it is easier to organize storage transfers if the submatrices all have the same dimensions.

5.9 SUBSTRUCTURE METHODS

Substructure methods have been applied extensively in the field of structural analysis. They are applicable to any problem requiring the solution of sets of linear equations which have a sparse coefficient matrix.

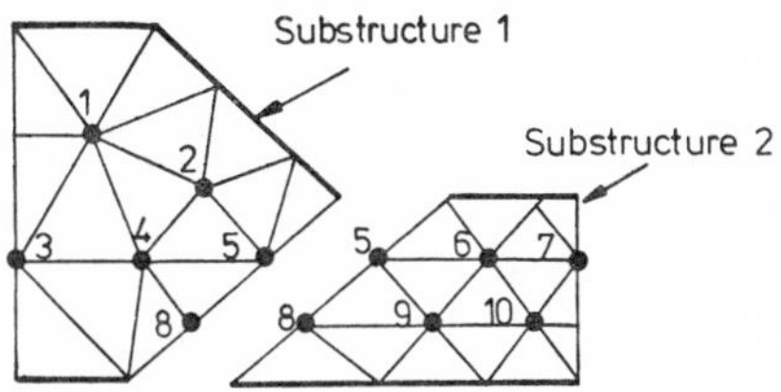

Figure 5.25 A substructure splitting of the finite element map of Figure 2.13

Although substructure techniques would normally be used for large-order problems, the finite element idealization of the heat transfer problem (Figure 2.13) will be used to illustrate the procedure. Figure 5.25 shows the finite element map separated into two substructures in such a way that individual finite elements are not dissected. It will be noticed that there are three types of node: those in substructure 1 only, those common to both substructures and those in substructure 2 only. This grouping is adopted in order to separate the temperature variables for the nodes into subvectors $\mathbf{x}_1$, $\mathbf{x}_2$ and $\mathbf{x}_3$ respectively. Since there is no direct coupling between the nodes of the first and third groups, the heat transfer equations have a submatrix form

$$\begin{bmatrix} \mathbf{A}_{11} & \mathbf{A}_{21}^T & \\ \mathbf{A}_{21} & \mathbf{A}_{22} & \mathbf{A}_{32}^T \\ & \mathbf{A}_{32} & \mathbf{A}_{33} \end{bmatrix} \begin{bmatrix} \mathbf{x}_1 \\ \mathbf{x}_2 \\ \mathbf{x}_3 \end{bmatrix} = \begin{bmatrix} \mathbf{b}_1 \\ \mathbf{b}_2 \\ \mathbf{b}_3 \end{bmatrix} \tag{5.19}$$

On the other hand, the equations for the two substructures may be specified separately as

$$\begin{bmatrix} \mathbf{A}_{11} & \mathbf{A}_{21}^T \\ \mathbf{A}_{21} & \mathbf{A}'_{22} \end{bmatrix} \begin{bmatrix} \mathbf{x}_1 \\ \mathbf{x}_2 \end{bmatrix} = \begin{bmatrix} \mathbf{b}_1 \\ \mathbf{b}'_2 + \boldsymbol{\gamma}_2 \end{bmatrix} \tag{5.20}$$

and

$$\begin{bmatrix} \mathbf{A}''_{22} & \mathbf{A}_{32}^T \\ \mathbf{A}_{32} & \mathbf{A}_{33} \end{bmatrix} \begin{bmatrix} \mathbf{x}_2 \\ \mathbf{x}_3 \end{bmatrix} = \begin{bmatrix} \mathbf{b}''_2 - \boldsymbol{\gamma}_2 \\ \mathbf{b}_3 \end{bmatrix} \tag{5.21}$$

where $\boldsymbol{\gamma}_2$ is a vector of unknown heat flows from substructure 1 to substructure 2 at the common nodes. The submatrices $\mathbf{A}'_{22}$ and $\mathbf{A}''_{22}$ will sum to $\mathbf{A}_{22}$ and the right-hand vectors $\mathbf{b}'_2$ and $\mathbf{b}''_2$ will sum to $\mathbf{b}_2$. It may easily be verified that the substructure equations (5.20) and (5.21) are together equivalent to the full system equations (5.19).

In substructure techniques the equations for an individual substructure are solved as far as possible, i.e. until just equations linking the common variables remain. This smaller set of equations is then added to the appropriate equations in the linking substructure(s), so eliminating the unknowns introduced at the common boundaries. The equations for the last substructure to be considered will therefore have no unknowns on the right-hand side and may be solved completely. As an example, consider the set of equations (5.20) for substructure 1. It is possible to perform a partial reduction by the Choleski submatrix method of the previous section, the resulting partially reduced equations being

$$\begin{bmatrix} \mathbf{L}_{11}^T & \mathbf{L}_{21}^T \\ & \mathbf{A}_{22}''' \end{bmatrix} \begin{bmatrix} \mathbf{x}_1 \\ \mathbf{x}_2 \end{bmatrix} = \begin{bmatrix} \mathbf{h}_1 \\ \mathbf{b}_2''' - \boldsymbol{\Upsilon}_2 \end{bmatrix} \tag{5.22}$$

Here the notation is similar to that used in section 5.8, with $\mathbf{A}_{22}''' = \mathbf{A}_{22}' - \mathbf{L}_{21}\mathbf{L}_{21}^T$ and $\mathbf{b}_2''' = \mathbf{b}_2' - \mathbf{L}_{21}\mathbf{h}_1$. The second submatrix equation links only the common variables. Adding the computed values of $\mathbf{A}_{22}'''$ and $\mathbf{b}_2'''$ to $\mathbf{A}_{22}''$ and $\mathbf{b}_2''$, the equations for substructure 2 may be modified to

$$\begin{bmatrix} \mathbf{A}_{22}^* & \mathbf{A}_{32}^T \\ \mathbf{A}_{32} & \mathbf{A}_{33} \end{bmatrix} \begin{bmatrix} \mathbf{x}_2 \\ \mathbf{x}_3 \end{bmatrix} = \begin{bmatrix} \mathbf{b}_2^* \\ \mathbf{b}_3 \end{bmatrix} \tag{5.23}$$

since $\mathbf{A}_{22}'' + \mathbf{A}_{22}''' = \mathbf{A}_{22} - \mathbf{L}_{21}\mathbf{L}_{21}^T = \mathbf{A}_{22}^*$ and $\mathbf{b}_2'' + \mathbf{b}_2''' = \mathbf{b}_2 - \mathbf{L}_{21}\mathbf{h}_1 = \mathbf{b}_2^*$. Equations (5.23) may be solved for $\mathbf{x}_2$ and $\mathbf{x}_3$, and $\mathbf{x}_1$ may be obtained by backsubstitution in the first submatrix equation of (5.22).

Efficient decomposition of substructure equations

The above substructure solution has been expressed in submatrix form. Indeed there is a clear equivalence with the submatrix solution of the previous section. However, there is no need to split the coefficient matrix for an individual substructure into three submatrices for storage purposes. It is possible, for instance, to store it in variable bandwidth form. Choosing the order of the variables for the heat transfer problem to be $\mathbf{x}_1 = \{T_3, T_1, T_4, T_2\}$, $\mathbf{x}_2 = \{T_8, T_5\}$ and $\mathbf{x}_3 = \{T_9, T_6, T_{10}, T_7\}$, the variable bandwidth stores are as shown in Figure 5.26. For substructure 1 the first four steps of a Choleski-type Gaussian elimination may then be carried out, resulting in the formation of the non-zero components of $\mathbf{L}_{11}$ and $\mathbf{L}_{21}$ in the first four columns and modifying columns 5 and 6 to represent the symmetric matrix $\mathbf{A}_{22}'''$. Corresponding operations need to be carried out with the right-hand vector. At this stage the computed values of $\mathbf{A}_{22}'''$ and $\mathbf{b}_2'''$ can be added to the values of $\mathbf{A}_{22}''$ and $\mathbf{b}_{21}''$, making it possible to proceed with the decomposition of substructure 2.

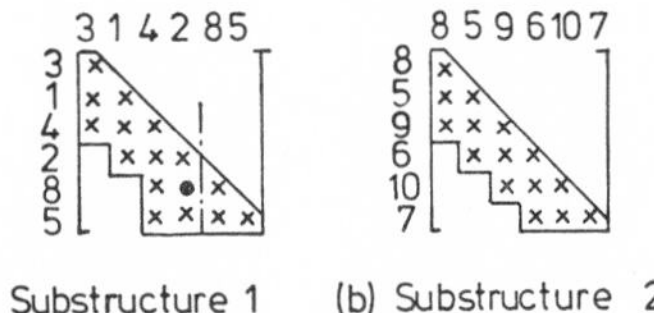

Figure 5.26 Variable bandwidth stores for the substructures shown in Figure 5.25

Notes on substructure techniques

(a) It is possible to replace the Choleski decomposition of the substructure equations with an $\mathbf{LDL}^T$ decomposition.

(b) In the example studied, the computational requirement for the substructure solution is effectively the same as that required by a variable bandwidth solution of the full set of equations with the variables specified in the same order. This is generally true where a structure is divided into any number of substructures, provided that the substructures are linked in series (i.e. substructure i is only linked to substructures $i - 1$ and $i + 1$).

(c) Nodes in a substructure which are common to a subsequent substructure must be listed last. However, nodes which are common to an earlier substructure need not be listed first. If the nodes in each substructure are to be listed in order of elimination then it may be necessary to give common nodes two node numbers, one for each substructure as illustrated in Figure 5.27(b).

(d) In some cases it is possible to obtain a more efficient solution by a substructure technique than by other methods. Figure 5.27 shows two node ordering schemes for a finite element problem, the second of which employs substructures. Table 5.6 gives a comparison of the storage space required for the coefficient matrices and the number of multiplications and divisions required for decomposition. The reason why the substructure method is more

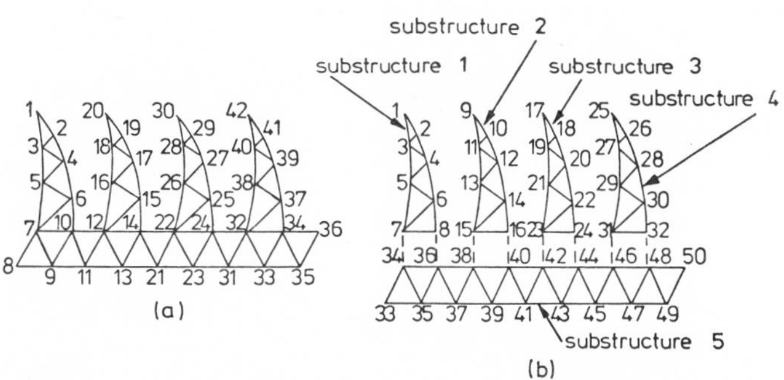

Figure 5.27 Alternative schemes for the solution of a finite element problem using variable bandwidth store

Table 5.6 Comparison of decompositions for schemes shown in Figure 5.27

	Scheme (a)	Scheme (b) using substructures	Scheme (b) with identical substructures
Storage locations required	160	135	72
No. of multiplications and divisions	368	202	112

efficient in this case is because it takes advantage of the tree structure of the finite element map.

(e) An efficient technique for substructure splitting of large sets of equations is to arrange the substructures to be subsets in a nested dissection of the graph (George, 1973).

(f) Substructure methods may be of particular advantage when some substructures are identical, since a partial decomposition of one substructure may also apply to other substructures. For instance, if substructures 1, 2, 3 and 4 of Figure 5.27(b) are identical, the storage space and computational requirement for the coefficient matrices are reduced to the values shown in the last column of Table 5.6.

5.10 ON THE USE OF BACKING STORE

There are a variety of ways in which the elimination techniques described in the previous sections may be implemented with backing store to allow much larger systems to be solved than are possible using only the main store. The amount of backing store at the disposal of the user is usually large and is unlikely to restrict the size of problem that can be solved. However, the amount of main store available usually imposes a restriction. For instance, with the first submatrix scheme shown in section 5.8, in which standard matrix inversion, multiplication, subtraction and transpose operations are performed with $b \times b$ submatrices, it will be necessary to hold three $b \times b$ matrices in the main store simultaneously during the implementation of the matrix multiplication $\mathbf{C}_{21}^T = \mathbf{A}_{11}^{-1}\mathbf{A}_{21}^T$. Thus the size of the submatrices must be restricted.

Although submatrix methods may appear to be very well suited to use with backing store it is difficult to develop efficient submatrix methods which are sufficiently versatile to accept a variety of non-zero element patterns for the coefficient matrix and which are not, at the same time, very complicated to implement. A number of alternative backing store procedures are outlined below.

A diagonal band backing store algorithm

In section 5.3 it was seen that each step of the reduction phase of diagonal band elimination involves only a $b \times b$ triangle of stored elements of the coefficient matrix. Thus, if the triangle spanning from column k to row $k + b - 1$ is held within the main store, the i-th reduction step may be performed. Column k will then contain the Choleski coefficients $l_{k,k}, l_{k+1,k}, \ldots, l_{k+b-1,k}$ (or, alternatively, $d_{k,k}, l_{k+1,k}, \ldots, l_{k+b-1,k}$ if an $\mathbf{LDL}^T$ decomposition is carried out), which must be transferred to the backing store. It is then possible to establish the triangular store for the next reduction step by shifting each of the stored elements $a_{ij}^{(k+1)}$ forward by one row and one column and inserting the next row of $\mathbf{A}$ into the last row of the triangular store. After decomposition, $\mathbf{L}$ will be stored by columns in the backing store. It may be recalled to the main store column by column to perform the right-hand side reduction, and then, with the order of recall reversed, to perform the backsubstitution. In both of these operations less main store will be required for coefficients of **L** and so some storage space will be available for at least part of the right-hand vector or matrix.

The storage limit is reached when the $b \times b$ triangle completely fills the available main store. Thus there is a restriction on the maximum bandwidth which may be handled. If the $b \times b$ triangle does not completely fill the available main store the extra storage space may be used during the decomposition to store extra rows of **A**. This will allow several reduction steps to be completed between storage transfers, so yielding an economy in the number of data transfers.

Variable bandwidth backing store algorithms

The scheme outlined above may be extended to variable bandwidth storage which does not contain re-entrant rows. The storage limit will be reached when the largest triangle just fits into the available main store.

For the general case, in which the variable bandwidth store is permitted to contain re-entrant rows, it is possible to divide the store into segments, each containing an integral number of consecutive rows for data transfer purposes. In order to perform the reduction it must be possible to contain two such segments in the main store simultaneously. If the number of rows in each segment always exceeds the maximum number of stored elements in any particular row (as illustrated in Figure 5.28(a)), the reduction for all of the elements of segment q can be carried out with just segments $q - 1$ and q in the main store. Thus the

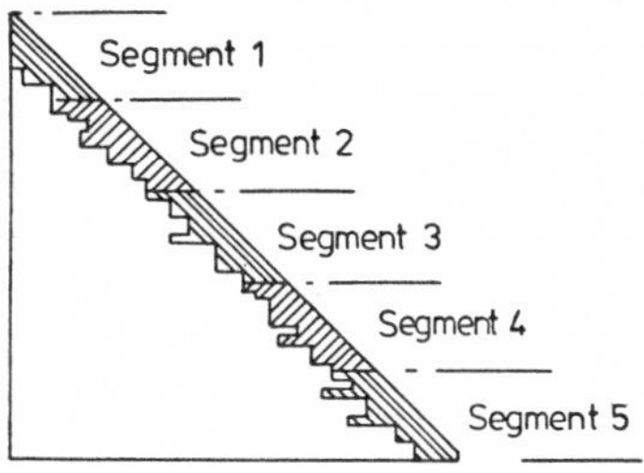

(a) allowing a simple reduction scheme

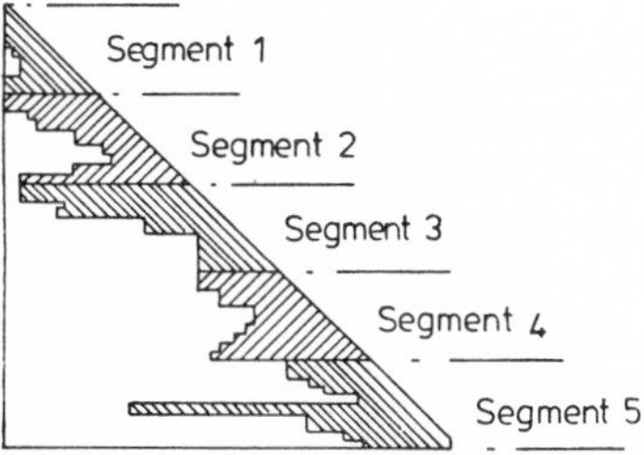

(b) requiring a more complex reduction scheme

Figure 5.28 Segmental arrangement of variable bandwidth store

Table 5.7 Use of main store for decomposition of variable bandwidth matrix shown in Figure 5.28(b)

Segments in main store		
Passive q	Active p	Operation
–	1	Reduction for segment 1
1	2	Reduction for segment 2
1, 2	3	Reduction for segment 3
3	4	Reduction for segment 4
2, 3, 4	5	Reduction for segment 5

complete reduction may be performed with the main store containing, in sequence, segments 1 and 2, segments 2 and 3, segments 3 and 4, etc. For right-hand side operations the segments can be brought into the main store one at a time.

On the other hand, if the number of rows in each segment is less than the number of stored elements in individual rows, the reduction for segment q is only likely to be accomplished by bringing into the main store more than one of the previous segments. For example, a reduction using the storage pattern shown in Figure 5.28(b) would entail bringing segments into the main store in the sequence given in Table 5.7. The organization of this scheme has been discussed by Jennings and Tuff (1971). The only storage restriction is that an individual segment should contain at least one complete row of the variable bandwidth store. However, when few rows are contained in each segment it will be necessary to carry out a large number of data transfers in order to perform the reduction.

Substructure operations with backing store

In section 5.9 it was found that the equations for an individual substructure can be constructed with the coefficient matrix stored in any way suitable for elimination (e.g. using a variable bandwidth store), provided that variables common to succeeding substructures are listed last. Substructure techniques are simple to organize with backing store provided that the main store is sufficiently large to hold the coefficient matrix for each substructure together with the corresponding right-hand vector or matrix.

Where the substructures are linked in series, the solution procedure may be summarized as follows:

(a) Set up the equations for substructure 1 in the main store.

(b) Perform the reduction as far as possible.

(c) Transfer the completed coefficients of **L** and the corresponding reduced right-hand side elements to backing store.

(d) Shift the remaining stored elements into their correct position for substructure 2.

(e) Complete the construction of the equations for substructure 2.

(f) Continue to process each substructure in turn. When the last substructure has been processed the reduction will be complete.

(g) The full solution can then be obtained by backsubstitution using the information held in the backing store.

Irons' frontal solution

Irons' (1970) frontal solution has been developed specifically for the solution of finite element equations. It may be considered as an extension of the substructure method in which each substructure is comprised of a single finite element together with frontal nodes, the latter being necessary to obtain a series linkage of the substructures.

Figure 5.29(a) shows a region which has been divided into five trapezoidal and five triangular finite elements such that there are fourteen node points. It will be assumed that one variable is associated with each node and that the node and

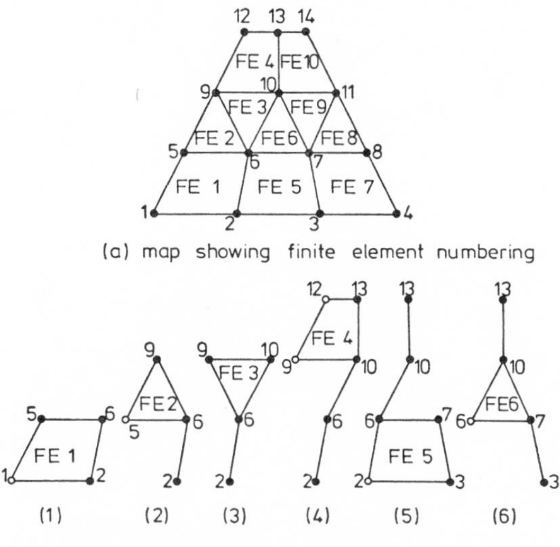

Figure 5.29 A frontal solution scheme for a finite element analysis

variable numbering schemes coincide. The finite elements are processed in the order in which they are numbered, the substructures corresponding to the first six stages of the reduction being as shown in Figure 5.29(b). Thus after three reduction stages, partially solved equations linking the four variables 2, 6, 9 and 10 will be present in the main store. The contributions of the fourth finite element are then added so that the resulting equations now link the six variables of the fourth substructure, namely, 2, 6, 9, 10, 12 and 13. The reduction for variables 9 and 12 is then carried out and the appropriate information transferred to backing store. The incomplete equations for the other variables are then carried forward to the next stage in the reduction.

Significant features of Irons' implementation of this technique are:

(a) The nodes may be numbered arbitrarily, the efficiency of the solution depending on the order in which the finite elements are numbered.

(b) In moving from one substructure to the next the coefficients corresponding to common nodes retain their position in the main store.

(c) As a consequence of (b), the variables to be reduced are not necessarily listed first in the substructure. Therefore the reduction for these variables and the transfer of elements to backing store may leave gaps in the list of active nodes and corresponding gaps in the triangle of stored coefficients. When new nodes are introduced they are entered into gaps in the active node list if such gaps have been left by previous reductions. Hence, for the substructure containing the largest number of nodes, the coefficient matrix will be stored in a full triangle.

(d) The total number of multiplications and divisions required to perform a complete solution of the finite element equations is the same as for a variable bandwidth solution, provided that the order of elimination of the variables is the same for both schemes (e.g. the frontal scheme of Figure 5.29 will be comparable with the variable bandwidth scheme having a node ordering of 1, 5, 9, 12, 2, 6, 3, 4, 8, 7, 10, 11, 13, 14).

5.11 UNSYMMETRIC BAND ELIMINATION

If a sparse unsymmetric matrix is diagonally dominant, the diagonal dominance will be retained if corresponding rows and columns are interchanged. Hence it will generally be possible to arrange the matrix so that the non-zero elements lie within a diagonal band store as shown in Figure 5.30(a). The band could be eccentric about the leading diagonal, with elements on row i stored between and including column $i - b_l + 1$ and $i + b_u - 1$ (b_l and b_u could be called the *lower* and *upper bandwidths* respectively). To perform decomposition $\mathbf{A} \rightarrow \mathbf{A}^F$ (equation 4.14) of an $n \times n$ matrix within this store requires approximately $n(b_l - 1)b_u$ multiplications and divisions. It is also possible to adopt a variable bandwidth storage scheme in which the lower triangle is stored by rows and the upper triangle by columns (Figure 5.30(b)).

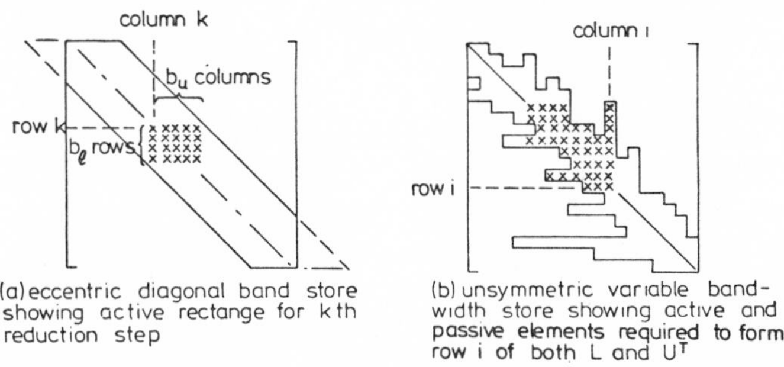

Figure 5.30 Unsymmetric band elimination schemes which do not allow for pivot selection

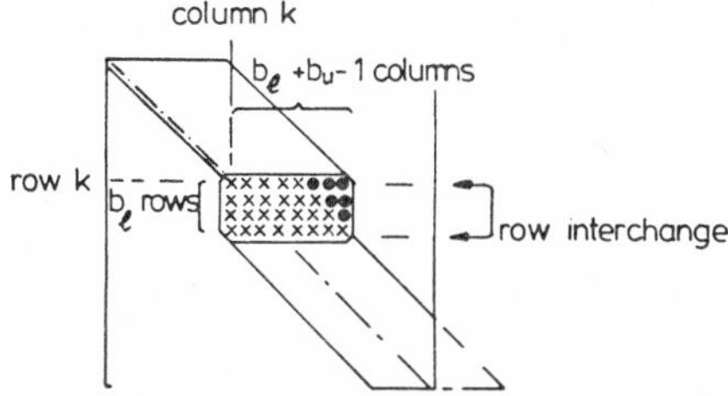

Figure 5.31 Unsymmetric diagonal band Gaussian elimination with row interchange

Unfortunately, many unsymmetric matrices will not have the property of diagonal dominance and so, for general unsymmetric matrix decomposition, storage schemes will have to be chosen which permit pivotal selection. Although it is not possible to employ a full pivoting strategy, a partial pivoting strategy may be carried out within a band store. Consider a standard Gaussian elimination with row interchange applied to a set of equations whose coefficient matrix has lower and upper bandwidths b_l and b_u respectively. Before the k-th reduction step, the possible interchange of rows k and $k + b_l - 1$ could result in elements within row k up to column $k + b_l + b_u - 2$ becoming non-zero. This means that the k-th reduction step could involve all of the elements within the $b_l \times (b_l + b_u - 1)$ rectangle shown in Figure 5.31. Although new non-zero elements could arise to the right of the original band, the full set of non-zero elements may be contained within an array with dimensions $n \times (b_l + b_u - 1)$ at all stages in the elimination. Figure 5.32 shows the contents of an array for a 9×9 matrix with $b_l = b_u = 3$ after the second reduction step. It may be noticed that when an element is eliminated from the left of a row all of the elements in the row must be displaced one position to the left after modification.

Provided that the corresponding interchange and forward-substitution operations on the right-hand side can be performed at the same time as the reduction of the

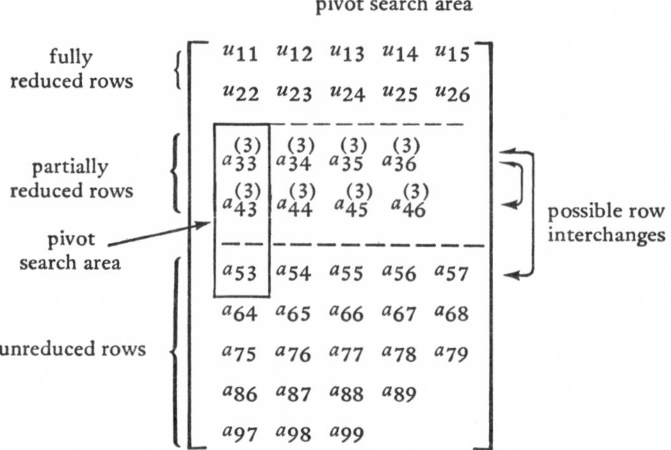

Figure 5.32 Unsymmetric band storage for a 9 × 9 matrix with $b_l = b_u = 3$, showing elements stored after two reduction steps

coefficient matrix, the multipliers will not need to be retained. However, if it is required to keep a record of the multipliers (which constitute the matrix **L** of the **LU** decomposition), it will be necessary to allocate extra storage space. Care must be taken in representing **L** since row interchanges during the reduction also interchange rows of **L**.

There is plenty of scope for rearranging a general sparse unsymmetric matrix into a convenient diagonal band form since it is not necessary that row and column interchanges should correspond. The main priority in rearranging the matrix should be to make b_l as small as possible.

5.12 UNSYMMETRIC ELIMINATION IN A PACKED STORE

Several schemes have been developed for solving equations with general large sparse unsymmetric coefficient matrices by using packed storage schemes (see Curtis and Reid, 1971; Tewarson, 1967; Tinney and Walker, 1967). The following is a brief outline to indicate some possible alternatives.

Adding new elements

When the coefficient matrix is stored in packed form, some extra storage space must be allocated for elements which become non-zero during the elimination process. Consider the case of standard Gaussian elimination for which the matrix is stored in a row-wise systematic packing. Reduction step k consists of scanning the elements of the matrix from rows $k + 1$ to n, modifying each row which contains a non-zero element in the pivotal column. Whenever a row has been modified it may contain more non-zero elements than it had originally. Hence the elements in other rows

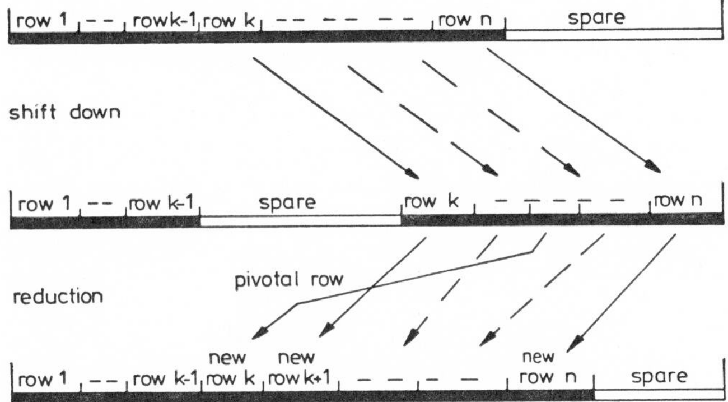

Figure 5.33 An allocation scheme for the primary array of a row-wise packed store during the k-th reduction step of Gaussian elimination

will have to be shifted to make room for the expanded rows. One way in which this can be performed is to begin by moving rows $k + 1$ to n in reverse order to the bottom of the available store and then modifying the appropriate rows as they are shifted back. Unless the choice of pivots is restricted to elements in row k, it may be necessary to re-order the rows. This may be accomplished by shifting up the pivotal row before the reduction commences, as shown in Figure 5.33.

The large amount of storage shifting operations inherent in this technique can be avoided by any of the three following methods:

(a) A random packing scheme may be used, so permitting all of the extra non-zero elements arising during the elimination to be added to the end of the list of stored information. Since the elements have to be scanned systematically during the elimination it is necessary to include address links.

(b) Extra storage space may be included at the end of each row so that extra non-zero elements can usually be inserted in a row without having to shift other rows.

(c) A compact elimination may be chosen, such that each row is fully reduced in one operation. It is only possible to use row pivot selection with this technique.

Pivot selection

A particular advantage of Gaussian elimination carried out in a packed store is that pivots may be chosen with the object of minimizing the growth in the number of non-zero elements. This objective is likely to be achieved by choosing particular pivots such that the pivotal row and the active part of the pivotal column have small numbers of non-zero elements. Integer arrays may be used to monitor the number of non-zero elements in each row and in the active part of each column so as to expedite the search for a good pivot. However, in order to prevent undue loss of accuracy, it is also necessary to ensure that a strong pivot is chosen. Hence the choice of pivot will normally require a compromise between these two principles. Whereas full pivoting will give the best opportunity of restricting the growth in the number of non-zero elements, partial pivoting procedures will require less computation to select the pivots.

General comments

At the moment insufficient methods have been programmed and comparative tests performed to give a clear guide as to which techniques are best to adopt. It is also difficult to assess whether or not packed storage is generally preferable to band storage for elimination involving large sparse unsymmetric matrices.

BIBLIOGRAPHY

Brameller, A., Allan, R. N., and Hamam, Y. M. (1976). *Sparsity, Its Practical Application to Systems Analysis*, Pitman, London. (Describes a random packing decomposition for symmetric matrices.)

Cheung, Y. K., and Khatua, T. P. (1976). 'A finite element solution program for large structures'. *Int. J. for Num. Methods in Engng.*, **10**, 401–412.

Curtis, A. R., and Reid, J. K. (1971). 'The solution of large sparse unsymmetric systems of linear equations'. *IFIP Conference Proceedings*, Ljubljana, Yugoslavia.

Cuthill, E. (1972). 'Several strategies for reducing the bandwidth of matrices'. In D. J. Rose and R. A. Willoughby (Eds.), *Sparse Matrices and Their Applications*, Plenum Press, New York.

Cuthill, E., and McKee, J. (1969). 'Reducing the bandwidth of sparse symmetric matrices'. *ACM Proceedings of 24th National Conference*, New York.

George, A. (1973). 'Nested dissection of a regular finite element mesh'. *SIAM J. Numer. Anal.*, **10**, 345–363.

Irons, B. M. (1970). 'A frontal solution program for finite element analysis'. *Int. J. for Num. Methods in Engng.*, **2**, 5–32.

Jennings, A. (1966). 'A compact storage scheme for the solution of linear simultaneous equations'. *Computer J.*, **9**, 281–285. (A variable bandwidth method.)

Jennings, A. (1971). 'Solution of variable bandwidth positive definite simultaneous equations'. *Computer J.*, **14**, 446. (A variable bandwidth algorithm.)

Jennings, A., and Tuff, A. D. (1971). 'A direct method for the solution of large sparse symmetric simultaneous equations'. In J. K. Reid (Ed.), *Large Sparse Sets of Linear Equations*, Academic Press, London. (A backing store version of the variable bandwidth method.)

Liu, W.-H., and Sherman, A. H. (1976). 'Comparative analysis of the Cuthill–McKee and the reverse Cuthill–McKee ordering algorithms for sparse matrices'. *SIAM J. Numer. Anal.*, **13**, 198–213.

Reid, J. K. (1974). 'Direct methods for sparse matrices'. In D. J. Evans (Ed.), *Software for Numerical Mathematics*, Academic Press, London.

Rosen, R., and Rubinstein, M. F. (1970). 'Substructure analysis by matrix decomposition'. *Proc. ASCE (J. of the Struct. Div.)*, **96**, 663–670.

Tewarson, R. P. (1967). 'Solution of a system simultaneous linear equations with a sparse coefficient matrix by elimination methods'. *BIT*, 7, 226–239.

Tewarson, R. P. (1973). *Sparse Matrices*, Academic Press, New York.

Tinney, W. F., and Walker, J. W. (1967). 'Direct solutions of sparse network equations by optimally ordered triangular factorization'. *Proc. IEEE*, **58**, 1801–1809.

Wilkinson, J. H., and Reinsch, C. (1971). *Handbook for Automatic Computation Vol. II, Linear Algebra*, Springer-Verlag, Berlin. (Gives algorithms for band symmetric and unsymmetric decompositions.)

Williams, F. W. (1973). 'Comparison between sparse stiffness matrix and substructure methods'. *Int. J. for Num. Methods in Engng.*, **5**, 383–394.

Zambardino, R. A. (1970). 'Decomposition of positive definite band matrices'. *Computer J.*, **13**, 421–422.

Chapter 6
Iterative Methods for Linear Equations

6.1 JACOBI AND GAUSS–SEIDEL ITERATION

In this chapter a variety of methods will be considered for solving linear simultaneous equations which are different in character from the elimination methods considered in the previous two chapters. They may be classified broadly into *stationary* and *gradient* methods, the stationary methods being presented first in the text. In each case trial values of variables are chosen, which are improved by a series of iterative corrections. It is the correction technique which distinguishes the methods. The earliest iterative methods are those of Jacobi and Gauss–Seidel.

Most of the iterative methods to be described will converge to the correct solution only if the coefficient matrix of the equations has a strong leading diagonal. In such cases it is convenient to assume that the original equations (and possibly the variables as well) have been scaled so that all of the leading diagonal coefficients are unity. The equations $Ax = b$ will then have the form

$$\begin{bmatrix} 1 & a_{12} & a_{13} & \cdots & a_{1n} \\ a_{21} & 1 & a_{23} & \cdots & a_{2n} \\ a_{31} & a_{32} & 1 & \cdots & a_{3n} \\ \cdot & \cdot & \cdot & \cdot & \cdot \\ a_{n1} & a_{n2} & a_{n3} & \cdots & 1 \end{bmatrix} \begin{bmatrix} x_1 \\ x_2 \\ x_3 \\ \cdot \\ x_n \end{bmatrix} = \begin{bmatrix} b_1 \\ b_2 \\ b_3 \\ \cdot \\ b_n \end{bmatrix} \tag{6.1}$$

where the coefficient matrix may be expressed as

$$\mathbf{A} = \mathbf{I} - \mathbf{L} - \mathbf{U} \tag{6.2}$$

in which

$$\mathbf{L} = \begin{bmatrix} 0 & & & & \\ -a_{21} & 0 & & & \\ -a_{31} & -a_{32} & 0 & & \\ \cdot & \cdot & \cdot & \cdot & \\ -a_{n1} & -a_{n2} & \cdots & -a_{n,n-1} & 0 \end{bmatrix}$$

and

$$\mathbf{U} = \begin{bmatrix} 0 & -a_{12} & -a_{13} & \cdots & -a_{1n} \\ & 0 & -a_{23} & \cdots & -a_{2n} \\ & & \cdot & \cdot & \cdot \\ & & & 0 & -a_{n-1,n} \\ & & & & 0 \end{bmatrix}$$

Jacobi iteration (or the method of simultaneous corrections)

If $x_i^{(k)}$ is the trial value of variable x_i after k iterations, then for Jacobi iteration the next approximation to x_i is

$$x_i^{(k+1)} = b_i - a_{i1}x_1^{(k)} - \cdots - a_{i,i-1}x_{i-1}^{(k)} - a_{i,i+1}^{(k)} - \cdots - a_{in}x_n^{(k)} \tag{6.3}$$

The full set of operations to perform one iteration step may be expressed in matrix form as

$$\mathbf{x}^{(k+1)} = \mathbf{b} + (\mathbf{L} + \mathbf{U})\mathbf{x}^{(k)} \tag{6.4}$$

where

$$\mathbf{x}^{(k)} = \{x_1^{(k)}, x_2^{(k)}, \ldots, x_n^{(k)}\}$$

Gauss–Seidel iteration (or the method of successive corrections)

If the variables are not corrected simultaneously but in the sequence $i = 1 \rightarrow n$, when $x_i^{(k+1)}$ is determined, revised values will already be available for $x_1, x_2, \ldots, x_{i-1}$. In Gauss–Seidel iteration these revised values are used instead of the original values, giving the general formula:

$$x_i^{(k+1)} = b_i - a_{i1}x_1^{(k+1)} - \cdots - a_{i,i-1}x_{i-1}^{(k+1)} - a_{i,i+1}x_{i+1}^{(k)} - \cdots - a_{in}x_n^{(k)} \tag{6.5}$$

In matrix form the iterative cycle is defined by

$$(\mathbf{I} - \mathbf{L})\mathbf{x}^{(k+1)} = \mathbf{b} + \mathbf{U}\mathbf{x}^{(k)} \tag{6.6}$$

A simple example

Consider the observational normal equations (2.30) obtained from the error minimization of a surveying network. Scaling these equations in the manner described by equation (4.47) with

$$\mathbf{R} = \mathbf{C} = \lceil 0.34711 \quad 0.27017 \quad 0.30861 \quad 0.32969 \rfloor$$

gives

$$\begin{bmatrix} 1 & -0.30009 & & -0.30898 \\ -0.30009 & 1 & -0.46691 & \\ & -0.46691 & 1 & -0.27471 \\ -0.30898 & & -0.27471 & 1 \end{bmatrix} \begin{bmatrix} x_1 \\ x_2 \\ x_3 \\ x_4 \end{bmatrix} = \begin{bmatrix} 5.32088 \\ 6.07624 \\ -8.80455 \\ 2.67600 \end{bmatrix} \tag{6.7}$$

(Note that the variables have also been scaled and hence are not the same as the variables of equation 4.47.) Applying Jacobi and Gauss–Seidel iterations to these equations using the initial trial vector $\mathbf{x}^{(0)} = \mathbf{0}$ give the results shown in Tables 6.1 and 6.2. Below the value of each set of trial variables is shown the Euclidean error norm

$$e_k = \| \mathbf{e}^{(k)} \|_E = \| \mathbf{x}^{(k)} - \mathbf{x} \|_E \tag{6.8}$$

computed using the solution $\mathbf{x} = \{8.4877 \quad 6.4275 \quad -4.7028 \quad 4.0066\}$.

Termination of the iterative process

In practice the true solution, $\mathbf{x}$, will not be available and hence the decision to terminate the process cannot be based on the error norm given above. Instead, either a vector difference norm or a residual norm may be adopted. If a vector difference norm is used a suitable criterion for termination would be when

$$\frac{\| \mathbf{x}^{(k)} - \mathbf{x}^{(k-1)} \|}{\| \mathbf{x}^{(k)} \|} \leqslant \text{tolerance} \tag{6.9}$$

It may be noticed that, with a tolerance of 1 and a trial vector $\mathbf{x}^{(0)} = \mathbf{0}$, this criterion will be satisfied at the first opportunity to apply it, namely at $k = 1$, but will be increasingly difficult to satisfy as the tolerance is reduced. The tolerance would normally be set to 0.001 or less in the hope of achieving an accuracy of three or more significant figures for the predicted values of the variables. If a residual norm is used a suitable criterion to terminate iteration would be if

$$\frac{\| \mathbf{r}^{(k)} \|}{\| \mathbf{b} \|} \leqslant \text{tolerance} \tag{6.10}$$

where $\mathbf{r}^{(k)}$ is the residual $\mathbf{b} - \mathbf{A}\mathbf{x}^{(k)}$. Again the tolerance will normally be set to 0.001 or less. The residual norm provides the more reliable of the two criteria since

Table 6.1 Jacobi iteration for equations (6.7)

k	0	1	2	3	4	5	–	19	20	21
x_1	0	5.3209	7.9711	6.9773	8.2727	7.7763	–	8.4840	8.4872	8.4860
x_2	0	6.0762	3.5621	6.0253	5.0795	6.2362	–	6.4265	6.4243	6.4271
x_3	0	−8.8046	−5.2324	−6.6191	−4.9745	−5.6050	–	−4.7075	−4.7035	−4.7050
x_4	0	2.6760	1.9014	3.7015	3.0135	3.8656	–	4.0059	4.0042	4.0063
e_k	12.3095	5.3609	3.6318	2.4916	1.7098	1.1732	–	0.0060	0.0041	0.0028

Table 6.2 Gauss–Seidel iteration for equations (6.7)

k	0	1	2	3	4	5	6	7	8	9
x_1	0	5.3209	8.5150	8.3846	8.3987	8.4471	8.4690	8.4786	8.4832	8.4855
x_2	0	7.6730	6.1933	6.2017	6.3380	6.3870	6.4074	6.4177	6.4228	6.4252
x_3	0	−5.2220	−5.1201	−4.8374	−4.7636	−4.7339	−4.7180	−4.7101	−4.7064	−4.7045
x_4	0	2.8855	3.9004	3.9378	3.9624	3.9856	3.9967	4.0018	4.0043	4.0055
e_k	12.3095	3.6202	0.4909	0.2906	0.1469	0.0685	0.0329	0.0160	0.0078	0.0038

with some slowly convergent iterations $\| \mathbf{e}^{(k)} \|$ may be much greater $\| \mathbf{x}^{(k)} - \mathbf{x}^{(k-1)} \|$.

6.2 RELAXATION TECHNIQUES

Hand relaxation

Relaxation techniques were used extensively for the hand solution of linear simultaneous equations before digital computers became available. They have been applied principally for solving field problems in which the governing partial differential equations were approximated to finite difference form. However, they have also been used for solving network and structural problems (e.g. the moment distribution method for structural frames). These methods are related to the Gauss–Seidel method but have the following differences.

Updating the residuals

The current values of all residuals are usually recorded on a graph of the mesh or network. As the value of the variable associated with a particular node is altered the residual for the node is set to zero (i.e. the node is *relaxed*), and the nodes which are linked to this node receive modifications (or *carry-over*).

Order of relaxation

At any stage in the relaxation it is possible to choose as the next relaxation node the one with the largest current residual.

Over-relaxation

The relaxation of finite difference meshes is often found to be slowly convergent. In such cases, after the residual at a node has been set to zero, it tends to build up again almost to its original value through carry-over from the relaxation of neighbouring nodes. Where this phenomenon is observed it is expedient to *over-relax*, i.e. to leave a negative value at each node which, at least partly, nullifies the effect of subsequent carry-over.

Successive over-relaxation (Frankel, 1950)

On a computer it is not efficient to search for the particular equation with the largest residual at every relaxation. However, the technique of over-relaxation is easy to implement and can often substantially improve the convergence rate.

The governing equation for successive over-relaxation (SOR) differs from the corresponding equation for Gauss–Seidel iteration only insofar as $x_i^{(k+1)} - x_i^{(k)}$ is

Table 6.3 SOR for equations (6.7) with $\omega = 1.16$

k	0	1	2	3	4	5	6	7
x_1	0	6.1722	9.6941	8.3398	8.4424	8.5137	8.4842	8.4868
x_2	0	9.1970	6.1177	6.3188	6.4880	6.4202	6.4253	6.4288
x_3	0	−5.2320	−4.8999	−4.5941	−4.7151	−4.7068	−4.7005	−4.7031
x_4	0	3.6492	4.4335	3.9200	4.0004	4.0157	4.0047	4.0065
e_k	12.3095	3.6659	1.3313	0.2302	0.0767	0.0288	0.0051	0.0016

scaled by the over-relaxation parameter ω, giving

$$x_i^{(k+1)} = \omega(b_i - a_{i1}x_1^{(k+1)} - \cdots - a_{i,i-1}x_{i-1}^{(k+1)}) + (1-\omega)x_i^{(k)} + \omega(-a_{i,i+1}x_{i+1}^{(k)} - \cdots - a_{in}x_n^{(k)}) \tag{6.11}$$

In matrix form one complete iteration step is represented by

$$(\mathbf{I} - \omega\mathbf{L})\mathbf{x}^{(k+1)} = \omega\mathbf{b} + [(1-\omega)\mathbf{I} + \omega\mathbf{U}]\,\mathbf{x}^{(k)} \tag{6.12}$$

Gauss–Seidel iteration is the special case of SOR with $\omega = 1$. The optimum value of ω lies almost invariably in the range $1 < \omega < 2$ (if the SOR formula were to be used with $\omega < 1$ then it should, logically, be called an under-relaxation). If equations (6.7) are solved by SOR using the parameter $\omega = 1.16$, as shown in Table 6.3, the convergence is faster than for Gauss–Seidel iteration.

6.3 GENERAL CHARACTERISTICS OF ITERATIVE METHODS

In all of the iterative methods the basic operation is very similar to the matrix premultiplication

$$\mathbf{x}^{(k+1)} = \mathbf{A}\mathbf{x}^{(k)} \tag{6.13}$$

Hence, if the coefficient matrix is fully populated with non-zero elements, the number of multiplications per iteration step is usually approximately n^2. In comparison with the number of multiplications required to solve the equations by elimination, it may be concluded that an iterative solution for one right-hand side is more efficient if satisfactory convergence can be achieved in less than $n/3$ iterations (or $n/6$ if the elimination solution can take advantage of symmetry).

If a set of equations has a sparse coefficient matrix, it may be stored by means of any of the sparse storage schemes, the most efficient and versatile schemes for general use being the various packing schemes of sections 3.8 to 3.12. Furthermore, if the equations have a repetitive form (whether the coefficient matrix is sparse or not), it is possible to avoid storing the coefficients explicitly. For instance, if Laplace's equation is to be solved over a rectangular field using a finite difference mesh corresponding to the network shown in Figure 5.8(a), the FORTRAN statement

```
X(I)=(1-OMEGA)*X(I)+OMEGA*0.25*(X(I-5)+X(I-1)+X(I+1)+X(I+5))
```

Table 6.4 On the computational efficiency of iterative solutions of the Laplace equation in finite difference form

Mesh size	No. of equations n	Semi-band-width	No. of non-zero elements per row c	No. of multiplications: For one iteration cn	No. of multiplications: For complete elimination $nb^2/2$	No. of iterations to break even $b^2/2c$
4 x 8	32	5	5	160	400	2½
12 x 24	288	13	5	1,440	24,400	17
40 x 80	3,200	41	5	16,000	2,690,000	168
4 x 6 x 12	288	25	7	2,000	90,000	45
9 x 13 x 26	3,042	118	7	21,000	21,000,000	1,000

Table 6.5 Merits of iterative methods as compared with elimination methods

Advantages	Disadvantages
1. Probably more efficient than elimination for very large-order systems.	1. Additional right-hand sides cannot be processed rapidly.
2. Implementation is simpler, particularly if backing store is needed.	2. The convergence, even if assured, can often be slow, and hence the amount of computation required to obtain a particular solution is not very predictable.
3. Advantage can be taken of a known approximate solution, if one exists.	3. The computer time and the accuracy of the the result depend on the judicious choice of parameters (e.g. the tolerance and the over-relaxation parameter for SOR).
4. Low accuracy solutions can be obtained rapidly.	4. If the convergence rate is poor, the results have to be interpreted with caution.
5. Where equations have a repetitive form their coefficients need not be individually stored (see text).	5. No particular advantage in computation time/iteration can be gained if the coefficient matrix is symmetric, whereas for an elimination method the computation time can be halved.
For sparse systems only:	
6. Less storage space is required for an iterative solution.	
7. The storage requirement is more easily defined in advance.	
8. The order of specification of the variables is not usually important.	

can be used for over-relaxation of all the inner mesh points. Corresponding statements can also be developed for the edge and corner node relaxations.

If the coefficient matrix of a set of equations is sparse, the number of multiplications per iteration is directly proportional to the sparsity. In Table 6.4 is shown a comparison between the number of multiplications for iterative and elimination solutions of Laplace's equation in finite difference form. It may be noticed that where the bandwidth of the equations is large it is even possible for slowly convergent iterative solutions to be competitive in efficiency with elimination solutions. Iterative methods are particularly attractive for the solution of large-order three-dimensional field problems.

Table 6.5 gives a summary of the advantages and disadvantages of using iterative methods in preference to elimination methods for solving linear equations.

6.4 THE ITERATION MATRIX

For a particular iterative method it would be useful to know:

(a) the range of problems for which convergence is assured,
(b) whether the convergence rate is likely to be rapid or not,
(c) whether the convergence rate is likely to be more rapid for the chosen method than for other iterative methods, and
(d) whether there are any means of accelerating the convergence rate if it is slow.

Let $\mathbf{e}^{(k)}$ be the vector of errors in the approximation $\mathbf{x}^{(k)}$ such that

$$\mathbf{x}^{(k)} = \mathbf{x} + \mathbf{e}^{(k)} \tag{6.14}$$

($\mathbf{x}$ being the exact solution). For stationary iterative methods the error propagation through one iteration step can be specified in the form

$$\mathbf{e}^{(k+1)} = \mathbf{M}\mathbf{e}^{(k)} \tag{6.15}$$

where $\mathbf{M}$, the *iteration matrix*, is a function of the coefficient matrix and does not vary from iteration to iteration. Substituting in equation (6.4) for $\mathbf{x}^{(k+1)}$ and $\mathbf{x}^{(k)}$ from equation (6.14) gives for Jacobi iteration

$$\mathbf{e}^{(k+1)} = (\mathbf{L} + \mathbf{U})\mathbf{e}^{(k)} \tag{6.16}$$

Hence the Jacobi iteration matrix is

$$\mathbf{M}_\mathrm{J} = \mathbf{L} + \mathbf{U} = \mathbf{I} - \mathbf{A} \tag{6.17}$$

Similarly equation (6.12) for SOR gives

$$(\mathbf{I} - \omega\mathbf{L})\mathbf{e}^{(k+1)} = [(1-\omega)\mathbf{I} + \omega\mathbf{U}]\,\mathbf{e}^{(k)} \tag{6.18}$$

Hence the SOR iteration matrix is

$$\mathbf{M}_\mathrm{SOR} = (\mathbf{I} - \omega\mathbf{L})^{-1}[(1-\omega)\mathbf{I} + \omega\mathbf{U}] \tag{6.19}$$

The eigenvalues of the iteration matrix

Consider the error propagation equation (6.15). If the iteration matrix $\mathbf{M}$ has a typical eigenvalue μ_i and corresponding eigenvector $\mathbf{p}_i$ then

$$\mathbf{M}\mathbf{p}_i = \mu_i\mathbf{p}_i \tag{6.20}$$

Expressing the initial error $\mathbf{e}^{(0)}$ as a linear combination of the eigenvectors according to

$$\mathbf{e}^{(0)} = \sum_{i=1}^{n} c_i\mathbf{p}_i \tag{6.21}$$

and applying the error propagation formula gives

$$\mathbf{e}^{(1)} = \sum_{i=1}^{n} c_i \mathbf{M}\mathbf{p}_i = \sum_{i=1}^{n} c_i \mu_i \mathbf{p}_i \tag{6.22}$$

and

$$\mathbf{e}^{(k)} = \sum_{i=1}^{n} c_i \mu_i^k \mathbf{p}_i \tag{6.23}$$

Thus the error vector will decay as k increases if the moduli of all of the eigenvalues of the iteration matrix are less than unity. Furthermore, the ultimate rate of convergence will be governed by the magnitude of the eigenvalue of largest modulus (otherwise known as the *spectral radius* of the iteration matrix).

Convergence of Jacobi iteration

The Jacobi iteration matrix is the simplest to investigate. Postmultiplying equation (6.17) by $\mathbf{p}_i$ and making use of the eigenvalue equation (6.20) gives

$$\mathbf{A}\mathbf{p}_i = (1 - \mu_i)\mathbf{p}_i \tag{6.24}$$

Hence $1 - \mu_i$ must be an eigenvalue of $\mathbf{A}$ with $\mathbf{p}_i$ as the corresponding eigenvector. Because this equation holds for all the eigenvalues of $\mathbf{M}_J$ it follows that each eigenvalue μ_i of $\mathbf{M}_J$ must be related to an eigenvalue λ_i of $\mathbf{A}$ according to

$$\mu_i = 1 - \lambda_i \tag{6.25}$$

Figure 6.1 shows the permissible regions of the Argand diagram for the eigenvalues

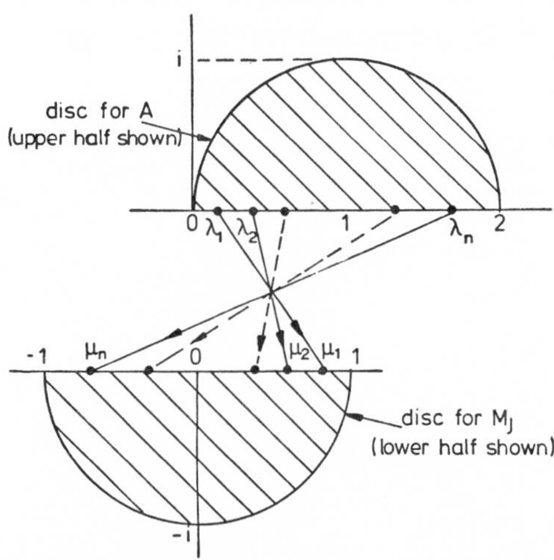

Figure 6.1 Jacobi iteration: permissible regions in the Argand diagram for eigenvalues of $\mathbf{M}_J$ and $\mathbf{A}$ to give convergence, showing transfer of real eigenvalues from $\mathbf{A}$ to $\mathbf{M}_J$

of **M** and **A**, and also the way in which a set of (real) eigenvalues of **A** transfer through equation (6.25) to a set of eigenvalues of $\mathbf{M}_J$.

Gerschgorin bounds for convergence of Jacobi iteration

If the Gerschgorin discs for **A** all lie within the permissible region for the eigenvalues shown in Figure 6.1 then Jacobi iteration must converge. Furthermore, since all of the Gerschgorin discs of **A** have their centres at the point (1, 0), the largest radius of any of the Gerschgorin discs provides a bound to the convergence rate. For example, the largest Gerschgorin radius for the coefficient matrix of equation (6.7) is 0.767. Thus the largest eigenvalue of $\mathbf{M}_J$ cannot be of magnitude greater than 0.767. Since $0.767^{26} \simeq 0.001$, Jacobi iteration applied to these equations should take no more than twenty-six iterations to gain three figures of accuracy in the predicted solution (see Table 6.1). Unfortunately this technique cannot in general give bounds for slowly convergent iterations since, in such cases, some of the Gerschgorin radii will be greater than or equal to unity.

Convergence of SOR

Substituting the eigenvalue equation (6.20) into the successive over-relaxation equation (6.19) yields

$$\mu_i(\mathbf{I} - \omega\mathbf{L})\mathbf{p}_i = [(1 - \omega)\mathbf{I} + \omega\mathbf{U}]\,\mathbf{p}_i \tag{6.26}$$

Hence

$$(\mu_i\mathbf{L} + \mathbf{U})\mathbf{p}_i = \frac{\mu_i + \omega - 1}{\omega}\,\mathbf{p}_i \tag{6.27}$$

It is not possible to derive from this equation a simple relationship between the eigenvalues of **A** and those of $\mathbf{M}_{SOR}$ which will hold for any matrix **A**.

Gerschgorin bounds for Gauss–Seidel iteration

Choosing $\omega = 1$ in equation (6.27) gives, for the Gauss–Seidel method,

$$\left(\mathbf{L} + \frac{1}{\mu_i}\mathbf{U}\right)\mathbf{p}_i = \mathbf{p}_i \tag{6.28}$$

If all of the Gerschgorin discs of $\mathbf{L} + \mathbf{U}$ have radius less than unity, it is possible to establish a maximum value for μ_i above which the Gerschgorin radii of $\mathbf{L} + (1/\mu_i)\mathbf{U}$ are too small for equation (6.28) to be satisfied. The bound for the convergence rate of Gauss–Seidel iteration obtained in this way will be better than the corresponding Jacobi bound. For instance, equation (6.7) gives

$$\mathbf{L} + \frac{1}{\mu_i} = \begin{bmatrix} 0 & -0.30009/\mu_i & & -0.30898/\mu_i \\ -0.30009 & 0 & -0.46691/\mu_i & \\ & -0.46691 & 0 & -0.27471/\mu_i \\ -0.30898 & & -0.27471 & 0 \end{bmatrix} \tag{6.29}$$

which has Gerschgorin column disc radii all less than unity for $\mu_i > 0.644$. Hence the largest eigenvalue of M_{GS} has a modulus less than 0.644. Since $0.644^{16} \simeq 0.0009$, a Gauss–Seidel iteration should take no more than sixteen iterations to gain three figures of accuracy in the predicted solution (see Table 6.2). As in the case of Jacobi's method it will not generally be possible to use this technique to give convergence bounds for slowly convergent Gauss–Seidel iterations.

Although Gauss–Seidel iteration normally converges faster than does Jacobi iteration when applied to the same set of equations, this is not a completely general rule. Fox (1964) quotes an example for which Jacobi iteration converges whereas Gauss–Seidel iteration does not.

6.5 CONVERGENCE WITH A SYMMETRIC POSITIVE DEFINITE COEFFICIENT MATRIX

If the coefficient matrix of a set of equations is known to be symmetric positive definite, the convergence of Jacobi iteration is assured, provided also that no eigenvalue of A is greater than or equal to 2. Sometimes this condition is known to be satisfied, e.g. where the Gerschgorin radii are all less than unity or when the matrix has property A (as described in section 6.6). However, in other cases convergence cannot be guaranteed. For instance, a symmetric positive definite coefficient matrix

$$\mathbf{A} = \begin{bmatrix} 1 & 0.8 & 0.8 \\ 0.8 & 1 & 0.8 \\ 0.8 & 0.8 & 1 \end{bmatrix} \tag{6.30}$$

has eigenvalues 0.2, 0.2 and 2.6, and consequently does not give a convergent Jacobi iteration.

The investigation of the convergence of SOR is complicated by the fact that the iteration matrix is unsymmetric and as a result may have complex conjugate pairs of eigenvalues. Let a typical eigenvalue μ of M_{SOR} have real and imaginary components θ and ϕ such that $\mu = \theta + i\phi$ and let its complex conjugate be μ^*. The right eigenvector **p** corresponding to μ will be complex. Premultiplying equation (6.27) by the Hermitian transpose of the eigenvector, and allowing for the symmetry of **A**, gives

$$\mathbf{p}^H(\mu\mathbf{L} + \mathbf{L}^T)\mathbf{p} = \frac{\mu + \omega - 1}{\omega}\mathbf{p}^H\mathbf{p} \tag{6.31}$$

the Hermitian transpose of which is

$$\mathbf{p}^H(\mu^*\mathbf{L}^T + \mathbf{L})\mathbf{p} = \frac{\mu^* + \omega - 1}{\omega}\mathbf{p}^H\mathbf{p} \tag{6.32}$$

Adding and subtracting these two equations yields respectively

$$\left.\begin{aligned}&(1+\theta)\mathbf{p}^H(\mathbf{L}+\mathbf{L}^T)\mathbf{p}+i\phi\mathbf{p}^H(\mathbf{L}-\mathbf{L}^T)\mathbf{p}=\frac{2(\theta+\omega-1)}{\omega}\mathbf{p}^H\mathbf{p}\\ &\text{and}\\ &i\phi\mathbf{p}^H(\mathbf{L}+\mathbf{L}^T)\mathbf{p}-(1-\theta)\mathbf{p}^H(\mathbf{L}-\mathbf{L}^T)\mathbf{p}=\frac{2i\phi}{\omega}\mathbf{p}^H\mathbf{p}\end{aligned}\right\}\tag{6.33}$$

It can be shown that both $\mathbf{p}^H\mathbf{p}$ and $\mathbf{p}^H\mathbf{A}\mathbf{p}$ are positive and real. If

$$k=\frac{\mathbf{p}^H\mathbf{A}\mathbf{p}}{\mathbf{p}^H\mathbf{p}}\tag{6.34}$$

then, by a generalization of the quadratic form (equation 1.134), it can be shown that

$$\lambda_1\leqslant k\leqslant\lambda_n\tag{6.35}$$

where λ_1 and λ_n are the minimum and maximum eigenvalues of $\mathbf{A}$ respectively. It may also be shown that the eigenvalues of $\mathbf{L}-\mathbf{L}^T$ are either imaginary or zero and that $\mathbf{p}^H(\mathbf{L}-\mathbf{L}^T)\mathbf{p}$ is either imaginary or zero. Specifically, if

$$l=\frac{i\mathbf{p}^H(\mathbf{L}-\mathbf{L}^T)\mathbf{p}}{\mathbf{p}^H\mathbf{p}}\tag{6.36}$$

then

$$|l|\begin{cases}=0 & \text{for } \mu \text{ and } \mathbf{p} \text{ real}\\ \leqslant\psi_n & \text{for } \mu \text{ and } \mathbf{p} \text{ complex}\end{cases}\tag{6.37}$$

where $\pm i\psi_n$ is the largest pair of eigenvalues of $\mathbf{L}-\mathbf{L}^T$. By substituting for k and l in equations (6.33) it can be shown that

$$\left.\begin{aligned}&(\alpha+k)\theta-l\phi=\alpha-k\\ &\text{and}\\ &l\theta+(\alpha+k)\phi=l\end{aligned}\right\}\tag{6.38}$$

where $\alpha=(2-\omega)/\omega$. Squaring and adding the two equations (6.38) yields

$$|\mu|^2=\theta^2+\phi^2=\frac{(\alpha-k)^2+l^2}{(\alpha+k)^2+l^2}\tag{6.39}$$

Since $k>0$, the modulus of every eigenvalue μ of $\mathbf{M}_{\mathrm{SOR}}$ must be less than unity if α is positive. However, α is positive only when ω satisfies $0<\omega<2$.

Hence successive over-relaxation converges for all cases where the coefficient matrix is symmetric positive definite provided that $0<\omega<2$ (an alternative proof is given by Varga, 1962).

Setting $l=0$ in equation (6.39) and using the limits for k given by equation (6.35), it can be proved that the real eigenvalues of $\mathbf{M}_{\mathrm{SOR}}$ must lie within the

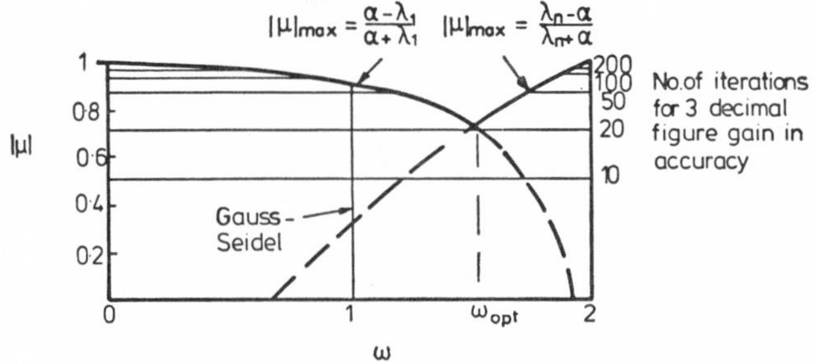

Figure 6.2 Bounds for moduli of real eigenvalues of M_{SOR} versus over-relaxation parameter: matrix **A** is symmetric with $\lambda_1 = 0.05$ and $\lambda_n = 1.95$

union of the two regions defined by

$$|\mu| \leqslant \frac{\alpha - \lambda_1}{\alpha + \lambda_1} \quad \text{and} \quad |\mu| \leqslant \frac{\lambda_n - \alpha}{\lambda_n + \alpha} \tag{6.40}$$

as illustrated in Figure 6.2. Whereas equation (6.39) does not preclude complex eigenvalues from lying outside these two regions, numerical tests seem to indicate that this possibility is remote. Thus, if conditions (6.40) can be taken as the bounds for the spectral radius of M_{SOR}, then the optimum over-relaxation parameter to use is the one corresponding to the intersection of the two boundaries, namely

$$\omega_{opt} = \frac{2}{1 + \sqrt{(\lambda_1\lambda_n)}} \tag{6.41}$$

which gives a bound for the convergence rate of

$$\mu_{opt} < \frac{\lambda_n^{1/2} - \lambda_1^{1/2}}{\lambda_n^{1/2} + \lambda_1^{1/2}} \tag{6.42}$$

Since the coefficient matrix of equation (6.7) has eigenvalues of 0.314, 0.910, 1.090 and 1.686 the optimum relaxation parameter is approximately 1.16. From the bound (6.42), $|\mu|_{opt} < 0.40$ and hence (since $0.40^8 \simeq 0.00065$) no more than eight iterations should be required to gain three figures of accuracy (see Table 6.3). Without determining the eigenvalues of the coefficient matrix it is possible to deduce that $\omega_{opt} < 1.22$ from the Gerschgorin bounds of the coefficient matrix.

Whereas Gerschgorin's theorem will always give a bound for the largest eigenvalue of the coefficient matrix it will not always give a bound for λ_1 which improves on $\lambda_1 > 0$. In cases where λ_1 is very small the optimum relaxation parameter will be just less than 2. Although the optimum convergence rate will be slow, it will be considerably better than the convergence rate for Gauss–Seidel iteration. Thus with $\lambda_1 = 0.001$ and $\lambda_n = 1.999$ the number of iterations required to gain three decimal figures of accuracy by Jacobi, Gauss–Seidel and optimum SOR iterations are likely to be approximately 7,000, 3,500 and 150 respectively.

6.6 MATRICES WITH PROPERTY A

Young (1954) has discovered a class of coefficient matrix for which the SOR convergence rate can be directly related to the convergence rate for Jacobi iteration. Matrices of this type are described as having *property A*. In Young's original presentation, the variables for the equations separate into two groups $\mathbf{x}_a$ and $\mathbf{x}_b$ in such a way that an individual variable in one group is only linked to variables in the other group. The equations will therefore have the submatrix form

$$\begin{bmatrix} \mathbf{I} & -\mathbf{R} \\ -\mathbf{S} & \mathbf{I} \end{bmatrix} \begin{bmatrix} \mathbf{x}_a \\ \mathbf{x}_b \end{bmatrix} = \begin{bmatrix} \mathbf{b}_a \\ \mathbf{b}_b \end{bmatrix} \tag{6.43}$$

The equation for a typical eigenvalue η and corresponding eigenvector $\{\mathbf{q}_a \quad \mathbf{q}_b\}$ of the Jacobi iteration matrix

$$\begin{bmatrix} \mathbf{0} & \mathbf{R} \\ \mathbf{S} & \mathbf{0} \end{bmatrix} \begin{bmatrix} \mathbf{q}_a \\ \mathbf{q}_b \end{bmatrix} = \eta \begin{bmatrix} \mathbf{q}_a \\ \mathbf{q}_b \end{bmatrix} \tag{6.44}$$

yields also

$$\begin{bmatrix} \mathbf{0} & \mathbf{R} \\ \mathbf{S} & \mathbf{0} \end{bmatrix} \begin{bmatrix} \mathbf{q}_a \\ -\mathbf{q}_b \end{bmatrix} = -\eta \begin{bmatrix} \mathbf{q}_a \\ -\mathbf{q}_b \end{bmatrix} \tag{6.45}$$

Hence the Jacobi iteration matrix must have non-zero eigenvalues occurring in pairs of the form $\pm\eta$.

SOR convergence with property A

Consider now successive over-relaxation. If μ is a typical eigenvalue of $\mathbf{M}_{SOR}$ with $\{\mathbf{p}_a \quad \mathbf{p}_b\}$ as the corresponding eigenvector, then, given that the coefficient matrix possesses property A, equation (6.27) becomes

$$\begin{bmatrix} \mathbf{0} & \mathbf{R} \\ \mu\mathbf{S} & \mathbf{0} \end{bmatrix} \begin{bmatrix} \mathbf{p}_a \\ \mathbf{p}_b \end{bmatrix} = \frac{\mu + \omega - 1}{\omega} \begin{bmatrix} \mathbf{p}_a \\ \mathbf{p}_b \end{bmatrix} \tag{6.46}$$

Hence

$$\begin{bmatrix} \mathbf{0} & \mathbf{R} \\ \mathbf{S} & \mathbf{0} \end{bmatrix} \begin{bmatrix} \mu^{1/2}\mathbf{p}_a \\ \mathbf{p}_b \end{bmatrix} = \frac{\mu + \omega - 1}{\omega\mu^{1/2}} \begin{bmatrix} \mu^{1/2}\mathbf{p}_a \\ \mathbf{p}_b \end{bmatrix} \tag{6.47}$$

Comparing equations (6.44) and (6.47) it may be seen that the eigenvalues for Jacobi and SOR iteration will be related according to

$$\frac{\mu + \omega - 1}{\omega\mu^{1/2}} = \eta \tag{6.48}$$

which gives

$$\mu^{1/2} = \frac{\omega\eta}{2} \pm \sqrt{\left\{\left(\frac{\omega\eta}{2}\right)^2 - \omega + 1\right\}} \tag{6.49}$$

For Gauss–Seidel iteration ($\omega = 1$), one-half of the eigenvalues are zero and the other half are real. The convergence rate is twice that for Jacobi iteration.

The optimum over-relaxation parameter is

$$\omega_{opt} = \frac{2}{1 + \sqrt{(1 - \eta^2_{max})}} \tag{6.50}$$

with

$$|\mu|_{opt} = \frac{1}{\eta^2_{max}} \{1 - \sqrt{(1 - \eta^2_{max})}\}^2 \tag{6.51}$$

For $\omega \leqslant \omega_{opt}$ the spectral radius of $\mathbf{M}_{SOR}$ is

$$|\mu|_{max} = \left[\frac{\omega\eta_{max}}{2} + \sqrt{\left\{\left(\frac{\omega\eta_{max}}{2}\right)^2 - \omega + 1\right\}}\right]^2 \tag{6.52}$$

For $\omega \geqslant \omega_{opt}$ the eigenvalues will be complex, except for those corresponding to $\eta = 0$. All of the eigenvalues will satisfy

$$|\mu| = \omega - 1 \tag{6.53}$$

Hence the graph of the spectral radius of $\mathbf{M}_{SOR}$ against ω will be as illustrated in Figure 6.3.

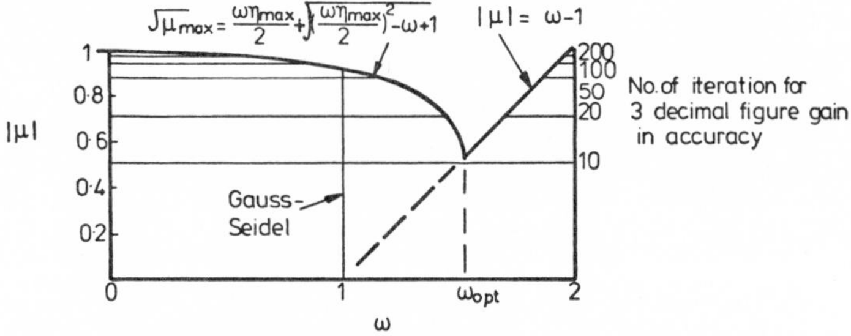

Figure 6.3 Spectral radius of $\mathbf{M}_{SOR}$ versus over-relaxation parameter, matrix $\mathbf{A}$ having property A with $\lambda_1 = 0.05$ and $\lambda_n = 1.95$ (i.e. $\eta = 0.95$)

In cases where the coefficient matrix is symmetric, positive definite and also has property A, the optimum convergence rate defined by equation (6.51) is twice the convergence rate indicated by the bound (6.42) for real eigenvalues. The optimum over-relaxation parameter is the same in both cases.

Ordering the variables for property A

If the graph of the coefficient matrix forms a rectangular mesh, property A can be obtained by separating the mesh points into two sets, simulating the black and white squares of a chess board and then numbering all of the nodes in one set before the nodes in the other set, as shown in Figure 6.4.

However, it is not necessary to order the variables in this way as several other more convenient ordering schemes also give rise to the property A characteristics.

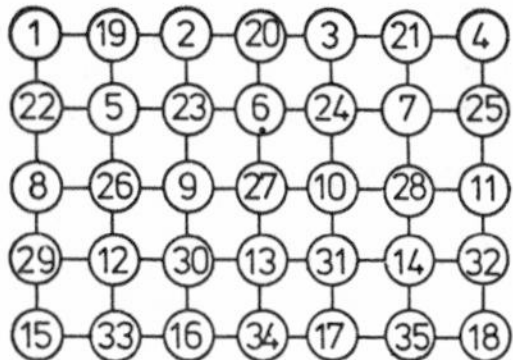

Figure 6.4 Chessboard node ordering to give property A

For instance, consider the equations arising from a 3 × 3 mesh in which the variables are ordered by rows. The scaled equations will have the form

$$\left[\begin{array}{ccc|ccc|ccc} 1 & a_{12} & & a_{14} & & & & & \\ a_{21} & 1 & a_{23} & & a_{25} & & & & \\ & a_{32} & 1 & & & a_{36} & & & \\ \hline a_{41} & & & 1 & a_{45} & & a_{47} & & \\ & a_{52} & & a_{54} & 1 & a_{56} & & a_{58} & \\ & & a_{63} & & a_{65} & 1 & & & a_{69} \\ \hline & & & a_{74} & & & 1 & a_{78} & \\ & & & & a_{85} & & a_{87} & 1 & a_{89} \\ & & & & & a_{96} & & a_{98} & 1 \end{array}\right] \left[\begin{array}{c} x_1 \\ x_2 \\ x_3 \\ \hline x_4 \\ x_5 \\ x_6 \\ \hline x_7 \\ x_8 \\ x_9 \end{array}\right] = \left[\begin{array}{c} b_1 \\ b_2 \\ b_3 \\ \hline b_4 \\ b_5 \\ b_6 \\ \hline b_7 \\ b_8 \\ b_9 \end{array}\right] \tag{6.54}$$

Hence, by equation (6.27) an eigenvalue μ of $\mathbf{M}_{\mathrm{SOR}}$ must satisfy

$$\left[\begin{array}{ccc|ccc|ccc} 0 & a_{12} & & a_{14} & & & & & \\ \mu a_{21} & 0 & a_{23} & & a_{25} & & & & \\ & \mu a_{32} & 0 & & & a_{36} & & & \\ \hline \mu a_{41} & & & 0 & a_{45} & & a_{47} & & \\ & \mu a_{52} & & \mu a_{54} & 0 & a_{56} & & a_{58} & \\ & & \mu a_{63} & & \mu a_{65} & 0 & & & a_{69} \\ \hline & & & \mu a_{74} & & & 0 & a_{78} & \\ & & & & \mu a_{85} & & \mu a_{87} & 0 & a_{89} \\ & & & & & \mu a_{96} & & \mu a_{98} & 0 \end{array}\right] \left[\begin{array}{c} p_1 \\ p_2 \\ p_3 \\ \hline p_4 \\ p_5 \\ p_6 \\ \hline p_7 \\ p_8 \\ p_9 \end{array}\right] = \frac{\mu + \omega - 1}{\omega} \left[\begin{array}{c} p_1 \\ p_2 \\ p_3 \\ \hline p_4 \\ p_5 \\ p_6 \\ \hline p_7 \\ p_8 \\ p_9 \end{array}\right] \tag{6.55}$$

from which it follows that

$$\left[\begin{array}{ccc|ccc|ccc} 0 & a_{12} & & a_{14} & & & & & \\ a_{21} & 0 & a_{23} & & a_{25} & & & & \\ & a_{32} & 0 & & & a_{36} & & & \\ \hline a_{41} & & & 0 & a_{45} & & a_{47} & & \\ & a_{52} & & a_{54} & 0 & a_{56} & & a_{58} & \\ & & a_{63} & & a_{65} & 0 & & & a_{69} \\ \hline & & & a_{74} & & & 0 & a_{78} & \\ & & & & a_{85} & & a_{87} & 0 & a_{89} \\ & & & & & a_{96} & & a_{98} & 0 \end{array}\right] \left[\begin{array}{c} \mu^2 p_1 \\ \mu^{3/2} p_2 \\ \mu p_3 \\ \hline \mu^{3/2} p_4 \\ \mu p_5 \\ \mu^{1/2} p_6 \\ \hline \mu p_7 \\ \mu^{1/2} p_8 \\ p_9 \end{array}\right] = \frac{\mu + \omega - 1}{\omega \mu^{1/2}} \left[\begin{array}{c} \mu^2 p_1 \\ \mu^{3/2} p_2 \\ \mu p_3 \\ \hline \mu^{3/2} p_4 \\ \mu p_5 \\ \mu^{1/2} p_6 \\ \hline \mu p_7 \\ \mu^{1/2} p_8 \\ p_9 \end{array}\right] \tag{6.56}$$

Thus equation (6.48) is satisfied for this node ordering scheme.

Specifically, where the graph of the coefficient matrix of a set of equations is a rectangular mesh, the numbering of the nodes either by rows, columns or diagonals, as illustrated in Figures 5.12 and 5.16, will give property A convergence characteristics. Such schemes are described as *consistently ordered.*

A simple example

The coefficient matrix for the equations (6.7) has the graph shown in Figure 6.5. Although this constitutes a rectangular mesh the node numbering is inconsistent with any of the property A schemes. If the variables have the consistent ordering $\{x_1 \;\; x_2 \;\; x_4 \;\; x_3\}$, the SOR iteration matrix for $\omega = 1.16$ has a spectral radius of 0.16 and hence accuracy should be gained at the rate of three decimal figures in four iterations. Table 6.6 shows the actual progress of SOR iterations. Although the

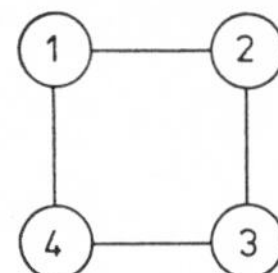

Figure 6.5 Graph of the coefficient matrix of equation (6.7)

Table 6.6 SOR for equation (6.7) with $\omega = 1.16$ and consistent ordering

k	0	1	2	3	4	5	6	7
x_1	0	6.1722	7.1996	8.2639	8.4330	8.4785	8.4859	8.4874
x_2	0	−10.2133	−5.9369	−5.0442	−4.7645	−4.7153	−4.7050	−4.7032
x_4	0	3.6653	5.7526	6.2727	6.3998	6.4220	6.4266	6.4273
x_3	0	2.0618	3.4629	3.9047	3.9837	4.0031	4.0059	4.0065
e_k	12.3095	6.2901	1.9833	0.4483	0.0899	0.0168	0.0030	0.0005

convergence rate is better than for the inconsistent ordering of Table 6.3, it falls short of the theoretical prediction given above. However, if iteration is continued, it is found that over a larger number of iterations the average convergence rate does conform to the property A prediction.

6.7 CHOICE OF RELAXATION PARAMETER

The main difficulty with using SOR lies in making a suitable choice of over-relaxation parameter. To determine the maximum or minimum eigenvalue of the coefficient matrix accurately may take longer than the SOR iteration itself. Thus, even when a formula is available for predicting the optimum over-relaxation parameter from the extreme eigenvalues of the coefficient matrix, it is not necessarily possible to make effective use of it to reduce the computing time.

For the case of equations with coefficient matrices having property A, Carré

(1961) has proposed a method of estimating the over-relaxation parameter as iteration proceeds.

From equations (6.14) and (6.15) it follows that

$$\mathbf{x}^{(k+1)} - \mathbf{x}^{(k)} = \mathbf{M}_{\text{SOR}}(\mathbf{x}^{(k)} - \mathbf{x}^{(k-1)}) \tag{6.57}$$

Hence the successive vector differences can be considered to be successive vectors in a power method iteration for the dominant eigenvalue of $\mathbf{M}_{\text{SOR}}$ (see section 10.1). Provided that the dominant eigenvalue is real, $\|\mathbf{x}^{(k+1)} - \mathbf{x}^{(k)}\|/\|\mathbf{x}^{(k)} - \mathbf{x}^{(k-1)}\|$ will provide an estimate of $\mu_{\max}$ which will improve as iteration proceeds. Using equation (6.48) to relate $\mu_{\max}$ to $\eta_{\max}$ and substituting into equation (6.50) yields the following estimate for the optimum over-relaxation parameter:

$$\omega' = \frac{2}{1 + \sqrt{\{1 - (\mu_{\max} + \omega - 1)^2/\omega^2\mu_{\max}\}}} \tag{6.58}$$

Iteration is commenced with a low relaxation parameter (say $\omega = 1$), and after about every ten iterations the over-relaxation parameter is re-evaluated according to equation (6.58). Initial estimates of ω_{opt} will tend to be low and hence the over-relaxation parameter will increase at each re-evaluation. It is suggested that if the prediction ω' is very close to ω then iteration should be continued without any further re-evaluation until the tolerance criterion is satisfied.

6.8 DOUBLE SWEEP AND PRECONDITIONING METHODS

Many modifications and extensions of the classical stationary iterative methods have been developed. The next three sections (sections 6.8 to 6.10) briefly describe some of these.

Double sweep successive over-relaxation (SSOR)

Aitken (1950) proposed a double sweep Gauss–Seidel iteration for use with his δ^2 acceleration process in cases where the coefficient matrix is symmetric. The method has been generalized to a double sweep successive over-relaxation by Sheldon (1955).

Each iteration of SSOR consists of a forward and a backward sweep of the element equations. If $\mathbf{x}^{(k+1/2)}$ represents the intermediate variables after the k-th iteration then, since the forward sweep is a conventional SOR iteration,

$$(\mathbf{I} - \omega\mathbf{L})\mathbf{x}^{(k+1/2)} = \omega\mathbf{b} - [(1-\omega)\mathbf{I} + \omega\mathbf{L}^T]\mathbf{x}^{(k)} \tag{6.59}$$

In the second sweep the variables are modified in reverse order according to

$$\begin{aligned} x_i^{(k+1)} = \omega(b_i - a_{i1}x_1^{(k+1/2)} - \cdots - a_{i,i-1}x_{i-1}^{(k+1/2)}) + (1-\omega)x^{(k+1/2)} \\ + \omega(-a_{i,i+1}x_{i+1}^{(k+1)} - \cdots - a_{in}x_n^{(k+1)}) \end{aligned} \tag{6.60}$$

Hence

$$(\mathbf{I} - \omega\mathbf{L}^T)\mathbf{x}^{(k+1)} = \omega\mathbf{b} + [(1-\omega)\mathbf{I} + \omega\mathbf{L}]\mathbf{x}^{(k+1/2)} \tag{6.61}$$

The iteration matrix for the full iteration cycle is

$$\left.\begin{aligned} \mathbf{M}_{\text{SSOR}} &= (\mathbf{I}-\omega\mathbf{L}^T)^{-1}[(1-\omega)\mathbf{I}+\omega\mathbf{L}](\mathbf{I}-\omega\mathbf{L})^{-1}[(1-\omega)\mathbf{I}+\omega\mathbf{L}^T] \\ &= (\mathbf{I}-\omega\mathbf{L}^T)^{-1}(\mathbf{I}-\omega\mathbf{L})^{-1}[(1-\omega)\mathbf{I}+\omega\mathbf{L}][(1-\omega)\mathbf{I}+\omega\mathbf{L}^T] \end{aligned}\right\} \tag{6.62}$$

Since $\mathbf{M}_{\text{SSOR}}$ may be transformed into a symmetric matrix by a similarity transformation (see section 8.9); it must have only real eigenvalues. If μ is a typical eigenvalue of $\mathbf{M}_{\text{SSOR}}$ and $\mathbf{p}$ is the corresponding right eigenvector it can be shown that

$$\left.\begin{aligned} &\mu = \frac{\mathbf{u}^T\mathbf{u}}{\mathbf{v}^T\mathbf{v}} \geqslant 0 \\ \text{and} \\ &1-\mu = \frac{\omega(2-\omega)\mathbf{p}^T\mathbf{A}\mathbf{p}}{\mathbf{v}^T\mathbf{v}} \end{aligned}\right\} \tag{6.63}$$

where $\mathbf{u} = [(1-\omega)\mathbf{I}+\mathbf{L}^T]\mathbf{p}$ and $\mathbf{v} = (\mathbf{I}-\omega\mathbf{L}^T)\mathbf{p}$. It follows from equations (6.63) that, if the coefficient matrix is symmetric positive definite, double sweep SOR is convergent for $0 < \omega < 2$.

In practice double sweep SOR is not as efficient as normal successive over-relaxation. However, the fact that the eigenvalues of the iteration matrix are real enable acceleration methods to be applied more easily (sections 6.11 and 6.12).

Evans' (1967) preconditioning method

This method may be applied to equations $\mathbf{Ax} = \mathbf{b}$ where the coefficient matrix is symmetric positive definite and has unit leading diagonal elements. The preconditioned equations are

$$\mathbf{By} = \mathbf{d} \tag{6.64}$$

where $\mathbf{B} = (\mathbf{I}-\omega\mathbf{L})^{-1}\mathbf{A}(\mathbf{I}-\omega\mathbf{L})^{-T}$ and $\mathbf{L}$ is defined for equation (6.2). This set of equations is equivalent to $\mathbf{Ax} = \mathbf{b}$ if

$$\left.\begin{aligned} &\mathbf{y} = (\mathbf{I}-\omega\mathbf{L}^T)\mathbf{x} \\ \text{and} \\ &\mathbf{d} = (\mathbf{I}-\omega\mathbf{L})^{-1}\mathbf{b} \end{aligned}\right\} \tag{6.65}$$

However, the eigenvalue spectrum of matrix $\mathbf{B}$ will be different to that of $\mathbf{A}$. Hence a Jacobi iteration of the transformed equations using the formula

$$\mathbf{y}^{(k+1)} = \mathbf{d} + (\mathbf{I}-\mathbf{B})\mathbf{y}^{(k)} \tag{6.66}$$

will have a different convergence rate to that of Jacobi iteration applied directly to the original equations.

In order to take advantage of any possible improvement in the convergence rate it is necessary to implement the iterative cycle without having to form matrix $\mathbf{B}$

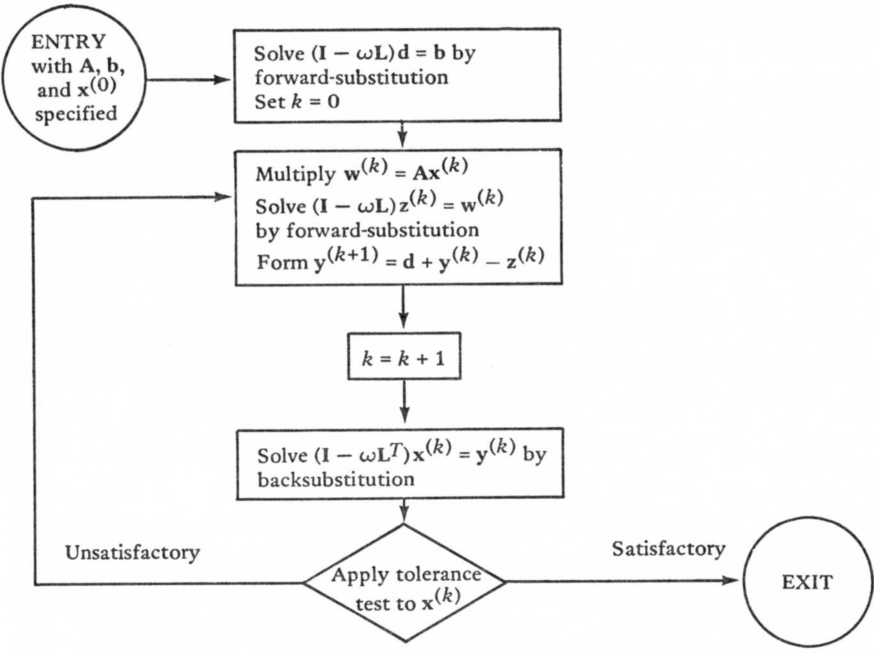

Figure 6.6 A procedure for unaccelerated iteration with preconditioned equations. (Note: A must have unit elements on the leading diagonal)

explicitly. Figure 6.6 is a flow diagram representing the complete solution process in which the tolerance test is applied to the vector $\mathbf{x}^{(k)}$. It may be deduced that one preconditioned iteration involves approximately twice as much computation as the corresponding standard iteration.

If η is a typical eigenvalue of $\mathbf{B}$ with $\mathbf{p}$ as the corresponding eigenvector, and if $\mathbf{v} = (\mathbf{I} - \omega\mathbf{L})^{-T}\mathbf{p}$, it follows from equation (6.64) that

$$\eta\mathbf{p}^T\mathbf{p} = \mathbf{v}^T\mathbf{A}\mathbf{v} \tag{6.67}$$

Hence η must always be positive. Substituting in equation (6.67) for $\mathbf{p}$ gives

$$\eta = \frac{\mathbf{v}^T\mathbf{A}\mathbf{v}}{\mathbf{v}^T(\mathbf{I} - \omega\mathbf{L})(\mathbf{I} - \omega\mathbf{L}^T)\mathbf{v}} = \frac{\mathbf{v}^T\mathbf{A}\mathbf{v}}{(1-\omega)\mathbf{v}^T\mathbf{v} + \omega\mathbf{v}^T\mathbf{A}\mathbf{v} + \omega^2\mathbf{v}^T\mathbf{L}\mathbf{L}^T\mathbf{v}} \tag{6.68}$$

When $\omega = 1$ this expression reduces to

$$\eta = \frac{\mathbf{v}^T\mathbf{A}\mathbf{v}}{\mathbf{v}^T\mathbf{A}\mathbf{v} + \mathbf{v}^T\mathbf{L}\mathbf{L}^T\mathbf{v}} \leqslant 1 \tag{6.69}$$

Hence the method is convergent for $\omega = 1$, with the iteration matrix $\mathbf{M}_{PC} = \mathbf{I} - \mathbf{B}$ having all of its eigenvalues real and non-negative. Although in practice the optimum value of ω is often greater than 1, the choice of this parameter is not so critical as it is with SOR. A choice of $\omega = 1$ normally gives a near optimum convergence rate.

The method of preconditioning is intended for use with acceleration methods or the conjugate gradient method, both of which are discussed later in the chapter. Evans' preconditioning method is very closely related to SSOR, being effectively the same at $\omega = 1$ and at other values of ω when accelerated.

6.9 BLOCK RELAXATION

Jacobi iteration and the relaxation methods so far described involve adjustment of the predicted values for the variables one at a time and are classified as point iterative methods. Block relaxation methods involve the simultaneous coupled adjustment of the predicted values for several variables at a time. Although it is possible to extend both Jacobi iteration and SOR to block form, only successive block over-relaxation will be considered since it yields better convergence rates.

Successive block over-relaxation (SBOR)

Each set of variables whose predictions are to be simultaneously adjusted is listed as a subvector. With corresponding segmentation of the right-hand vector and the coefficient matrix, the full set of equations are

$$\begin{bmatrix} \mathbf{A}_{11} & \mathbf{A}_{12} & \cdots & \mathbf{A}_{1n} \\ \mathbf{A}_{21} & \mathbf{A}_{22} & \cdots & \mathbf{A}_{2n} \\ \cdot & \cdot & \cdot & \cdot \\ \mathbf{A}_{n1} & \mathbf{A}_{n2} & \cdots & \mathbf{A}_{nn} \end{bmatrix} \begin{bmatrix} \mathbf{x}_1 \\ \mathbf{x}_2 \\ \cdot \\ \mathbf{x}_n \end{bmatrix} = \begin{bmatrix} \mathbf{b}_1 \\ \mathbf{b}_2 \\ \cdot \\ \mathbf{b}_n \end{bmatrix} \tag{6.70}$$

where $\mathbf{A}_{11}$, $\mathbf{A}_{22}$, etc., are square. In successive block over-relaxation the simultaneous adjustment at the k-th iteration step of the estimates for all of the variables $\mathbf{x}_i$ is such that

$$\begin{aligned} \mathbf{A}_{ii}\mathbf{x}_i^{(k+1)} = \omega(\mathbf{b}_i - \mathbf{A}_{i1}\mathbf{x}_1^{(k+1)} - \cdots - \mathbf{A}_{i,i-1}\mathbf{x}_{i-1}^{(k+1)}) + (1-\omega)\mathbf{A}_{ii}\mathbf{x}_i^{(k)} \\ + \omega(-\mathbf{A}_{i,i+1}\mathbf{x}_{i+1}^{(k)} - \cdots - \mathbf{A}_{in}\mathbf{x}_n^{(k)}) \end{aligned} \tag{6.71}$$

The adjustment is implemented by evaluating the right-hand side of equations (6.71) and then solving these equations for $\mathbf{x}_i^{(k+1)}$. It is expedient to determine the triangular factors of all of the matrices of the form $\mathbf{A}_{ii}$ before iteration commences so that only right-hand side operations need to be performed within the iteration cycle. Advantage may be taken of sparseness in the off-diagonal submatrices $\mathbf{A}_{ij}$ $(i \neq j)$ and of band structure in the submatrices $\mathbf{A}_{ii}$, in order to reduce the amount of computation.

Equivalent relaxation procedures

Equation (6.71) may be written in the alternative form

$$\begin{aligned} \mathbf{x}_i^{(k+1)} = \omega(\bar{\mathbf{b}}_i - \bar{\mathbf{A}}_{i1}\mathbf{x}_1^{(k+1)} - \cdots - \bar{\mathbf{A}}_{i,i-1}\mathbf{x}_{i-1}^{(k)}) + (1-\omega)\mathbf{x}_i^{(k)} \\ + \omega(-\bar{\mathbf{A}}_{i,i+1}\mathbf{x}_{i+1}^{(k)} - \cdots - \mathbf{A}_{in}\mathbf{x}_n^{(k)}) \end{aligned} \tag{6.72}$$

where $\bar{b}_i = A_{ii}^{-1} b_i$ and $\bar{A}_{ij} = A_{ii}^{-1} A_{ij}$. This corresponds to the standard SOR prediction for the variables x_i using equations whose right-hand subvector is $\bar{b}_i$ and whose coefficients contain submatrices $\bar{A}_{ij}$ in off-diagonal positions. Hence SBOR is equivalent to standard SOR applied to modified equations $\bar{A}x = \bar{b}$ having the submatrix form

$$\begin{bmatrix} I & \bar{A}_{12} & \cdots & \bar{A}_{1n} \\ \bar{A}_{21} & I & \cdots & \bar{A}_{2n} \\ \cdot & \cdot & \cdot & \cdot \\ \bar{A}_{n1} & \bar{A}_{n2} & \cdots & I \end{bmatrix} \begin{bmatrix} x_1 \\ x_2 \\ \cdot \\ x_n \end{bmatrix} = \begin{bmatrix} \bar{b}_1 \\ \bar{b}_2 \\ \cdot \\ \bar{b}_n \end{bmatrix} \tag{6.73}$$

Property A characteristics for SBOR

In equations (6.73) there are no coefficients linking variables in the same group. Consequently it is possible to prove that property A characteristics apply to these equations (and hence to the original SBOR equations) provided that the non-null coefficient submatrices conform to one of the allowable property A patterns. For instance, property A characteristics are obtained with SBOR if the coefficient matrix has a tridiagonal pattern of submatrices. Equations which do not exhibit property A characteristics with SOR may be ordered in such a way that property A characteristics are obtained with SBOR, thus extending the usefulness of the property A concept.

SBOR with symmetric positive definite matrices

If A is a symmetric positive definite coefficient matrix, the submatrices of the form A_{ii} must also be symmetric positive definite. If the Choleski factorization of A_{ii} is $L_{ii}L_{ii}^T$, it is possible to show that an SBOR iteration is equivalent to an SOR iteration

$$\left.\begin{aligned} &\bar{A}\bar{x} = \bar{b} \\ &\text{where} \\ &\bar{A} = \bar{L}^{-1} A \bar{L}^{-T}, \quad \bar{x} = \bar{L}^T x, \quad \bar{b} = \bar{L}^{-1} b \\ &\text{and} \\ &\bar{L} = \lceil L_{11} L_{22} \ldots L_{nn} \rfloor \end{aligned}\right\} \tag{6.74}$$

Since $\bar{A}$ must be symmetric positive definite if SBOR is applied to equations which have a symmetric positive definite coefficient matrix, convergence is assured provided that $0 < \omega < 2$.

In general, SBOR converges more rapidly than SOR applied to the same set of equations. The improvement in convergence rate will be most pronounced when the largest off-diagonal coefficients occur within the submatrices A_{ii}.

6.10 SLOR, ADIP AND SIP

Successive line over-relaxation (SLOR)

The simple and repetitive structure of finite difference equations is utilized in certain iterative techniques, of which three will be discussed. Successive line over-relaxation is a form of block relaxation in which the variables associated with an individual line of a rectangular mesh constitute a subset. For five-point difference equations such as those derived from Laplace's equation using a rectangular mesh and row-wise numbering of the mesh nodes, there is a tridiagonal pattern of non-null submatrices. Furthermore, the submatrices which require to be decomposed are themselves tridiagonal. This means that the decomposition can be carried out with only a small amount of computation and without any of the zero coefficients becoming non-zero. Where the mesh has rectangular boundaries the off-diagonal elements in the lower triangle of the coefficient matrix lie along two subdiagonals as illustrated in Figure 6.7. Hence if the coefficient matrix is symmetric, it may be stored in three one-dimensional arrays, one for the diagonal elements, one for the row-wise mesh coupling coefficients and one for the column-wise mesh coupling coefficients. The decomposition of the tridiagonal submatrices may be carried out using the first two of these arrays. If the mesh does not have a rectangular boundary the coefficient matrix will still have a tridiagonal pattern of non-zero submatrices. For instance, the heat transfer equations (2.52) have the block structure

$$
\left[\begin{array}{cc|ccc|ccccc}
3 & & & & & & & & & \\
-1 & 4 & & & & & & & & \\
\hline
-1 & & 3 & & & & & & & \\
 & -1 & -1 & 4 & & & \text{symmetric} & & & \\
 & & & -1 & 4 & & & & & \\
\hline
 & & -1 & & & 3 & & & & \\
 & & & -1 & & -1 & 4 & & & \\
 & & & & -1 & & -1 & 4 & & \\
 & & & & & & & -1 & 4 & \\
 & & & & & & & & -1 & 3
\end{array}\right]
\left[\begin{array}{c}
T_1 \\ T_2 \\ \hline T_3 \\ T_4 \\ T_5 \\ \hline T_6 \\ T_7 \\ T_8 \\ T_9 \\ T_{10}
\end{array}\right]
=
\left[\begin{array}{c}
T_L \\ 2T_L \\ \hline \\ \\ 2T_L \\ \hline T_H \\ T_H \\ T_H \\ T_L + T_H \\ T_L + T_H
\end{array}\right]
\tag{6.75}
$$

In this case the coefficient matrix may also be stored using three one-dimensional arrays. However, elements stored in the mesh column-wise coupling array will not all be from the same subdiagonal of the matrix.

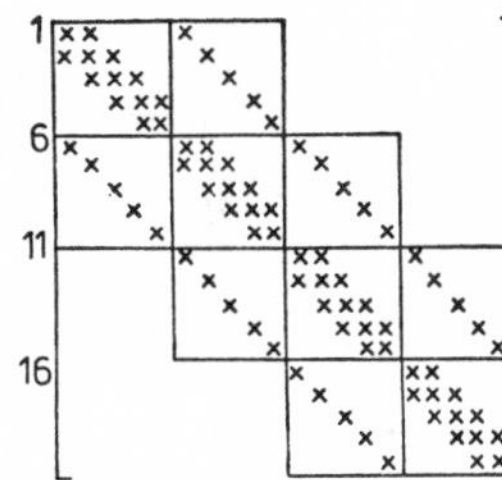

Figure 6.7 Five-point finite difference equations for a 4 × 5 grid: pattern of non-zero elements in the coefficient matrix

Alternating direction implicit procedure (ADIP)

Use is made of the interchangeability of the roles of mesh rows and columns in the alternating direction implicit procedure developed by Peaceman and Rachford (1955). The coefficient matrix of the equations is split according to

$$\mathbf{A} = \mathbf{H} + \mathbf{V} \tag{6.76}$$

where **H** and **V** contain the components of the difference equations in the directions of the mesh rows and columns respectively. For a set of five-point Laplace difference equations with row-wise listing of the mesh nodes **H** and **V** are both symmetric positive definite. Furthermore, they are both easily stored and decomposed without any of the zero elements becoming non-zero (**H** is a tridiagonal matrix and **V** would be a tridiagonal matrix if the nodes were listed by mesh columns). The matrices **H** and **V** are employed in two alternating sweeps to produce an ADIP iteration step as follows:

$$\left.\begin{aligned} &(\mathbf{H} + \omega\mathbf{I})\mathbf{x}^{(k+\frac{1}{2})} = \mathbf{b} - (\mathbf{V} - \omega\mathbf{I})\mathbf{x}^{(k)} \\ &\text{and} \\ &(\mathbf{V} + \omega\mathbf{I})\mathbf{x}^{(k+1)} = \mathbf{b} - (\mathbf{H} - \omega\mathbf{I})\mathbf{x}^{(k+\frac{1}{2})} \end{aligned}\right\} \tag{6.77}$$

The parameter ω may either be given a fixed value for the whole iterative sequence or else changed from iteration to iteration using a strategy developed by Wachspress and Harbetler (1960).

Strongly implicit procedure (SIP)

In ADIP some of the off-diagonal coefficients are retained on the left-hand side of the equations in the first sweep, while the rest are retained in the second sweep. In contrast, the iteration step of the strongly implicit procedure (Stone, 1968) is executed with all of the off-diagonal coefficients retained on the left-hand side of the equations. If **N** is a matrix of additional elements, a basic SIP iteration is given by

$$(\mathbf{A} + \mathbf{N})\mathbf{x}^{(k+1)} = \mathbf{b} + \mathbf{N}\mathbf{x}^{(k)} \tag{6.78}$$

If **A** is symmetric, **N** will also be symmetric and the modified coefficient matrix may be decomposed into Choleski triangular factors according to

$$\mathbf{A} + \mathbf{N} = \mathbf{L}\mathbf{L}^T \tag{6.79}$$

The matrix **L** has non-zero elements only in positions corresponding to non-zero elements of **A**. These elements satisfy the normal formulae for the Choleski decomposition of **A**. Thus the elements in the first two columns of **L** corresponding to a matrix **A**, which has the pattern of non-zero elements shown in Figure 6.7, may be computed according to

$$\left.\begin{aligned} l_{11} &= \sqrt{(a_{11})}, & l_{21} &= a_{21}/l_{11}, & l_{61} &= a_{61}/l_{11} \\ l_{22} &= \sqrt{(a_{22} - l_{21}^2)}, & l_{32} &= a_{32}/l_{22}, & l_{72} &= a_{72}/l_{22} \end{aligned}\right\} \tag{6.80}$$

However, in the Choleski decomposition of the unmodified coefficient matrix **A** it is necessary to introduce a non-zero element l_{62} such that

$$l_{61}l_{21} + l_{62}l_{22} = 0 \tag{6.81}$$

which in turn introduces further non-zero elements into **L** as the elimination proceeds. In the SIP decomposition n_{62} is given the value $l_{61}l_{21}$ so that l_{62} will be zero. If this procedure for preventing additional non-zero elements from occurring in **L** is repeated wherever necessary throughout the decomposition, then matrix **N** will contain elements as shown in Figure 6.8.

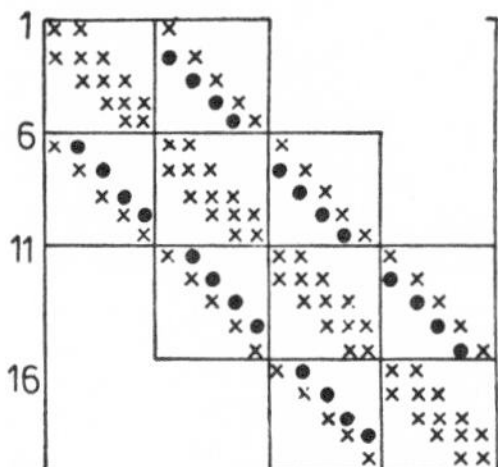

Figure 6.8 SIP solution of five-point finite difference equations for a 4 × 5 grid

Whereas ADIP is particularly effective for certain types of finite difference equations, SIP has been claimed to be effective for a wider class of problems (Weinstein, 1969). Unfortunately there is not much theoretical evidence available concerning the convergence of SIP.

6.11 CHEBYSHEV ACCELERATION

Richardson (Young, 1954) has shown that the Jacobi iteration method can be improved by converting it into a non-stationary process in which the predictions for the variables are extrapolated by a different amount at each iteration. The technique can be used to modify any stationary iterative process for which the iteration matrix has real eigenvalues provided that approximate values of the extreme eigenvalues of the iteration matrix are available. The following acceleration method using Chebyshev polynomials may be considered to be an extension of Richardson's method.

Scope for variation of the predictions

Consider a stationary iterative process characterized by an iteration matrix **M** whose typical (real) eigenvalue is μ_i and corresponding eigenvector is $\mathbf{p}_i$. Let the initial error vector be expressed as a linear function of these eigenvectors according to

$$\mathbf{e}^{(0)} = \sum_{i=1}^{n} c_i \mathbf{p}_i \tag{6.82}$$

After k iterations of the stationary process the error vector $\mathbf{e}^{(k)}$ is

$$\mathbf{e}^{(k)} = \mathbf{M}^k \mathbf{e}^{(0)} = \sum_{i=1}^{n} \mu_i^k c_i \mathbf{p}_i \tag{6.83}$$

Thus each eigenvector component has been factored by the corresponding value of μ_i^k. This process is illustrated in Figure 6.9 for $k = 5$ and $0 \leqslant \mu \leqslant 0.95$.

Consider an alternative predicted solution $\bar{\mathbf{x}}^{(k)}$ obtained by forming a linear combination of the vector iterates according to

$$\bar{\mathbf{x}}^{(k)} = d_0 \mathbf{x}^{(0)} + d_1 \mathbf{x}^{(1)} + \cdots + d_k \mathbf{x}^{(k)} \tag{6.84}$$

For such a prediction to give the correct solution when $\mathbf{x}^{(0)} = \mathbf{x}^{(1)} = \cdots = \mathbf{x}^{(k)} = \mathbf{x}$ it is necessary that

$$d_0 + d_1 + \cdots + d_k = 1 \tag{6.85}$$

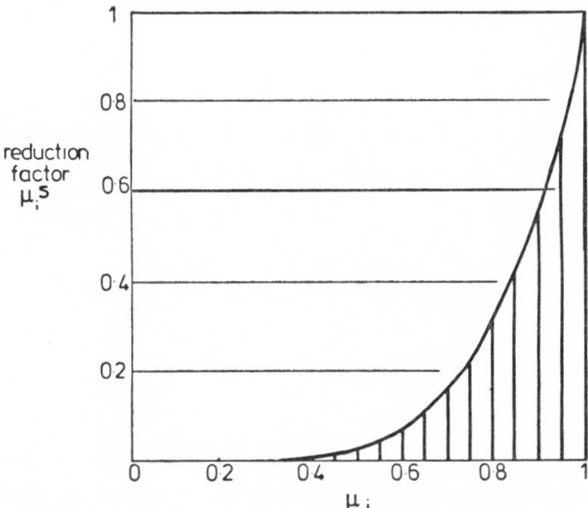

Figure 6.9 Effect of five stationary iterations on the eigenvector components of the trial vector, for $0 \leqslant \mu \leqslant 0.95$

The error vector associated with $\bar{x}^{(k)}$ can be shown to be

$$\bar{e}^{(k)} = \sum_{i=1}^{n} t_i^{(k)} c_i \mathbf{p}_i \tag{6.86}$$

where

$$t_i^{(k)} = d_0 + d_1\mu_i + \cdots + d_k\mu_i^k \tag{6.87}$$

The scope for the choice of $d_0, d_1, \ldots, d_k$ is such that $t_i^{(k)}$ may be any k-th order polynomial in μ_i which passes through the point $(\mu_i, t_i^{(k)}) = (1, 1)$.

Minimax error control

In order to obtain the most effective reduction of all the error components, the coefficients $d_0, d_1, \ldots, d_k$ should be chosen in such a way that the maximum value of t_i is minimized. If it is assumed that the eigenvalues are distributed throughout the region $\mu_{min} \leqslant \mu \leqslant \mu_{max}$, this minimax principle is achieved by using the Chebyshev polynomials $T_k(y)$. The optimum function $t_i^{(k)}$ may be shown to be

$$t_i^{(k)} = \frac{T_k(y_i)}{T_k(y')} \tag{6.88}$$

where

$$y_i = \frac{2\mu_i - \mu_{max} - \mu_{min}}{\mu_{max} - \mu_{min}} \tag{6.89}$$

and y' is the value of y_i at $\mu_i = 1$.

Figure 6.10 shows $t_i^{(5)}$ plotted as a function of μ_i for the case where $\mu_{min} = 0$ and $\mu_{max} = 0.95$. It has been evaluated from the fifth-order Chebyshev polynomial

$$T_5(y) = 16y^5 - 20y^3 + 5y \tag{6.90}$$

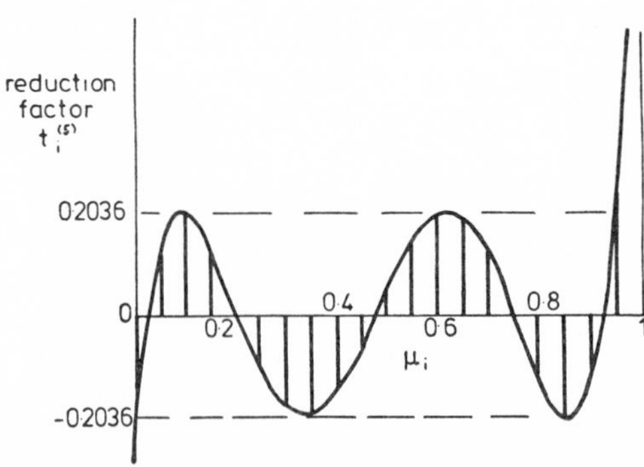

Figure 6.10 Effect of fifth-order Chebyshev acceleration on the components of the trial vector, for $0 \leqslant \mu \leqslant 0.95$

Because $t_i^{(5)} \leqslant 0.2036 \simeq 0.95^{31}$ the error of the accelerated vector $\bar{\mathbf{x}}^{(5)}$ is likely to be of the same order of magnitude as $\mathbf{x}^{(31)}$, showing a considerable improvement over the basic stationary iteration.

A recurrence relationship for generating vectors

The most convenient and numerically stable way of generating accelerated vectors $\bar{\mathbf{x}}^{(k)}$ is by using a recurrence relationship. The Chebyshev functions satisfy the recurrence relationship

$$T_{k+1}(y) = 2yT_k(y) - T_{k-1}(y) \tag{6.91}$$

Hence, from equation (6.88),

$$T_{k+1}(y')t_i^{(k+1)} = 2y_iT_k(y')t_i^{(k)} - T_{k-1}(y')t_i^{(k-1)} \tag{6.92}$$

and consequently

$$T_{k+1}(y') \sum_{i=1}^{n} t_i^{(k+1)} c_i \mathbf{p}_i = 2T_k(y') \sum_{i=1}^{n} \left(\frac{2\mu_i - \mu_{\max} - \mu_{\min}}{\mu_{\max} - \mu_{\min}}\right) t_i^{(k)} c_i \mathbf{p}_i - T_{k-1}(y') \sum_{i=1}^{n} t_i^{(k-1)} c_i \mathbf{p}_i \tag{6.93}$$

However, from equation (6.86),

$$\sum_{i=1}^{n} t_i^{(k)} c_i \mathbf{p}_i = \bar{\mathbf{e}}^{(k)} = \bar{\mathbf{x}}^{(k)} - \mathbf{x} \tag{6.94}$$

Also

$$\sum_{i=1}^{n} \mu_i t_i^{(k)} c_i \mathbf{p}_i = \mathbf{M}\bar{\mathbf{e}}^{(k)} = \mathbf{z}^{(k)} - \mathbf{x} \tag{6.95}$$

where $\mathbf{z}^{(k)}$ has been obtained from $\bar{\mathbf{x}}^{(k)}$ by one application of the basic iteration cycle. On substituting equations (6.94) and (6.95) into equation (6.93) it can be shown that the coefficients of $\mathbf{x}$ cancel leaving

$$T_{k+1}(y')\bar{\mathbf{x}}^{(k+1)} = 2T_k(y') \left[\left(\frac{2}{\mu_{\max} - \mu_{\min}}\right)\mathbf{z}^{(k)} - \left(\frac{\mu_{\max} + \mu_{\min}}{\mu_{\max} - \mu_{\min}}\right)\bar{\mathbf{x}}^{(k)}\right] - T_{k-1}(y')\bar{\mathbf{x}}^{(k-1)} \tag{6.96}$$

Since $T_{k+1}(y')$ is non-zero a recurrence formula has been obtained which may be used to generate accelerated vectors from lower order accelerated vectors. Computation of a sequence of accelerated vectors may be initiated by setting $\bar{\mathbf{x}}^{(0)} = \mathbf{x}^{(0)}$, $\bar{\mathbf{x}}^{(1)} = \mathbf{x}^{(1)}$, $T_0(y') = 1$ and $T_1(y') = y'$. It may be noted from equation (6.96) that only three vectors need to be stored at any given time during the evaluation of the sequence.

Notes on Chebyshev acceleration

(a) If accurate estimates of the extreme eigenvalues of the iteration matrix are available, Chebyshev acceleration can improve the convergence rate of a stationary iterative process provided that the eigenvalues of the iteration matrix are real.

(b) If Chebyshev acceleration is used with underestimates of the limiting eigenvalues, the convergence rate is still likely to be improved although to a much lesser extent than if the estimates are accurate.

(c) If the iteration matrix has an eigenvalue very close to unity, then even a Chebyshev accelerated iteration will take a large number of iterations to obtain an accurate solution.

6.12 DYNAMIC ACCELERATION METHODS

The effective application of Chebyshev acceleration requires prior knowledge of the extreme eigenvalues of the iteration matrix. Such information may not be readily available in the case of slowly convergent iterations where acceleration methods would be most useful. Acceleration methods which make use of the successive iterates to provide the means of acceleration may be classified as *dynamic acceleration methods*. Carré's method of revising the over-relaxation parameter, when SOR is applied to equations with property A coefficient matrices (section 6.7), is an example of a dynamic acceleration method.

Aitken's (1950) δ^2 acceleration

If an iterative process has a slow convergence rate, successive errors will generally exhibit an exponential decay in the later stages of the iteration. (An exception to this rule is when the largest eigenvalues of the iteration matrix are complex.) In such cases Aitken's δ^2 process can be used to predict the asymptotic limit to which the predictions for each variable is tending. If three successive estimates for x_i are $x_i^{(k)}$, $x_i^{(k+1)}$ and $x_i^{(k+2)}$, the δ^2 prediction is given by

$$x_i^* = \frac{x_i^{(k)} x_i^{(k+2)} - (x_i^{(k+1)})^2}{x_i^{(k)} - 2x_i^{(k+1)} + x_i^{(k+2)}} \tag{6.97}$$

Alternatively,

$$\left.\begin{aligned} x_i^* &= x_i^{(k+2)} + s_i^*(x_i^{(k+2)} - x_i^{(k+1)}) \\ &\text{where} \\ s_i^* &= \frac{\mu_i^*}{1-\mu_i^*} = \frac{x_i^{(k+1)} - x_i^{(k+2)}}{x_i^{(k)} - 2x_i^{(k+1)} + x_i^{(k+2)}} \end{aligned}\right\} \tag{6.98}$$

Hence μ_i^* is the ratio of successive differences $(x_i^{(k+1)} - x_i^{(k+2)})/(x_i^{(k)} - x_i^{(k+1)})$ and s_i^* is the amount by which the last step is extrapolated. Thus s_i^* can be described as an acceleration parameter.

If Aitken acceleration is applied at the wrong time the denominator of one or more of the expressions (6.97) could be zero or very small. In such circumstances the acceleration method will either fail to yield a prediction or else give a predicted value which is grossly in error. The 'wrong time' may be interpreted as either too soon, before an exponential decay is properly established, or too late when rounding errors affect the predictions. It is therefore not very suitable for computer implementation.

Modified Aitken acceleration (Jennings, 1971)

When the iteration matrix is symmetric Aitken's prediction can be modified by using a common acceleration parameter for all of the variables as follows:

$$\bar{\mathbf{x}} = \mathbf{x}^{(k+2)} + s(\mathbf{x}^{(k+2)} - \mathbf{x}^{(k+1)}) \tag{6.99}$$

where

$$s = \frac{\bar{\mu}}{1-\bar{\mu}} = \frac{(\mathbf{x}^{(k)} - \mathbf{x}^{(k+1)})^T(\mathbf{x}^{(k+1)} - \mathbf{x}^{(k+2)})}{(\mathbf{x}^{(k)} - \mathbf{x}^{(k+1)})^T(\mathbf{x}^{(k)} - 2\mathbf{x}^{(k+1)} + \mathbf{x}^{(k+2)})} \tag{6.100}$$

In order to investigate the effect of applying this acceleration method, let the error vector $\mathbf{e}^{(0)}$ be expressed as a linear combination of the eigenvectors $\mathbf{p}_i$ of the iteration matrix normalized such that $\mathbf{p}_i^T\mathbf{p}_i = 1$. Thus

$$\mathbf{x}^{(0)} - \mathbf{x} = \mathbf{e}^{(0)} = \sum_{i=1}^{n} a_i\mathbf{p}_i \tag{6.101}$$

Hence

$$\mathbf{x}^{(k)} - \mathbf{x} = \mathbf{M}^k\mathbf{e}^{(0)} = \sum_{i=1}^{n} a_i\mu_i^k\mathbf{p}_i \tag{6.102}$$

By substituting equation (6.102) into equation (6.100) it may be shown that

$$s = \frac{\sum_{i=1}^{n} \mu_i z_i}{\sum_{i=1}^{n} (1-\mu_i) z_i} \tag{6.103}$$

where

$$z_i = \mu_i^{2k}(1-\mu_i)^2 a_i^2 \geqslant 0 \tag{6.104}$$

If the basic iterative process is convergent, $\mu_i < 1$, and hence the denominator of s must be positive. (The case where all z_i are zero can be discounted since, in this case, an accurate solution would have already been obtained.) From equations

(6.100) and (6.103) it follows that

$$\sum_{i=1}^{n} (\bar{\mu} - \mu_i) z_i = 0 \tag{6.105}$$

and hence $\bar{\mu}$ must lie between the extreme eigenvalues of M. The error vector after acceleration is given by

$$\bar{\mathbf{e}} = \bar{\mathbf{x}} - \mathbf{x} = \sum_{i=1}^{n} (\mu_i + s\mu_i - s)\mu_i^{k+1} a_i \mathbf{p}_i = \sum_{i=1}^{n} \frac{\mu_i - \bar{\mu}}{1 - \bar{\mu}} \mu_i^{k+1} a_i \mathbf{p}_i \tag{6.106}$$

If acceleration is applied after so many iterations that all but one value of z_i is effectively zero, then $\bar{\mathbf{e}}$ can be shown to be zero. This acceleration method is therefore particularly effective when only one eigenvalue of the iteration matrix has modulus close to unity. It is possible to apply the acceleration after two or more iterations and to repeat the acceleration frequently. Thus it is satisfactory to use in computer implementation of iterative methods.

Double acceleration

The above procedure may be extended to produce a double-acceleration method which uses five successive vector iterates. Defining

$$t_j = (\mathbf{x}^{(k+2)} - \mathbf{x}^{(k+3)})^T (\mathbf{x}^{(k+j)} - \mathbf{x}^{(k+j+1)}) \tag{6.107}$$

double acceleration gives the prediction

$$\bar{\bar{\mathbf{x}}} = \frac{A\mathbf{x}^{(k+4)} - B\mathbf{x}^{(k+3)} + C\mathbf{x}^{(k+2)}}{A - B + C} \tag{6.108}$$

where

$$\left.\begin{aligned} A &= \begin{vmatrix} t_0 & t_1 \\ t_1 & t_2 \end{vmatrix}, \quad B = \begin{vmatrix} t_0 & t_1 \\ t_2 & t_3 \end{vmatrix}, \quad C = \begin{vmatrix} t_1 & t_2 \\ t_2 & t_3 \end{vmatrix}, \\ &\text{and} \\ A - B + C &= \begin{vmatrix} t_0 - t_1 & t_1 - t_2 \\ t_1 - t_2 & t_2 - t_3 \end{vmatrix} \end{aligned}\right\} \tag{6.109}$$

The denominator $A - B + C$ can be shown to be positive provided that M is symmetric and at least two coefficients a_i are non-zero. If double accelerations are applied after every four iterations of an iterative process in which the extreme eigenvalues of the iteration matrix are 0 and $1 - e$ (where e is small), the number of iterations will be reduced to about one-eighth of the number required without acceleration.

6.13 GRADIENT METHODS

The process of solving a set of n simultaneous equations is equivalent to that of finding the position of the minimum of an error function defined over an

n-dimensional space. In each step of a gradient method a trial set of values for the variables is used to generate a new set corresponding to a lower value of the error function. In most gradient methods, successive error vectors cannot be generated by means of an iteration matrix and hence they are classified as non-stationary processes.

Error function minimization

If $\bar{\mathbf{x}}$ is the trial vector, the corresponding vector of residuals is

$$\mathbf{r} = \mathbf{b} - \mathbf{A}\bar{\mathbf{x}} \tag{6.110}$$

If the coefficient matrix of the equations is symmetric and positive definite, its inverse will also be symmetric and positive definite. Hence the positive error function such that

$$h^2 = \mathbf{r}^T\mathbf{A}^{-1}\mathbf{r} \tag{6.111}$$

must have a real value for all possible vectors $\bar{\mathbf{x}}$ except the correct solution $\bar{\mathbf{x}} = \mathbf{x}$. In the latter case $\mathbf{r} = \mathbf{0}$ and hence $h = 0$. Substituting for $\mathbf{r}$ from equation (6.110)

$$h^2 = \bar{\mathbf{x}}^T\mathbf{A}\bar{\mathbf{x}} - 2\mathbf{b}^T\bar{\mathbf{x}} + \mathbf{b}^T\mathbf{A}^{-1}\mathbf{b} \tag{6.112}$$

showing that h^2 is quadratic in the variables $\bar{\mathbf{x}}$.

The vector $\mathbf{x}^{(k)}$ may be represented by a point in n-dimensional space. The equation

$$\bar{\mathbf{x}} = \mathbf{x}^{(k)} + \alpha\mathbf{d}^{(k)} \tag{6.113}$$

defines a line through the point whose direction is governed by the vector $\mathbf{d}^{(k)}$ (a graphical interpretation for $n = 2$ is given in Figure 6.11). The parameter α is proportional to the distance of $\bar{\mathbf{x}}$ from $\mathbf{x}^{(k)}$. Substituting for $\bar{\mathbf{x}}$ in equation (6.112)

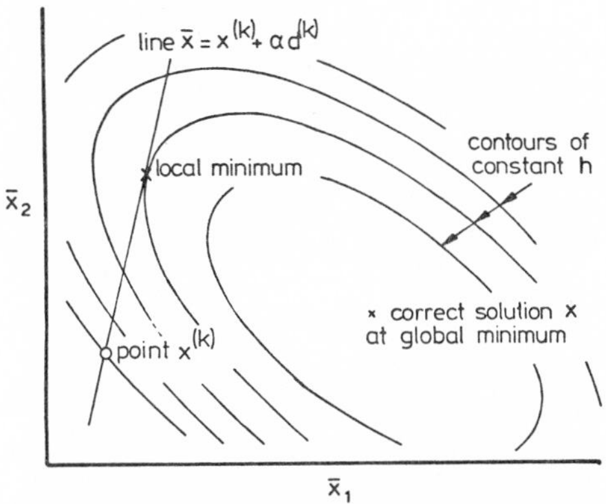

Figure 6.11 Graphical interpretation of local minimization of the error function h

gives

$$b^2 = \alpha^2 [\mathbf{d}^{(k)}]^T \mathbf{A}\mathbf{d}^{(k)} - 2\alpha[\mathbf{d}^{(k)}]^T \mathbf{r}^{(k)} + [\mathbf{x}^{(k)}]^T \mathbf{A}\mathbf{x}^{(k)} + \mathbf{b}^T\mathbf{A}^{-1}\mathbf{b} \qquad (6.114)$$

Therefore the function b^2 varies quadratically with α having a local minimum at which

$$\frac{\partial(b^2)}{\partial\alpha} = 2[\mathbf{d}^{(k)}]^T [\alpha \mathbf{A}\mathbf{d}^{(k)} - \mathbf{r}^{(k)}] = 0 \qquad (6.115)$$

Both the methods of steepest descent and conjugate gradients use the position of this local minimum to define the next trial vector, i.e.

$$\left.\begin{aligned} &\mathbf{x}^{(k+1)} = \mathbf{x}^{(k)} + \alpha_k \mathbf{d}^{(k)} \\ &\text{where} \\ &\alpha_k = \frac{[\mathbf{d}^{(k)}]^T \mathbf{r}^{(k)}}{[\mathbf{d}^{(k)}]^T \mathbf{A}\mathbf{d}^{(k)}} \end{aligned}\right\} \qquad (6.116)$$

The methods differ only in the choice of the direction vectors $\mathbf{d}^{(k)}$.

The method of steepest descent

In the method of steepest descent, $\mathbf{d}^{(k)}$ is chosen to be the direction of maximum gradient of the function at the point $\mathbf{x}^{(k)}$. This can be shown to be proportional to $\mathbf{r}^{(k)}$. The $(k+1)$-th step of the process can be accomplished by means of the following algorithm:

$$\left.\begin{aligned} \mathbf{u}^{(k)} &= \mathbf{A}\mathbf{r}^{(k)} \\ \alpha_k &= [\mathbf{r}^{(k)}]^T \mathbf{r}^{(k)} / [\mathbf{r}^{(k)}]^T \mathbf{u}^{(k)} \\ \mathbf{x}^{(k+1)} &= \mathbf{x}^{(k)} + \alpha_k \mathbf{r}^{(k)} \\ \mathbf{r}^{(k+1)} &= \mathbf{r}^{(k)} - \alpha_k \mathbf{u}^{(k)} \end{aligned}\right\} \qquad (6.117)$$

(Before the first step can be undertaken it is necessary to compute $\mathbf{r}^{(0)} = \mathbf{b} - \mathbf{A}\mathbf{x}^{(0)}$.)

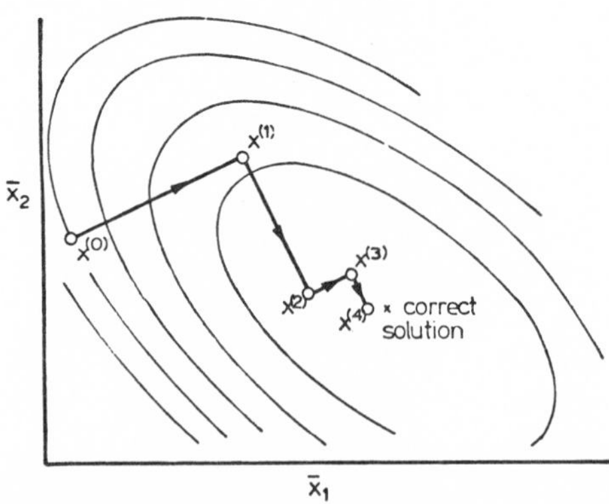

Fig. 6.12 Convergence of the method of steepest descent

Figure 6.12 illustrates the convergence of the method of steepest descent graphically for a problem involving only two variables.

The method of conjugate gradients (Hestenes and Stiefel, 1952)

In the method of conjugate gradients the direction vectors are chosen to be a set of vectors $\mathbf{p}^{(0)}$, $\mathbf{p}^{(1)}$, etc., which represent, as nearly as possible, the directions of steepest descent of points $\mathbf{x}^{(0)}$, $\mathbf{x}^{(1)}$, etc., respectively, but with the overriding condition that they be mutually conjugate. Here the term *conjugate* means that the vectors are orthogonal with respect to $\mathbf{A}$ and hence satisfy the condition

$$[\mathbf{p}^{(i)}]^T\mathbf{A}\mathbf{p}^{(j)} = 0 \quad \text{for } i \neq j \tag{6.118}$$

Step $k + 1$ of the procedure may be accomplished by means of the following algorithm:

$$\left.\begin{aligned} \mathbf{u}^{(k)} &= \mathbf{A}\mathbf{p}^{(k)} \\ \alpha_k &= [\mathbf{r}^{(k)}]^T\mathbf{r}^{(k)}/[\mathbf{p}^{(k)}]^T\mathbf{u}^{(k)} \\ \mathbf{x}^{(k+1)} &= \mathbf{x}^{(k)} + \alpha_k\mathbf{p}^{(k)} \\ \mathbf{r}^{(k+1)} &= \mathbf{r}^{(k)} - \alpha_k\mathbf{u}^{(k)} \\ \beta_k &= [\mathbf{r}^{(k+1)}]^T\mathbf{r}^{(k+1)}/[\mathbf{r}^{(k)}]^T\mathbf{r}^{(k)} \\ \mathbf{p}^{(k+1)} &= \mathbf{r}^{(k+1)} + \beta_k\mathbf{p}^{(k)} \end{aligned}\right\} \tag{6.119}$$

with the initial values for $k = 0$ being

$$\mathbf{p}^{(0)} = \mathbf{r}^{(0)} = \mathbf{b} - \mathbf{A}\mathbf{x}^{(0)} \tag{6.120}$$

It can be shown by induction that the following orthogonal relationships are satisfied:

$$\left.\begin{aligned} [\mathbf{r}^{(i)}]^T\mathbf{p}^{(j)} &= 0 \quad \text{for } i > j \\ [\mathbf{r}^{(i)}]^T\mathbf{r}^{(j)} &= 0 \quad \text{for } i \neq j \end{aligned}\right\} \tag{6.121}$$

The conjugate property, equation (6.118), is also satisfied. The parameters α_k and β_k may alternatively be specified by

$$\alpha_k = \frac{[\mathbf{p}^{(k)}]^T\mathbf{r}^{(k)}}{[\mathbf{p}^{(k)}]^T\mathbf{u}^{(k)}} \quad \text{and} \quad \beta_k = -\frac{[\mathbf{r}^{(k+1)}]^T\mathbf{u}^{(k)}}{[\mathbf{p}^{(k)}]^T\mathbf{u}^{(k)}} \tag{6.122}$$

Because of the orthogonal relationships, the correct solution is theoretically obtained after a total of n steps. Hence the method should not, strictly speaking, be classified as iterative. Table 6.7 shows the n step convergence when the method is used to solve equation (6.7). However, it is not advisable to assume that convergence will always be obtained at $k = n$. In some cases convergence to an acceptable accuracy will be obtained for lower values of k, and in other cases rounding errors will affect the computation to such an extent that more than n steps will need to be performed. The method should therefore be programmed as an iterative method,

Table 6.7 The conjugate gradient method applied to equation (6.7)

k	0	1	2	3	4
x_1	0	4.3208	7.4446	8.3457	8.4876
x_2	0	4.9342	4.5856	5.8057	6.4275
x_3	0	−7.1497	−6.6948	−5.0983	−4.7029
x_4	0	2.1730	2.2546	4.2461	4.0066
r_1	5.3209	3.1522	−0.0510	0.0294	0.0000
r_2	6.0762	−0.8996	0.5988	0.3945	0.0000
r_3	−8.8046	1.2459	0.6506	0.1709	0.0000
r_4	2.6760	−0.1261	0.8825	−0.3920	0.0000
p_1	5.3209	3.5893	0.4047	0.1172	
p_2	6.0762	−0.4005	0.5480	0.5135	0
p_3	−8.8046	0.5227	0.7170	0.3266	
p_4	2.6760	0.0937	0.8944	−0.1978	
$\mathbf{r}^T\mathbf{r}$	149.914	12.3137	1.5633	0.3394	0
$\mathbf{p}^T\mathbf{A}\mathbf{p}$	184.613	14.1485	0.7021	0.2803	0
α	0.8120	0.8703	2.2266	1.2108	—
β	0.0821	0.1270	0.2171	—	—

with the process terminating when the residual vector tolerance criterion (6.10) is satisfied.

6.14 ON THE CONVERGENCE OF GRADIENT METHODS

It may be shown that in n-dimensional space the surfaces of constant h are ellipsoids with principal axes of length $(h^2/\lambda_1)^{1/2}$, $(h^2/\lambda_2)^{1/2}$, . . . , $(h^2/\lambda_n^{1/2})$ where $\lambda_1, \lambda_2, \ldots, \lambda_n$ are the eigenvalues of the coefficient matrix. The convergence characteristics of gradient methods are dependent on the shapes of these surfaces, and hence on the eigenvalues of the coefficient matrix. For instance, if the magnitude of all of the eigenvalues are almost the same, then the surfaces become spheroids and the convergence by either of the given methods will be very rapid.

Convergence of the method of steepest descent

Let the eigensolution of the coefficient matrix be such that $\mathbf{AQ} = \mathbf{Q\Lambda}$ (as in equation 1.112). Expressing $\mathbf{r}^{(k)}$ as a linear combination of the eigenvectors of $\mathbf{A}$ according to

$$\mathbf{r}^{(k)} = \mathbf{Q}\mathbf{s}^{(k)} \tag{6.123}$$

and substituting in the steepest descent expression for α_k, it can be shown that

$$\sum_{i=1}^{n} (1 - \alpha_k\lambda_i)(s_i^{(k)})^2 = 0 \tag{6.124}$$

Since the terms under the summation cannot all have the same sign,

$$\frac{1}{\lambda_n} \leqslant \alpha_k \leqslant \frac{1}{\lambda_1} \tag{6.125}$$

Substituting equation (6.123) in the expression for $\mathbf{r}^{(k+1)}$ gives

$$\mathbf{r}^{(k+1)} = \mathbf{Q}\mathbf{s}^{(k+1)} = (\mathbf{I} - \alpha_k \mathbf{A})\mathbf{Q}\mathbf{s}^{(k)} = \mathbf{Q}(\mathbf{I} - \alpha_k \mathbf{\Lambda})\mathbf{s}^{(k)} \tag{6.126}$$

and, since $\mathbf{Q}$ is non-singular,

$$\mathbf{s}^{(k+1)} = (\mathbf{I} - \alpha_k \mathbf{\Lambda})\mathbf{s}^{(k)} \tag{6.127}$$

showing that in the $(k + 1)$-th step of the method of steepest descent the i-th eigenvector component of the residual $s_i^{(k)}$ is multiplied by $(1 - \alpha_k \lambda_i)$. Since α_k must satisfy the inequalities (6.125), the largest positive and negative factors are for s_1 and s_n respectively. Hence it will be the extreme eigenvector components which govern the convergence rate.

A steady convergence rate is obtained if $s_1^{(0)} = s_n^{(0)} = s$ and $s_2^{(0)} = s_3^{(0)} = \cdots = s_{n-1}^{(0)} = 0$. The two non-zero eigenvector components are reduced in absolute magnitude by the same factor $(\lambda_n - \lambda_1)/(\lambda_1 + \lambda_n)$ at each iteration, with α_k having the constant value of $2/(\lambda_1 + \lambda_n)$. If $\lambda_1 + \lambda_n = 2$ this steady convergence rate corresponds to that of Jacobi iteration. If $\lambda_1 + \lambda_n \neq 2$ the correspondence is with extrapolated Jacobi iteration. Hence it cannot be expected that the method of steepest descent will converge significantly more rapidly than Jacobi iteration.

Convergence of the method of conjugate gradients

In both of the gradient methods it can be shown that the reduction in the error function during step $k + 1$ is such that

$$b_k^2 - b_{k+1}^2 = \alpha_k [\mathbf{r}^{(k)}]^T \mathbf{r}^{(k)} \tag{6.128}$$

It can also be shown that, for $k > 0$, the value of α_k for the conjugate gradient method is greater than the corresponding value for the method of steepest descent. Hence a conjugate gradient step other than the first will always give a greater reduction in error than a corresponding step of the steepest descent method starting from the same point $\mathbf{x}^{(k)}$.

After two steps of the conjugate gradient method the residuals are related to the initial residuals by

$$\mathbf{r}^{(2)} = [(\mathbf{I} - \alpha_1 \mathbf{A})(\mathbf{I} - \alpha_0 \mathbf{A}) - \alpha_1 \beta_0 \mathbf{A}]\,\mathbf{r}^{(0)} \tag{6.129}$$

Hence, if the residuals are expressed in the eigenvector component form of equation (6.123), it can be shown that

$$s_i^{(2)} = [(1 - \alpha_1 \lambda_i)(1 - \alpha_0 \lambda_i) - \alpha_1 \beta_0 \lambda_i]\, s_i^{(0)} \tag{6.130}$$

Similar relationships may be obtained for other values of k, establishing the general formula:

$$s_i^{(k)} = f_k(\lambda_i)\, s_i^{(0)} \tag{6.131}$$

where $f_k(\lambda)$ is a k-th order polynomial in λ.

The effect of choosing conjugate directions is to make the particular k-th order polynomial the one which results in the greatest possible reduction in the error function. Hence the conjugate gradient method is effectively a dynamic acceleration of the method of steepest descent. As such it has been shown to be at least as efficient as Chebyshev acceleration of Jacobi iteration (Reid, 1971). The convergence rate is not principally dependent on the values of the limiting eigenvalues (Stewart, 1975), but on the grouping of the full spectrum of eigenvalues. For instance, if the eigenvalues are in three groups such that

$$\lambda_i = \begin{cases} & \lambda_a(1+\epsilon_i) \\ \text{or} & \lambda_b(1+\epsilon_i) \\ \text{or} & \lambda_c(1+\epsilon_i) \end{cases} \tag{6.132}$$

where λ_a, λ_b and λ_c are well separated and $|\epsilon_i| \ll 1$, then the residual obtained after three steps should be small. Equations which will give slow convergence by the conjugate gradient method are those for which the coefficient matrix has a large number of small but distinct eigenvalues.

6.15 APPLICATION OF THE METHOD OF CONJUGATE GRADIENTS

Use as a direct method

Because the conjugate gradient method should yield the correct solution after n steps, it was first considered as an alternative to elimination methods for the solution of any set of linear equations for which the coefficient matrix is symmetric and positive definite. However, if the coefficient matrix is fully populated with non-zero coefficients, the number of multiplications required to perform n steps is in excess of n^3. Hence a conjugate gradient solution would entail approximately six times the amount of computation necessary for a solution by elimination.

The only operation in the conjugate gradient procedure which involves the coefficient matrix is the multiplication $\mathbf{u}^{(k)} = \mathbf{A}\mathbf{p}^{(k)}$. If the coefficient matrix is sparse, only its non-zero elements need to be stored, thus yielding a considerable saving in both storage space and computation time. If $\mathbf{A}$ has an average of c non-zero elements per row, each step of the conjugate gradient method requires approximately $(c+5)n$ multiplications. Hence, in theory, approximately $(c+5)n^2$ multiplications are required to obtain the exact solution. Comparisons with the number of multiplications required for band elimination solutions ($nb^2/2$) reveal that, for most problems, the elimination procedure will be much more efficient (this is true for all of the examples given in Table 6.4).

Use as an iterative method

If the conjugate gradient method is applied to large-order sets of equations satisfactory convergence is normally obtained in much fewer than n steps. In such cases it is the convergence rate of the method which is of significance rather than its terminal property, and in effect it is being used as an iterative method rather than as a direct method. If the convergence rate is particularly good the conjugate gradient method may be more efficient than elimination in providing a satisfactory solution.

Initial scaling

In the application of the conjugate gradient method it is unnecessary for the coefficient matrix $\mathbf{A}$ to have unit elements on the leading diagonal. However, it will be found that, if the equations are symmetrically scaled in such a way that the leading diagonal elements all have the same value, the convergence rate will usually be improved.

Transformation of the equations

The conjugate gradient method may be modified by applying it to transformations of the original equations. One possible transformation is Evans' preconditioning (equation 6.64). The conjugate gradient method may be applied to the preconditioned equations without forming the matrix $\mathbf{B} = (\mathbf{I} - \omega\mathbf{L})^{-1}\mathbf{A}(\mathbf{I} - \omega\mathbf{L})^{-T}$ explicitly, by a similar technique to that used in the basic preconditioning method shown in Figure 6.6. Another possible transformation is the symmetric equivalent of the block relaxation transformation given by equations (6.74). In this case it is unnecessary to form the off-diagonal blocks of $\bar{\mathbf{A}} = \bar{\mathbf{L}}^{-1}\mathbf{A}\bar{\mathbf{L}}^{-T}$ explicitly. However, advantage may be taken of the fact that the diagonal blocks are unit matrices.

Some numerical examples

Table 6.8 gives a comparison of the numerical efficiency of various conjugate gradient solutions with those of variable bandwidth elimination and optimum SOR. The equations to which the methods have been applied are derived from four problems as follows:

1. A finite difference formulation of Laplace's equation using a square mesh.
2. A finite element plate-bending analysis (the governing partial differential equation being biharmonic).
3. A box beam analysis using the same rectangular bending elements as in (2).
4. A structural analysis of a three-dimensional rectangular building frame having two storeys and 4 x 4 bays.

Table 6.8 Comparisons of numerical efficiency for the conjugate gradient method showing numbers of multiplications (k = 1,000). Numbers of iterations, shown in brackets, are to achieve a tolerance of $\|x^{(k)} - x\|_E / \|x\|_E < 0.00005$. (From a Ph.D. dissertation by G. M. Malik, Queen's University, Belfast, 1976)

	Problem no.	1	2	3	4
Properties of coefficient matrix	Order n	300	273	320	300
	Average semi-bandwidth, b	19	21	41	52
	Average no. of non-zero elements per row, c	5	15	15	9
	$\frac{\lambda_n}{\lambda_1}$	182	12,900	247	185,000
Gauss–Seidel		$749k(436)$	$>2{,}148k(>500)$	$2{,}184k(485)$	$>1{,}480k(>500)$
Optimum SOR		$79k(46)$ $\omega_{\text{opt}} = 1.75$	$>2{,}148k(>500)$	$225k(50)$ $\omega_{\text{opt}} = 1.74$	$>1{,}480k(>500)$
Conjugate gradient method	Unscaled	$154k(48)$	$1{,}183k(209)$	$384k(63)$	$794k(178)$
	Scaled	$140k(43)$	$734k(129)$	$334k(54)$	$489k(109)$
	Block	$117k(33)$	$556k(91)$	$293k(39)$	$404k(50)$
	Preconditioned	$62k(12)$ $\omega_{\text{opt}} = 1.6$	$544k(54)$ $\omega_{\text{opt}} = 1.0$	$240k(22)$ $\omega_{\text{opt}} = 1.2$	$420k(56)$ $\omega_{\text{opt}} = 1.0$
Variable bandwidth elimination		$69k$	$90k$	$305k$	$451k$

The fourth example was chosen because, for this type of problem, the coefficient matrix has a large ratio λ_n/λ_1 and hence is badly conditioned. Although this large ratio made SOR ineffective, the convergence of the conjugate gradient solutions was satisfactory. The conjugate gradient solution for problem 2 was the least satisfactory, particularly when compared with the variable bandwidth solution.

Solution of unsymmetric equations

It is always possible to transform a set of equations $\mathbf{Ax} = \mathbf{b}$, with a non-singular unsymmetric coefficient matrix, into the form

$$\left.\begin{aligned} &\bar{\mathbf{A}}\mathbf{x} = \bar{\mathbf{b}} \\ &\text{where} \\ &\bar{\mathbf{A}} = \mathbf{A}^T\mathbf{A} \quad \text{and} \quad \bar{\mathbf{b}} = \mathbf{A}^T\mathbf{b} \end{aligned}\right\} \tag{6.133}$$

Hence $\bar{\mathbf{A}}$ is symmetric and positive definite. The conjugate gradient method may be applied to the transformed equations without forming $\bar{\mathbf{A}}$ explicitly by using the following algorithm:

$$\left.\begin{aligned} \mathbf{u}^{(k)} &= \mathbf{A}\mathbf{p}^{(k)} \\ \alpha_k &= [\bar{\mathbf{r}}^{(k)}]^T\bar{\mathbf{r}}^{(k)}/[\mathbf{u}^{(k)}]^T\mathbf{u}^{(k)} \\ \mathbf{x}^{(k+1)} &= \mathbf{x}^{(k)} + \alpha_k\mathbf{p}^{(k)} \\ \mathbf{r}^{(k+1)} &= \mathbf{r}^{(k)} - \alpha_k\mathbf{u}^{(k)} \\ \bar{\mathbf{r}}^{(k+1)} &= \mathbf{A}^T\mathbf{r}^{(k+1)} \\ \beta_k &= [\bar{\mathbf{r}}^{(k+1)}]^T\bar{\mathbf{r}}^{(k+1)}/[\bar{\mathbf{r}}^{(k)}]^T\bar{\mathbf{r}}^{(k)} \\ \mathbf{p}^{(k+1)} &= \bar{\mathbf{r}}^{(k+1)} + \beta_k\mathbf{p}^{(k)} \end{aligned}\right\} \tag{6.134}$$

starting with $\mathbf{r}^{(0)} = \mathbf{b} - \mathbf{A}\mathbf{x}^{(0)}$, $\bar{\mathbf{r}}^{(0)} = \mathbf{p}^{(0)} = \mathbf{A}^T\mathbf{r}_0$.

Although this transformation makes the gradient methods universally applicable for the solution of linear equations, it must be noted that, if $\mathbf{A}$ is a near singular matrix, $\mathbf{A}^T\mathbf{A}$ will be very much more ill conditioned than $\mathbf{A}$. For example, the matrix

$$\mathbf{A} = \begin{bmatrix} 1 & 1 \\ 0.99 & 1 \end{bmatrix} \tag{6.135}$$

has an eigenvalue ratio $\lambda_2/\lambda_1 \simeq 400$, while $\mathbf{A}^T\mathbf{A}$ has an eigenvalue ratio $\lambda_2/\lambda_1 \simeq 160{,}000$. For large-order sets of equations, this will mean that there is a strong risk of encountering poor convergence rates.

The above transformation may also be applied in the case of other iterative methods. However, most iterative methods, including SOR, can only be implemented if $\bar{\mathbf{A}}$ is computed explicitly.

BIBLIOGRAPHY

Aitken, A. C. (1950). 'On the iterative solution of a system of linear equations'. *Proc. Roy. Soc. Edinburgh*, 63, 52–60. (Double-sweep Gauss–Seidel iteration.)

Allen, D. N. de G. (1954). 'Relaxation Methods in Engineering and Science'. McGraw-Hill, New York.

Carré, B. A. (1961). 'The determination of the optimum accelerating factor for successive over-relaxation'. *Computer J.*, 4, 73–78.

Evans, D. J. (1967). 'The use of preconditioning in iterative methods for solving linear equations with symmetric positive definite matrices'. *J. Inst. Maths. Applics.*, 4, 295–314.

Evans, D. J. (1973). 'The analysis and application of sparse matrix algorithms in the finite element method'. In J. R. Whiteman (Ed.), *The Mathematics of Finite Elements and Applications*, Academic Press, London. (The preconditioned conjugate gradient method.)

Fox, L. (1964). *An Introduction to Numerical Linear Algebra*, Clarendon Press, Oxford. (Chapter 8 on iterative and gradient methods.)

Frankel, S. P. (1950). 'Convergence rates of iterative treatments of partial differential equations'. *Math. Tables Aids Comput.*, 4, 65–77. (Successive over-relaxation.)

Golub, G. H., and Varga, R. S. (1961). 'Chebyshev semi-iterative methods, successive over-relaxation iterative methods and second order Richardson iterative methods'. *Num. Math.*, **3**, 147–156 and 157–168.

Hestenes, M. R., and Stiefel, E. (1952). 'Methods of conjugate gradients for solving linear systems'. *J. Res. Nat. Bur. Standards*, **49**, 409–436.

Jennings, A. (1971). 'Accelerating the convergence of matrix iterative processes'. *J. Inst. Maths. Applics.*, **8**, 99–110.

Martin, D. W., and Tee, G. J. (1961). 'Iterative methods for linear equations with symmetric positive definite matrix'. *The Computer Journal*, **4**, 242–254. (A comparative study.)

Peaceman, D. W., and Rachford, H. H. (1955). 'The numerical solution of parabolic and elliptic differential equations'. *SIAM Journal*, **3**, 28–41. (Alternating direction implicit procedure.)

Reid, J. K. (1971). 'On the method of conjugate gradients for the solution of large sparse systems of linear equations'. In J. K. Reid (Ed.) *Large Sparse Sets of Linear Equations*, Academic Press, London.

Schwarz, H. R., Rutishauser, H., and Stiefel, E. (1973). *Numerical Analysis of Symmetric Matrices*, English translation: Prentice-Hall, Englewood Cliffs, New Jersey. (Chapter 2 on relaxation methods.)

Sheldon, J. W. (1955). 'On the numerical solution of elliptic difference equations'. *Math. Tables Aids Comput.*, **12**, 174–186. (Double-sweep SOR.)

Stewart, G. W. (1975). 'The convergence of the method of conjugate gradients at isolated extreme points of the spectrum'. *Num. Math.*, **24**, 85–93.

Stone, H. L. (1968). 'Iterative solution of implicit approximations of multi-dimensional partial differential equations'. *SIAM J. Numer. Anal.*, **5**, 530–558. (Strongly implicit procedure.)

Varga, R. S. (1962). *Matrix Iterative Analysis*, Prentice-Hall, Englewood Cliffs, New Jersey.

Varga, R. S. (1972). 'Extensions of the successive overrelaxation theory with applications to finite element approximations'. In J. J. H. Millar (Ed.), *Topics in Numerical Analysis*, *Proc. Royal Irish Academy*, Academic Press, London.

Wachspress, E. L., and Harbetler, G. J. (1960). 'An alternating-direction-implicit iteration technique'. *SIAM Journal*, **8**, 403–424.

Weinstein, H. G. (1969). 'Iteration procedure for solving systems of elliptic partial differential equations'. In *IBM Sparse Matrix Proceedings*, IBM Research Center, Yorktown Heights, New York. (On the SIP method.)

Young, D. (1954). 'Iterative methods for solving partial difference equations of elliptic type'. *Trans. Amer. Math. Soc.*, **76**, 92–111. (Successive over-relaxation.)

Chapter 7
Some Matrix Eigenvalue Problems

7.1 COLUMN BUCKLING

The characteristics of an eigenvalue problem are that the equations are homogeneous and hence always accommodate a trivial solution. However, at certain critical values of a parameter the equations also accommodate finite solutions for which the relative values of the variables are defined, but not their absolute values. These critical values are called *eigenvalues.*

Consider the column shown in Figure 7.1(a) which is subject to an axial compressive force P. Assume that the column is free to rotate at both ends and that the only unconstrained deflection at either end is the vertical deflection of the top (represented diagrammatically by the roller unit). The question to be investigated is whether, under ideal conditions, the column could deflect laterally.

Assuming that the column is in equilibrium with a lateral deflection $y(x)$ (Figure 7.1b) for which y is small compared with the length l. It can be shown that, if the column has a bending stiffness EI, the following differential equation and end conditions must be satisfied:

$$\left.\begin{aligned} &\frac{\mathrm{d}^2y}{\mathrm{d}x^2} + \frac{Py}{EI} = 0 \\ &y = 0 \text{ at } x = 0 \\ &y = 0 \text{ at } x = l \end{aligned}\right\} \tag{7.1}$$

If the bending stiffness is uniform the general solution of the differential equation

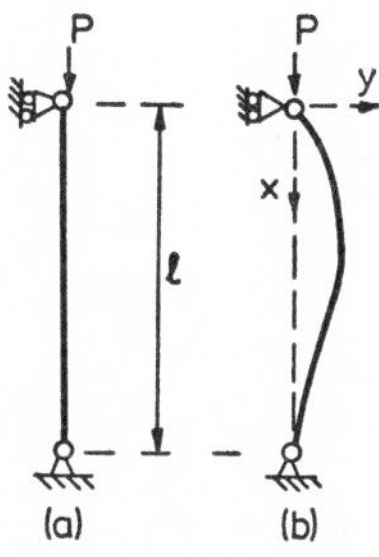

Figure 7.1 Axially loaded column without and with lateral deflection

is

$$y = a \sin\left\{\sqrt{\left(\frac{P}{EI}\right)}x\right\} + b \cos\left\{\sqrt{\left(\frac{P}{EI}\right)}x\right\} \tag{7.2}$$

and in order to satisfy the end conditions

$$\left.\begin{aligned} &b = 0 \\ \text{and} \quad &a \sin\left\{\sqrt{\left(\frac{Pl^2}{EI}\right)}\right\} = 0 \end{aligned}\right\} \tag{7.3}$$

Examination of the last condition reveals that either $a = 0$, in which case the column does not deflect laterally, or

$$P = \frac{n^2\pi^2 EI}{l^2} \tag{7.4}$$

where n is a positive integer, in which case it can adopt any value subject only to the small deflection assumption.

Thus in the buckling equations (7.1) the trivial solution $y = 0$ exists, but also at the buckling loads given by equation (7.4) non-trivial solutions exist of the form

$$y = a \sin\left(\frac{n\pi x}{l}\right) \tag{7.5}$$

The shapes of the buckling modes associated with the three lowest buckling loads are illustrated in Figure 7.2. In practice the structural engineer is rarely interested in any but the lowest buckling load as it would be impossible to load the column beyond this point unless extra lateral restraint is supplied.

The buckling problem becomes a matrix eigenvalue problem if a numerical rather than an analytical solution is sought. For instance, if a finite difference approximation is used, having a spacing between displacement nodes of b

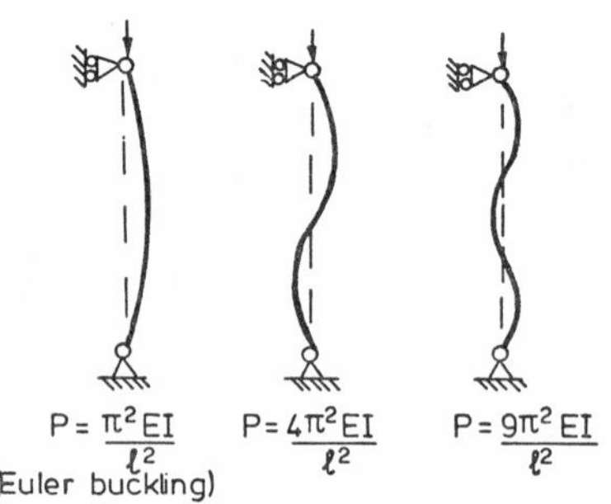

Figure 7.2 The lowest three buckling loads of a pin-ended column

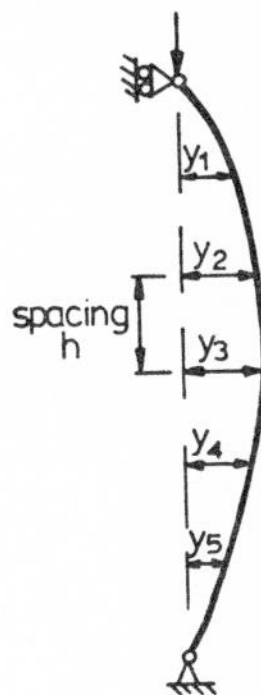

Figure 7.3 Finite difference representation for column buckling

(Figure 7.3), the differential equation of (7.1) is converted to

$$\frac{1}{h^2}(-y_{i-1} + 2y_i - y_{i+1}) = \frac{P}{EI} y_i \tag{7.6}$$

If $h = l/6$, so that $y_0 = y_6 = 0$ specifies the end conditions, and $\lambda = Ph^2/EI$, the full set of finite difference equations appears in standard matrix eigenvalue form as

$$\begin{bmatrix} 2 & -1 & & & \\ -1 & 2 & -1 & & \\ & -1 & 2 & -1 & \\ & & -1 & 2 & -1 \\ & & & -1 & 2 \end{bmatrix} \begin{bmatrix} y_1 \\ y_2 \\ y_3 \\ y_4 \\ y_5 \end{bmatrix} = \lambda \begin{bmatrix} y_1 \\ y_2 \\ y_3 \\ y_4 \\ y_5 \end{bmatrix} \tag{7.7}$$

Since the matrix is symmetric and positive definite, the eigenvalues must all be real and positive. The five eigenvalues computed from this matrix are 0.2679, 1.0000, 2.0000, 3.0000 and 3.7320, which give loads less than the lowest five buckling loads by amounts of 2.3, 8.8, 18.9, 31.6 and 45.5 per cent. respectively, the accuracy falling off for the higher modes as the finite difference approximation becomes less valid.

Although the Euler classical solution is available, and hence there is no need to perform a numerical solution of this problem, the numerical solution can easily be extended to configurations for which no classical solutions exist, such as columns with various non-uniform bending stiffnesses or with axial load varying with x.

7.2 STRUCTURAL VIBRATION

Figure 7.4(a) shows a cantilever beam which is supporting three concentrated masses. This could represent, for instance, a very simple idealization of an aircraft wing.

Subject to static lateral forces p_1, p_2 and p_3 acting at the mass positions, the

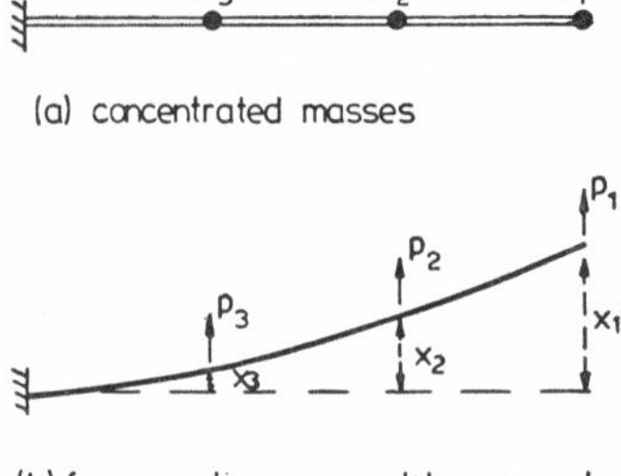

(a) concentrated masses

(b) forces acting on cantilever and corresponding displacements

Figure 7.4 Vibration of a cantilever cantilever with concentrated masses

cantilever will yield deflections x_1, x_2 and x_3 at the mass points of

$$\left.\begin{aligned} &\begin{bmatrix} x_1 \\ x_2 \\ x_3 \end{bmatrix} = \begin{bmatrix} f_{11} & f_{12} & f_{13} \\ f_{21} & f_{22} & f_{23} \\ f_{31} & f_{32} & f_{33} \end{bmatrix} \begin{bmatrix} p_1 \\ p_2 \\ p_3 \end{bmatrix} \\ &\text{or} \\ &\mathbf{x} = \mathbf{F}\mathbf{p} \end{aligned}\right\} \tag{7.8}$$

where **F** is a flexibility matrix of the cantilever, provided that the cantilever is loaded within the linear range (Figure 7.4(b)). In the absence of any externally applied force, the force acting on a particular mass to cause it to accelerate and the corresponding force acting on the cantilever at that point must sum to zero. Hence each force acting on the cantilever will be the *reversed inertia force* appropriate to the particular mass according to Newton's second law of motion, i.e.

$$\left.\begin{aligned} &\begin{bmatrix} p_1 \\ p_2 \\ p_3 \end{bmatrix} = -\begin{bmatrix} m_1 & & \\ & m_2 & \\ & & m_3 \end{bmatrix} \begin{bmatrix} \ddot{x}_1 \\ \ddot{x}_2 \\ \ddot{x}_3 \end{bmatrix} \\ &\text{or} \\ &\mathbf{p} = -\mathbf{M}\ddot{\mathbf{x}} \end{aligned}\right\} \tag{7.9}$$

where **M** is a diagonal matrix of masses and $\ddot{\mathbf{x}}$ is a column vector of accelerations.

Substituting for **p** in the second equation of (7.8) gives

$$\mathbf{F}\mathbf{M}\ddot{\mathbf{x}} + \mathbf{x} = \mathbf{0} \tag{7.10}$$

This is a set of homogeneous equations with the trivial solution that $\mathbf{x} = \mathbf{0}$. However, the general solution is obtained by assuming that the beam oscillates according to

$$\left.\begin{aligned} x_i &= \bar{x}_i \sin(\omega t + \epsilon) \\ \text{i.e.} \quad \mathbf{x} &= \sin(\omega t + \epsilon)\bar{\mathbf{x}} \\ \text{and} \quad \ddot{\mathbf{x}} &= -\omega^2 \sin(\omega t + \epsilon)\bar{\mathbf{x}} \end{aligned}\right\} \tag{7.11}$$

where $\bar{\mathbf{x}}$ is a column vector of maximum displacements of mass points. Substituting the above in equation (7.10) gives

$$\mathbf{FM}\bar{\mathbf{x}} = \frac{1}{\omega^2}\bar{\mathbf{x}} \tag{7.12}$$

Thus the possible values of $1/\omega^2$ are given by the eigenvalues of the *dynamical matrix* **FM**. Since ω is the circular frequency, the lowest frequency of vibration corresponds to the largest eigenvalue of the dynamical matrix.

The eigenvalue solution implies that steady vibration at one of several frequencies may take place if the beam has been previously disturbed in a suitable way. Theoretically the beam will continue vibrating indefinitely, but in reality, friction, air resistance and hysteresis of the material cause the vibration to damp down. The eigenvectors define the maximum values of each of the variables (which occur at the same instant of time in each oscillation). Only the relative values of the elements in each eigenvector are physically meaningful, because the absolute amplitude and phase angle ϵ of the vibration will depend on the nature and magnitude of the preceding disturbance.

As an example, consider that Figure 7.4 represents a uniform cantilever of total length 15 m which has a bending stiffness of 426 x 10^6 Nm^2 and masses m_1, m_2 and m_3 of 200, 400 and 400 kg respectively placed at 5 m spacing. From the bending stiffness it is possible to evaluate the static flexibility equations as

$$\left.\begin{aligned} x_1 &= 2.641p_1 + 1.369p_2 + 0.391p_3 \\ x_2 &= 1.369p_1 + 0.782p_2 + 0.245p_3 \\ x_3 &= 0.391p_1 + 0.245p_2 + 0.098p_3 \end{aligned}\right\} \tag{7.13}$$

the coefficients being in units of 10^{-6} m/N. Hence in these units

$$\mathbf{F} = \begin{bmatrix} 2.641 & 1.369 & 0.391 \\ 1.369 & 0.782 & 0.245 \\ 0.391 & 0.245 & 0.098 \end{bmatrix} \tag{7.14}$$

Multiplying the flexibility and mass matrices gives

$$\mathbf{FM} = \begin{bmatrix} 528.2 & 547.6 & 156.4 \\ 273.8 & 312.8 & 98.0 \\ 78.2 & 98.0 & 39.1 \end{bmatrix} \tag{7.15}$$

and, since 1 N = 1 kgm/sec^2, the elements of **FM** are in 10^{-6} sec^2 units.

Eigenvalue analysis of matrix (7.15) gives

$$\left.\begin{aligned} \lambda_1 &= 849.2 & q_1 &= \{1 \quad 0.5400 \quad 0.1619\} \\ \lambda_2 &= 26.83 & q_2 &= \{1 \quad -0.7061 \quad -0.7333\} \\ \lambda_3 &= 4.056 & q_3 &= \{0.4521 \quad -0.7184 \quad 1\} \end{aligned}\right\} \tag{7.16}$$

Because the units of λ are the same as of the elements of **FM**, the relationship $\lambda = 1/\omega^2$ gives the circular frequencies of the cantilever as $\omega_1 = 34.32$ rad/sec, $\omega_2 = 193.1$ rad/sec and $\omega_3 = 496.5$ rad/sec. The corresponding vibration modes obtained from the three eigenvectors are shown in Figure 7.5.

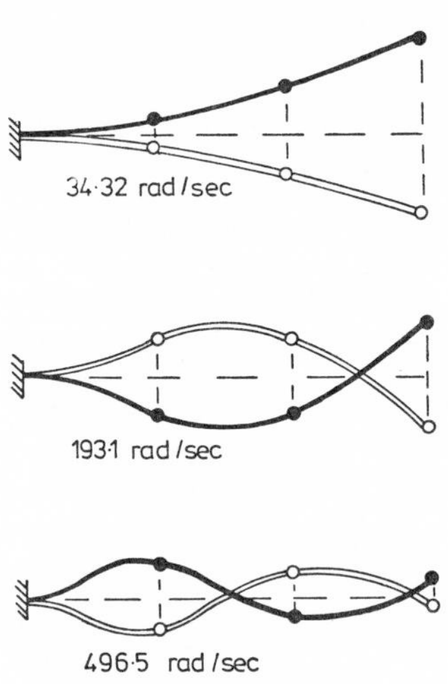

Figure 7.5 Vibration modes of the cantilever

It will be noticed that, although the matrices **F** and **M** are both symmetric, the matrix requiring eigensolution is not necessarily symmetric. This may be remedied by using the transformation

$$\bar{y}_i = \sqrt{(m_i)}\bar{x}_i \tag{7.17}$$

in which case the following symmetric form is obtained

$$\begin{bmatrix} \sqrt{(m_1)} & & \\ & \sqrt{(m_2)} & \\ & & \sqrt{(m_3)} \end{bmatrix} \begin{bmatrix} & & \\ & \mathbf{F} & \\ & & \end{bmatrix} \begin{bmatrix} \sqrt{(m_1)} & & \\ & \sqrt{(m_2)} & \\ & & \sqrt{(m_3)} \end{bmatrix} \begin{bmatrix} \bar{y}_1 \\ \bar{y}_2 \\ \bar{y}_3 \end{bmatrix} = \frac{1}{\omega^2} \begin{bmatrix} \bar{y}_1 \\ \bar{y}_2 \\ \bar{y}_3 \end{bmatrix} \tag{7.18}$$

In practice a concentrated mass system will be an idealization of a problem in which mass is distributed either evenly or unevenly throughout the system. The error caused by this idealization will most affect the highest computed frequencies, and so the number of concentrated masses should normally be chosen to be much larger than the number of required frequencies of vibration in order to give results of acceptable accuracy.

7.3 LINEARIZED EIGENVALUE PROBLEMS

In the vibration analysis of more complex structures two factors affect the form of the equations obtained:

(a) It is normally more convenient to represent the load–deflection characteristics of the structure as a stiffness matrix rather than a flexibility matrix. If $\mathbf{K}$ is the stiffness matrix associated with a set of displacements $\mathbf{x}$ then

$$\mathbf{p} = \mathbf{K}\mathbf{x} \tag{7.19}$$

It follows that, if the same displacements are chosen as for equation (7.8),

$$\mathbf{K} = \mathbf{F}^{-1} \tag{7.20}$$

By choosing sufficient displacements to make the structure *kinematically determinate*, it is not necessary to invert the flexibility matrix. Instead, the stiffness matrix can be constructed directly by adding the member contributions, the process having similarities with the construction of the coefficient matrices of equations (2.9) and (2.61). Equation (7.11) becomes

$$\mathbf{M}\ddot{\mathbf{x}} + \mathbf{K}\mathbf{x} = \mathbf{0} \tag{7.21}$$

and the eigenvalue equation (7.12) yields

$$\mathbf{M}\bar{\mathbf{x}} = \frac{1}{\omega^2}\mathbf{K}\bar{\mathbf{x}} \tag{7.22}$$

The stiffness matrix will normally be symmetric, positive definite and sparse, although for a structure which is free to move bodily, such as an aircraft or ship, the stiffness matrix will also have zero eigenvalues and hence only be positive semidefinite.

(b) In cases where the concentrated mass idealization is not very suitable, assumptions of distributed mass may be made which lead to a symmetric, positive definite mass matrix. This matrix usually has a sparse form corresponding roughly to the sparse form of the stiffness matrix.

In the buckling analysis of more complex structures, equations of the form

$$\mathbf{G}\mathbf{x} = \frac{1}{\theta}\mathbf{K}\mathbf{x} \tag{7.23}$$

may be derived. $\mathbf{K}$ is the structural stiffness matrix associated with the lateral deflections involved in the buckling deformation. The matrix **G** represents the modification to the stiffness matrix due to the presence of the primary loading (i.e. the axial load in the case of a column, or in-plane forces for a plate), the load parameter θ having the effect of scaling the whole primary loading. Whereas the matrix **K** will normally be symmetric, positive definite and sparse, the matrix **G** is likely to be symmetric and sparse, but may not be positive definite.

Both equations (7.22) and (7.23) are of the form given by

$$\mathbf{Ax} = \lambda\mathbf{Bx} \tag{7.24}$$

and may be described as linearized eigenvalue problems.

7.4 SOME PROPERTIES OF LINEARIZED EIGENVALUE PROBLEMS

The following properties appertain to the equations $\mathbf{Ax} = \lambda\mathbf{Bx}$.

(a) The roles of **A** and **B** can be interchanged. If this is done the eigenvalues will be changed to their reciprocals but the eigenvectors will remain unaltered, viz.:

$$\mathbf{Bx} = \frac{1}{\lambda}\mathbf{Ax} \tag{7.25}$$

(b) If **B** is non-singular the equations reduce to the standard eigenvalue form

$$(\mathbf{B}^{-1}\mathbf{A})\mathbf{x} = \lambda\mathbf{x} \tag{7.26}$$

(c) If **A** and **B** are real and symmetric, the eigenvalues will be real and the eigenvectors will be orthogonal with respect to both **A** and **B**.

Premultiplying equation (7.24) by $\mathbf{x}^H$ gives

$$\mathbf{x}^H\mathbf{Ax} = \lambda\mathbf{x}^H\mathbf{Bx} \tag{7.27}$$

and forming the Hermitian transpose of equation (7.27) gives

$$\mathbf{x}^H\mathbf{Ax} = \lambda^*\mathbf{x}^H\mathbf{Bx} \tag{7.28}$$

where λ^* is the complex conjugate of λ. From these two equations, $\lambda = \lambda^*$, and hence λ must be real (it can also be shown that the eigenvectors are real).

The condition that the eigenvectors are orthogonal with respect to **A** and **B** is defined according to

$$\mathbf{x}_i^T\mathbf{Ax}_j = \mathbf{x}_i^T\mathbf{Bx}_j = 0 \quad \text{for } i \neq j \tag{7.29}$$

This may be deduced from the eigenvalue equations

$$\left.\begin{aligned} &\mathbf{Ax}_i = \lambda_i\mathbf{Bx}_i \\ \text{and} \\ &\mathbf{Ax}_j = \lambda_j\mathbf{Bx}_j \end{aligned}\right\} \tag{7.30}$$

by premultiplying them by $\mathbf{x}_j^T$ and $\mathbf{x}_i^T$ respectively, and transposing the first equation to give

$$\left.\begin{aligned} \mathbf{x}_i^T\mathbf{A}\mathbf{x}_j = \lambda_i\mathbf{x}_i^T\mathbf{B}\mathbf{x}_j \\ \mathbf{x}_i^T\mathbf{A}\mathbf{x}_j = \lambda_j\mathbf{x}_i^T\mathbf{B}\mathbf{x}_j \end{aligned}\right\} \tag{7.31}$$

The orthogonality condition then follows immediately provided that $\lambda_i \neq \lambda_j$. When $\lambda_i = \lambda_j$ it is possible to choose eigenvectors which obey this property.

(d) If **A** and **B** are both real symmetric and **B** is positive definite, the equations may be transformed to standard symmetric eigenvalue form.

Since **B** is symmetric and positive definite it may be reduced by Choleski decomposition to

$$\mathbf{B} = \mathbf{L}\mathbf{L}^T \tag{7.32}$$

where **L** is a lower triangular matrix. Premultiplying equation (7.24) by $\mathbf{L}^{-1}$ gives

$$(\mathbf{L}^{-1}\mathbf{A}\mathbf{L}^{-T})(\mathbf{L}^T\mathbf{x}) = \lambda(\mathbf{L}^T\mathbf{x}) \tag{7.33}$$

Hence λ values satisfying equation (7.24) must be eigenvalues of the symmetric matrix $\mathbf{L}^{-1}\mathbf{A}\mathbf{L}^{-T}$, the eigenvectors of which take the form $\mathbf{L}^T\mathbf{x}$.

(e) If both **A** and **B** are real, symmetric and positive definite the eigenvalues will all be positive.

Premultiplying equation (7.24) by $\mathbf{x}^T$ gives

$$\mathbf{x}^T\mathbf{A}\mathbf{x} = \lambda\mathbf{x}^T\mathbf{B}\mathbf{x} \tag{7.34}$$

but as $\mathbf{x}^T\mathbf{A}\mathbf{x}$ and $\mathbf{x}^T\mathbf{B}\mathbf{x}$ must both be positive, λ must be positive.

(f) If **A** is singular $\lambda = 0$ must be a possible solution and if **B** is singular $\lambda \to \infty$ must be a solution.

If **A** is singular and **x** is the eigenvector corresponding to the zero eigenvalue, $\mathbf{A}\mathbf{x} = \mathbf{0}$. Hence $\lambda = 0$ must satisfy equation (7.24) with a non-null vector **x**. The condition for **B** being singular can be deduced by invoking property (a).

(g) If $\bar{\mathbf{B}} = \mathbf{B} + \mathbf{A}/\alpha$, where α is any appropriate scalar, the modified eigenvalue problem

$$\mathbf{A}\mathbf{x} = \bar{\lambda}\bar{\mathbf{B}}\mathbf{x} \tag{7.35}$$

has eigenvalues related to the original eigenvalues according to

$$\frac{1}{\bar{\lambda}} = \frac{1}{\lambda} + \frac{1}{\alpha} \tag{7.36}$$

This modification may be useful where both **A** and **B** are singular. In most cases $\bar{\mathbf{B}}$ will be non-singular, so allowing the problem to be solved by equation (7.26) or, where $\bar{\mathbf{B}}$ is symmetric and positive definite, by equation (7.33).

The above properties of $\mathbf{A}\mathbf{x} = \lambda\mathbf{B}\mathbf{x}$ can be used to transform a problem into

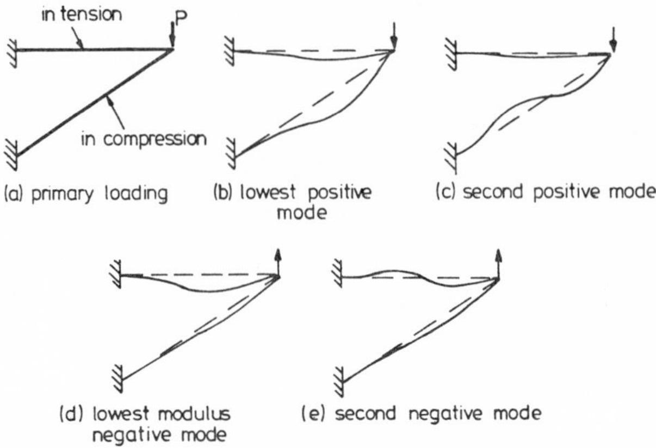

Figure 7.6 Buckling modes of a simple stiff-jointed frame

standard eigenvalue form, maintaining symmetry where possible. However, in cases where **A** and **B** are large and sparse, the matrix of the standard eigenvalue form is likely to be fully populated and its explicit derivation would be numerically cumbersome. Hence special consideration to problems of this form will be given in the discussion of numerical procedures.

Consider the vibration equation (7.22) in which both the mass and stiffness matrices are symmetric and positive definite. It follows that the eigenvalues $\lambda = 1/\omega^2$ will be real and positive, and hence no extraneous solutions should be obtained. In the case of the buckling equations (7.23) the eigenvalues must be real, but if **G** is not positive definite, negative as well as positive eigenvalues $\lambda = 1/\theta$ will be obtained. These negative eigenvalues will arise when the structure can buckle with the primary loading reversed. This could occur, for instance, with the stiff-jointed frame of Figure 7.6(a), the inclined member of which is loaded in compression and the horizontal member in tension under the specified (primary) load P. Figures 7.6(b) and (c) indicate the first two buckling modes with the primary load positive and Figures 7.6(d) and (e) indicate the first two buckling modes with primary load negative. The possible presence of unwanted negative eigenvalues due to reversed loading are of particular significance to vector iterative methods of determining the eigenvalues (Chapter 10).

7.5 DAMPED VIBRATION

A damped structural free vibration in which the damping forces are viscous (i.e. proportional to velocity) may be written in the form

$$\mathbf{A}\ddot{\mathbf{x}} + \mathbf{B}\dot{\mathbf{x}} + \mathbf{C}\mathbf{x} = \mathbf{0} \tag{7.37}$$

where **A** represents the mass matrix, **B** the damping matrix and **C** the stiffness matrix. (It may be noticed that, if the damping matrix is null, the equations are in the form of the undamped vibration equations 7.21.) The solution of equations of this form may be obtained using the substitution

$$\mathbf{x} = e^{\lambda t}\mathbf{y} \tag{7.38}$$

which gives the homogeneous equations

$$(\lambda^2\mathbf{A} + \lambda\mathbf{B} + \mathbf{C})\mathbf{y} = \mathbf{0} \tag{7.39}$$

The feasible values for λ satisfy the determinantal equation

$$|\lambda^2\mathbf{A} + \lambda\mathbf{B} + \mathbf{C}| = 0, \tag{7.40}$$

the characteristic equation of which is a real polynomial of order $2n$. The $2n$ possible roots of this polynomial may either be real or be twinned in complex conjugate pairs.

For a real root λ the associated characteristic motion is either an exponential convergence or divergence, as shown in Figure 7.7. For a complex conjugate pair $\lambda = \mu + i\omega$, $\lambda^* = \mu - i\omega$, the corresponding modal vectors take the form $\mathbf{y} = \mathbf{p} + i\mathbf{q}$ and $\bar{\mathbf{y}} = \mathbf{p} - i\mathbf{q}$. The characteristic motion associated with the pair of roots can be expressed with arbitrary coefficients a and b as

$$\mathbf{x} = (a + ib)e^{(\mu + i\omega)t}(\mathbf{p} + i\mathbf{q}) + (a - ib)e^{(\mu - i\omega)t}(\mathbf{p} - i\mathbf{q}) \tag{7.41}$$

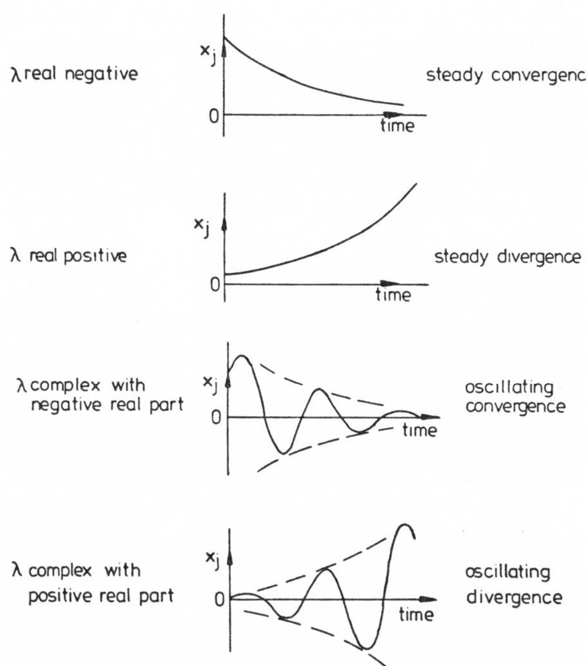

Figure 7.7 Movement of dispacement x_j for different types of characteristic motion

which may be reduced to

$$\left.\begin{aligned} &\mathbf{x} = e^{\mu t}(\sin(\omega t)\,\mathbf{u} + \cos(\omega t)\,\mathbf{v}) \\ &\text{where} \\ &\mathbf{u} = 2(a\mathbf{p} - b\mathbf{q}) \quad \text{and} \quad \mathbf{v} = -2(b\mathbf{p} - a\mathbf{q}) \end{aligned}\right\} \tag{7.42}$$

The motion represented by one of the variables x_i is therefore a sinusoidal oscillation of circular frequency ω whose amplitude is either decreasing or increasing exponentially with time according to $e^{\mu t}$, as shown in Figure 7.7. Other variables will have the same frequency and decay or attenuation of amplitude, but all of the variables will not be in phase since the vectors **u** and **v** cannot be proportional to each other.

In the case of a structural vibration which is damped by the presence of the surrounding air or a small amount of hysteresis in the material, the characteristic modes would consist of convergent oscillations. However, where damping is strong steadily convergent modes could be obtained.

7.6 DYNAMIC STABILITY

Where there is an external source of energy both steady and oscillating divergence are possible. To prove that the system under investigation is stable it is necessary to ensure that all of the eigenvalues occur in the negative half-plane of the Argand diagram. An indication of the degree of stability is given by the maximum real component in any of the roots.

Dynamic instability can occur with structures subject to wind loading. One form of structure which has to be designed in such a way that dynamic instability cannot occur is the suspension bridge, although most work in this field has been carried out by aircraft designers where the relative air speed produces, effectively, a very strong wind acting on the structure.

In the classical British method of flutter analysis the dynamic equations appear in the form of equation (7.37) in which

$$\left.\begin{aligned} \mathbf{B} &= \mathbf{B}_S + \rho V \mathbf{B}_A \\ \mathbf{C} &= \mathbf{C}_S + \rho V^2 \mathbf{C}_A \end{aligned}\right\} \tag{7.43}$$

where $\mathbf{B}_S$ and $\mathbf{C}_S$ are structural damping and stiffness matrices, $\mathbf{B}_A$ and $\mathbf{C}_A$ are aerodynamic damping and stiffness matrices, ρ is the air density and V is the air speed. For a particular flight condition of an aircraft **B** and **C** may be evaluated and the stability investigated through the solution of equation (7.37). It is possible for instability to occur at a certain flight speed even though higher flight speeds do not produce instability. Hence it may be necessary to carry out a thorough search of the whole speed and altitude range of an aircraft to ensure stability at all times. If an aircraft is designed to fly at sonic or supersonic speeds the aerodynamic matrices are also affected by the compressibility of the air. Because of the large number of analyses that need to be performed, it is normal to use as few variables as possible.

Often the normal modes of undamped structural vibration are used as coordinates, so that for the analysis of a complete aircraft only about thirty variables may be required.

7.7 REDUCTION OF THE QUADRATIC EIGENVALUE PROBLEM TO STANDARD FORM

If **A** is non-singular it is possible to transform the quadratic eigenvalue problem as specified by equation (7.39) to the form

$$(\lambda^2\mathbf{I} + \lambda\mathbf{A}^{-1}\mathbf{B} + \mathbf{A}^{-1}\mathbf{C})\mathbf{y} = \mathbf{0} \tag{7.44}$$

By introducing the additional variable $\mathbf{z} = \lambda\mathbf{y}$ this equation can be specified in standard eigenvalue form as

$$\begin{bmatrix} -\mathbf{A}^{-1}\mathbf{B} & -\mathbf{A}^{-1}\mathbf{C} \\ \mathbf{I} & \mathbf{0} \end{bmatrix}\begin{bmatrix} \mathbf{z} \\ \mathbf{y} \end{bmatrix} = \lambda\begin{bmatrix} \mathbf{z} \\ \mathbf{y} \end{bmatrix} \tag{7.45}$$

Furthermore, the roles of **A** and **C** may be reversed if $\bar{\lambda} = 1/\lambda$ is substituted for λ.

An alternative reduction which may be adopted, if for instance both **A** and **C** are singular, is obtained by firstly transforming the equations to the auxiliary parameter μ where

$$\mu = \frac{\theta + \lambda}{\theta - \lambda} \tag{7.46}$$

In this case $\lambda = \theta(\mu - 1)/(\mu + 1)$, giving

$$[\mu^2(\theta^2\mathbf{A} + \theta\mathbf{B} + \mathbf{C}) + 2\mu(\mathbf{C} - \theta^2\mathbf{A}) + \theta^2\mathbf{A} - \theta\mathbf{B} + \mathbf{C}]\,\mathbf{y} = \mathbf{0} \tag{7.47}$$

Thus, writing $\theta^2\mathbf{A} + \theta\mathbf{B} + \mathbf{C} = \mathbf{S}$, the standard form becomes

$$\begin{bmatrix} -2\mathbf{S}^{-1}(\mathbf{C} - \theta^2\mathbf{A}) & -\mathbf{S}^{-1}(\theta^2\mathbf{A} - \theta\mathbf{B} + \mathbf{C}) \\ \mathbf{I} & \mathbf{0} \end{bmatrix}\begin{bmatrix} \bar{\mathbf{z}} \\ \mathbf{y} \end{bmatrix} = \mu\begin{bmatrix} \bar{\mathbf{z}} \\ \mathbf{y} \end{bmatrix} \tag{7.48}$$

where $\bar{\mathbf{z}} = \mu\mathbf{y}$. A property of this eigenvalue equation is that stable modes are such that $|\mu| < 1$ whereas unstable modes have $|\mu| < 1$ if a positive choice has been made for θ. Hence the magnitude of the eigenvalue of largest modulus defines whether the system is stable or not. The parameter θ may be used to regulate the relative magnitudes of the contributions of **A**, **B** and **C** to **S**.

7.8 PRINCIPAL COMPONENT ANALYSIS

Principal component analysis is a means of extracting the salient features of a mass of data. Consider the data given in Table 7.1 which could represent the examination percentage scores of twenty students who each sit six subjects. Without some form of mathematical analysis it would be difficult to draw any reliable conclusions about the presence or absence of correlation in the results.

The first step in analysing this information is to subtract from each column its own average value. The resulting table may be represented by the data matrix

$$
\mathbf{X} = \begin{bmatrix}
-2.6 & -8.7 & -11.65 & -14.55 & -15 & -19.8 \\
-35.6 & -24.7 & -12.65 & -16.55 & -21 & -27.8 \\
7.4 & -2.7 & -3.65 & 9.45 & -3 & 1.2 \\
-3.6 & 16.3 & -6.65 & 3.45 & 10 & -5.8 \\
11.4 & 9.3 & -0.65 & -6.55 & 7 & 11.2 \\
-17.6 & 1.3 & -3.65 & -15.55 & 5 & -2.8 \\
-1.6 & 13.3 & -4.65 & 4.45 & 3 & 7.2 \\
4.4 & 15.3 & 7.35 & -7.55 & 12 & 10.2 \\
46.4 & 23.3 & 5.35 & 26.45 & 20 & 17.2 \\
-16.6 & 6.3 & 28.35 & 14.45 & 7 & 2.2 \\
-9.6 & -20.7 & -12.65 & -7.55 & -16 & -6.8 \\
-16.6 & -14.7 & -18.65 & -4.55 & -12 & -5.8 \\
-15.6 & -23.7 & -5.65 & -19.55 & -9 & -14.8 \\
17.4 & -5.7 & 5.35 & -4.55 & 11 & 10.2 \\
25.4 & -5.7 & 7.35 & 10.45 & 0 & -8.8 \\
13.4 & 18.3 & 13.35 & 13.45 & 22 & 16.2 \\
-3.6 & 9.3 & 2.35 & 9.45 & -6 & 7.2 \\
15.4 & -3.7 & 15.35 & 2.45 & -6 & -2.8 \\
-9.6 & 17.3 & 15.35 & 11.45 & 10 & 9.2 \\
-8.6 & -19.7 & -19.65 & -8.55 & -19 & 3.2
\end{bmatrix} \tag{7.49}
$$

In general, if there are n variables specified for each of m objects, $\mathbf{X}$ is of order $m \times n$. In the particular example the variables are the different examination subjects and the objects are the students.

The next step in analysing the information is to form the *covariance matrix* according to the formula

$$
\mathbf{C} = \frac{1}{m-1} \mathbf{X}^T \mathbf{X} \tag{7.50}
$$

The variance for variable i is equal to the i-th diagonal element, i.e.

$$
c_{ii} = \frac{1}{m-1} \sum_{k=1}^{m} x_{ki}^2 \tag{7.51}
$$

and the covariance for variables i and j is equal to the off-diagonal element

$$
c_{ij} (= c_{ji}) = \frac{1}{m-1} \sum_{k=1}^{m} x_{ki} x_{kj} \tag{7.52}
$$

Table 7.1 Examination percentages

Student	Maths.	English	Physics	Geography	French	Art
			Subject			
1	50	45	41	45	46	30
2	17	29	40	43	40	22
3	60	51	49	69	58	51
4	49	70	46	63	71	44
5	64	63	52	53	68	61
6	35	55	49	44	66	47
7	51	67	48	64	64	57
8	57	69	60	52	73	60
9	99	77	58	86	81	67
10	36	60	81	74	68	52
11	43	33	40	52	45	43
12	36	39	34	55	49	44
13	37	30	47	40	52	35
14	70	48	58	55	72	60
15	78	48	60	70	61	41
16	66	72	66	73	83	66
17	49	63	55	69	55	57
18	68	50	68	62	55	47
19	43	71	68	71	71	59
20	44	34	33	51	42	53
Average	52.6	53.7	52.65	59.55	61.0	49.8

The covariance gives a measure of the correlation between the data for variables i and j. For the given example the covariance matrix is

$$\frac{1}{m-1}\mathbf{X}^T\mathbf{X} = \begin{bmatrix} 336.15 & & & & & \\ 139.77 & 233.38 & & \text{symmetric} & & \\ 83.12 & 111.47 & 154.66 & & & \\ 137.76 & 130.54 & 93.89 & 154.05 & & \\ 134.53 & 168.42 & 104.84 & 94.95 & 161.58 & \\ 127.44 & 135.09 & 71.14 & 91.43 & 114.32 & 141.43 \end{bmatrix} \tag{7.53}$$

relevant features of which are as follows:

(a) The largest diagonal element occurs in the first row. Hence the mathematics results have the largest variance.

(b) All of the off-diagonal elements are positive, signifying that there is positive correlation between the results in all of the subjects, i.e. there is a tendency for students doing well/badly in one subject to do well/badly in the others.

(c) The largest off-diagonal element appears at position (5, 2), showing that the strongest correlation is between the results in English and French.

Whereas the covariance matrix and the information it yields can be considered as an end in itself, it may be difficult to interpret its significance, particularly when the number of variables is large. The eigensolution of this matrix produces more information which can often be valuable. If the full set of eigenvalues of the covariance matrix is expressed as a diagonal matrix $\mathbf{\Lambda} = [\lambda_1 \quad \lambda_2 \quad \ldots \quad \lambda_n]$, where $\lambda_1 \geqslant \lambda_2 \geqslant \cdots \geqslant \lambda_n$, and the corresponding eigenvectors are compounded into one matrix $\mathbf{Q} = [\mathbf{q}_1 \quad \mathbf{q}_2 \quad \ldots \quad \mathbf{q}_n]$, then

$$\left(\frac{1}{m-1}\mathbf{X}^T\mathbf{X}\right)\mathbf{Q} = \mathbf{Q}\mathbf{\Lambda} \tag{7.54}$$

Because $\mathbf{X}^T\mathbf{X}$ is symmetric and positive definite, its eigenvalues will be real and positive and $\mathbf{Q}$ will be an orthogonal matrix. Hence, premultiplying by $\mathbf{Q}^T$ gives

$$\frac{1}{m-1}\mathbf{Q}^T\mathbf{X}^T\mathbf{X}\mathbf{Q} = \mathbf{\Lambda} \tag{7.55}$$

Consider a matrix $\mathbf{Y}$ of order $m \times n$ such that

$$\left.\begin{aligned} &\mathbf{Y} = \mathbf{X}\mathbf{Q} \\ \text{i.e.} \\ &y_{ij} = \sum_{k=1}^{n} x_{ik}q_{kj} \end{aligned}\right\} \tag{7.56}$$

Since row i of $\mathbf{X}$ contains the performances of student i in all of the examinations, y_{ij} is the linear combination, as represented by the j-th eigenvector, of the examination performances of student i. Matrix $\mathbf{Y}$ can be interpreted as an alternative data matrix appertaining to a transformed set of variables. For the transformed variables the covariance matrix is

$$\frac{1}{m-1}(\mathbf{X}\mathbf{Q})^T(\mathbf{X}\mathbf{Q}) = \frac{1}{m-1}\mathbf{Q}^T\mathbf{X}^T\mathbf{X}\mathbf{Q} = \mathbf{\Lambda} \tag{7.57}$$

Hence the transformation has the effect of completely eliminating the correlation between the variables.

The principal components are the eigenvectors of the covariance matrix associated with the larger eigenvalues. An important property is that the total variance of the original variables (given by the trace of the covariance matrix) is equal to the total variance of the transformed variables (given by the sum of the eigenvalues):

$$\frac{1}{m-1}\sum_{i=1}^{n}\sum_{k=1}^{m} x_{ki}^2 = \sum_{i=1}^{n}\lambda_i = \frac{1}{m-1}\sum_{i=1}^{n}\sum_{k=1}^{m} y_{ki}^2 \tag{7.58}$$

The first principal component $\mathbf{q}_1$ is the linear combination of the variables for which the variance is maximum. The second principal component $\mathbf{q}_2$ also obeys this condition, but subject to the overriding condition that it is uncorrelated with $\mathbf{q}_1$. Higher order principal components may be defined in a similar way.

A complete eigensolution of the covariance matrix (7.53) is given in Table 7.2. The proportion of the total variance attributable to each of the transformed

Table 7.2 Eigensolution of covariance matrix (7.53)

i	1	2	3	4	5	6	Σ
$\lambda_i =$	799.15	178.35	87.38	58.85	41.56	15.96	1,181.25
$q_i =$	0.5181	0.8152	0.0567	0.1812	0.1200	0.1284	
	0.4743	−0.3890	−0.3735	−0.1487	0.4736	0.4877	
	0.3052	−0.3496	0.7366	0.3897	−0.1547	0.2574	
	0.3598	−0.0273	0.3711	−0.7896	−0.0014	−0.3296	
	0.4013	−0.2348	−0.2028	0.4073	0.1416	−0.7462	
	0.3506	−0.0774	−0.3686	0.0618	−0.8468	0.1186	

variables can be evaluated by dividing the eigenvalues by 1,181.25. The first eigenvector, having all of its elements of similar magnitude, describes the previously noted tendency for students performing well/badly in one subject to perform well/badly in all the others. This trend accounts for 67.7 per cent. of the total variance. The second eigenvector suggests a secondary tendency for the performance in mathematics to vary more than the performance in other subjects (15.1 per cent. of the total variance). The other eigenvectors together account for only 17.2 per cent. of the total variance and it is unlikely that any of them signify any important trends.

7.9 A GEOMETRICAL INTERPRETATION OF PRINCIPAL COMPONENT ANALYSIS

A data set appertaining to only two variables may be represented graphically. Each column of the matrix

$$\mathbf{X}^T = \begin{bmatrix} -6 & -4 & -4 & -2 & -2 & -2 & 0 & 2 & 4 & 6 & 8 \\ -4 & -6 & -2 & -6 & -3 & 3 & -2 & 7 & 2 & 3 & 8 \end{bmatrix} \tag{7.59}$$

may be plotted as a point in Cartesian coordinates, the full set of eleven columns yielding points as shown in Figure 7.8(a). The rows of $\mathbf{X}^T$ have been previously scaled so that their average is zero. Graphically this can be interpreted as shifting the origin to the centre of gravity of the group of points, with the centre of gravity evaluated on the assumption that each point carries the same weight. The covariance matrix and its eigensolution are

$$\left.\begin{aligned} &\frac{1}{m-1}\mathbf{X}^T\mathbf{X} = \begin{bmatrix} 20.0 & 17.2 \\ 17.2 & 24.0 \end{bmatrix} \\ &\lambda_1 = 39.3, \quad q_1 = \{0.8904 \quad 1\} \\ &\lambda_2 = \ \ 4.7, \quad q_2 = \{1 \quad -0.8904\} \end{aligned}\right\} \tag{7.60}$$

Using the eigenvectors to define transformed variables is equivalent to a rotation of the coordinate axes of the graph, as shown in Figure 7.8(b). In this problem there is a strong correlation between the two variables (only one point has variables of

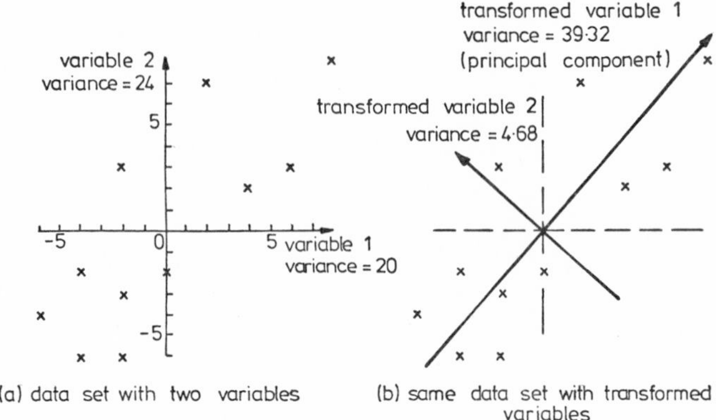

Figure 7.8 Geometrical interpretation of principal component analysis

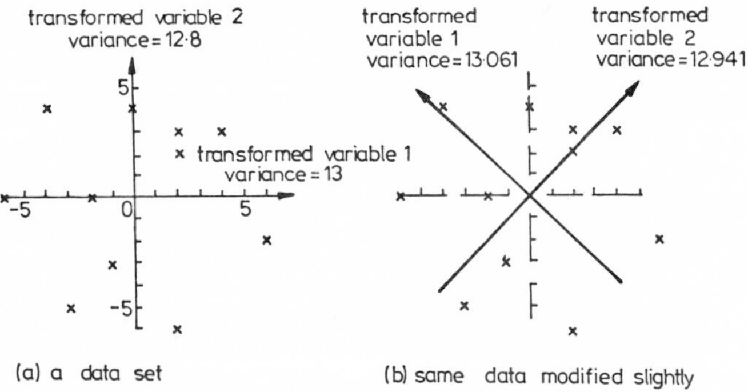

Figure 7.9 Geometrical example with no significant correlation

different sign). This correlation appears in the principal component which claims 89.3 per cent. of the total variance.

In contrast the data set

$$\mathbf{X}^T = \begin{bmatrix} -6 & -4 & -3 & -2 & -1 & 0 & 2 & 2 & 2 & 4 & 6 \\ 0 & 4 & -5 & 0 & -3 & 4 & -6 & 2 & 3 & 3 & -2 \end{bmatrix} \tag{7.61}$$

does not exhibit any correlation between the variables. As a result the transformed variables are coincident with the basic variables (Figure 7.9(a)). The covariance matrix and its eigensolution are

$$\left.\begin{aligned} &\frac{1}{m-1}\mathbf{X}^T\mathbf{X} = \begin{bmatrix} 13.0 & 0 \\ 0 & 12.8 \end{bmatrix} \\ &\lambda_1 = 13.0, \quad q_1 = \{1 \quad 0\} \\ &\lambda_2 = 12.8, \quad q_2 = \{0 \quad 1\} \end{aligned}\right\} \tag{7.62}$$

Not only are the eigenvalues close in magnitude to each other, but also the eigenvectors are sensitive to small changes in the data. For example, if 0.1 is added to element (2, 2) and subtracted from element (7, 2) of matrix **X**, the covariance matrix and its eigensolution will be modified to

$$\left.\begin{array}{l} \dfrac{1}{m-1}\mathbf{X}^T\mathbf{X} = \begin{bmatrix} 13.0 & -0.06 \\ -0.06 & 13.002 \end{bmatrix} \\ \lambda_1 = 13.061 \qquad q_1 = \{-0.983 \quad 1\ \} \\ \lambda_2 = 12.941 \qquad q_2 = \{\ 1 \quad -0.983\} \end{array}\right\} \tag{7.63}$$

Figure 7.9(b) shows the graphical interpretation of this slight modification. In this example the distribution of the points appears to be random since they do not show any significant correlation. The eigenvalues are almost equal, and because of this the eigenvectors are so sensitive to small changes in the data that they have no significant meaning.

One of the problems in principal component analysis is to distinguish correlations between the variables from the effects of random variations (or *noise*). From the previous discussion it appears that where eigenvalues are closely spaced the corresponding variance can probably be attributed to random variations, whereas if the dominant eigenvalue is well separated from the rest, or where a set of dominant eigenvalues are all well separate from the rest, their corresponding eigenvectors may contain significant correlations between the variables. Where data has been collected from a sufficiently large number of objects this may be divided into several sets. Hence if examination results similar to those in Table 7.1 are available for 600 students, three principal component analyses could be undertaken, each with the examination results from 200 of the students. Comparison of the results of the three analyses would then give a strong indication of which trends were consistently present throughout the data and which appeared to be random. Although in most applications only one or two dominant eigenvalues and corresponding eigenvectors will be required, in meteorology up to 50 components have been obtained from data sets involving over 100 variables for use in long-range weather forecasting (Craddock and Flood, 1969).

In some analyses the variables are not alike in character. For instance, if the objects are different factories, the variables could represent the number of employees, their average age, their average productivity, etc. In this case no physical interpretation can be placed on the relative values of the diagonal elements of the covariance matrix, and so the rows of $\mathbf{X}^T$ should be scaled so that the diagonal elements of the covariance matrix are equal to unit. A covariance matrix so modified is called a *correlation matrix.*

7.10 MARKOV CHAINS

Markov chains can be used for the response analysis of systems which behave spasmodically. Consider a system which can adopt any one of n possible states and which is assumed to pass through a sequence of transitions, each of which may change the state of the system according to a known set of probabilities. If the system is in state j, let the probability of changing to state i at the next transition be p_{ij}. Since the system must be in one of the states after the transition,

$$\sum_{i=1}^{n} p_{ij} = 1 \tag{7.64}$$

and, since no probability can be negative,

$$0 \leqslant p_{ij} \leqslant 1 \tag{7.65}$$

The full set of probabilities of moving from any one state to any other, or indeed of remaining static, form an $n \times n$ matrix $\mathbf{P} = [p_{ij}]$ known as a *stochastic* or *transition matrix*, each column of which must sum to unity (equation 7.64). (This matrix is often defined in transposed form.) The stochastic matrix is assumed to be independent of the transition number. Thus

$$\mathbf{P} = \begin{bmatrix} 0.4 & 0.2 & 0.3 \\ 0 & 0.2 & 0.3 \\ 0.6 & 0.6 & 0.4 \end{bmatrix} \tag{7.66}$$

could be a stochastic matrix represented by the directed graph shown in Figure 7.10.

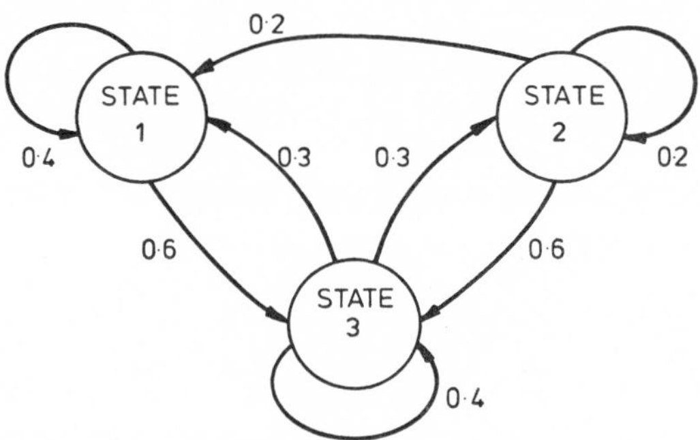

Figure 7.10 Probabilities for one transition expressed by stochastic matrix (7.66)

Defining also a sequence of probability vectors $\mathbf{x}^{(k)}$ whose typical element $x_i^{(k)}$ gives the probability of being in state i after transition k, it can be shown that

$$\sum_{i=1}^{n} x_i^{(k)} = 1 \tag{7.67}$$

and also

$$0 \leqslant x_i^{(k)} \leqslant 1 \tag{7.68}$$

Table 7.3 Markov chain for system described by stochastic matrix (7.66) and which is in state 1 initially

k	0	1	2	3	4	5	6	7	∞
x_1	1	0.4	0.34	0.316	0.3136	0.31264	0.312544	0.3125056	→ 0.3125
x_2	0	0	0.18	0.180	0.1872	0.18720	0.187488	0.1874880	→ 0.1875
x_3	0	0.6	0.48	0.504	0.4992	0.50016	0.499968	0.5000064	→ 0.5000
Σ	1	1	1	1	1	1	1	1	1

If the system described by the stochastic matrix (7.66) is initially in state 1, $\mathbf{x}^{(0)} = \{1 \quad 0 \quad 0\}$ and it follows from the definition of the stochastic matrix that $\mathbf{x}^{(1)} = \mathbf{P}\mathbf{x}^{(0)} = \{0.4 \quad 0 \quad 0.6\}$. Considering the probabilities for the next transition it can be deduced that $\mathbf{x}^{(2)} = \mathbf{P}\mathbf{x}^{(1)} = \{0.34 \quad 0.18 \quad 0.48\}$. In general the probability vector may be derived after any number of transitions according to the formula

$$\mathbf{x}^{(k+1)} = \mathbf{P}\mathbf{x}^{(k)} = \mathbf{P}^{k+1}\mathbf{x}^{(0)} \tag{7.69}$$

the sequence of such vectors being called a *Markov chain*. The Markov chain for this particular example is shown in Table 7.3 and is seen to be convergent to what may be described as a *long-run probability vector*.

The long-run probability vector ($\mathbf{q}$) is an interesting phenomenon since it must satisfy the equation

$$\mathbf{P}\mathbf{q} = \mathbf{q} \tag{7.70}$$

and hence must be a right eigenvector of the stochastic matrix for which the corresponding eigenvalue is unity. It is pertinent to ask several questions:

(a) Does every stochastic matrix have a long-run probability vector?
(b) Is the long-run probability vector independent of the initial starting vector?
(c) Is an eigenvalue analysis necessary in order to determine the long-run probability vector?
(d) How can the rate of convergence of the Markov chain be assessed?

Before discussing these questions a stochastic matrix will be developed for analysing the performance of a simple computer system.

7.11 MARKOV CHAINS FOR ASSESSING COMPUTER PERFORMANCE

Let a simple computer system be represented by a queue and two phases, the two phases being job loading and execution. It will be assumed that the computer cannot load while it is involved in execution. Hence once a job has finished loading it can always proceed immediately to the execution phase without waiting in a queue. The only queue is for jobs which arrive while another job is being either loaded or executed. If the queue is restricted to no more than two jobs, the seven possible states of the system, $s_1, \ldots, s_7$ are shown in Table 7.4.

Table 7.4 Possible states of a simple computer system

	State	s_1	s_2	s_3	s_4	s_5	s_6	s_7
Number of jobs in	queue	0	0	0	1	1	2	2
	loading phase	0	0	1	0	1	0	1
	execution phase	0	1	0	1	0	1	0

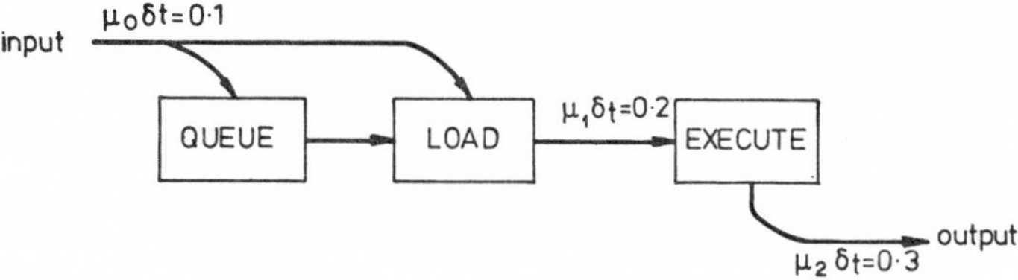

Figure 7.11 Probabilities of a change of state in time δt for a small computer system

In order to convert the problem into a Markov process it is necessary to designate a time interval δt for one transition. The possibility that more than one job movement occurs during any time interval can be precluded, provided that the latter is chosen to be sufficiently short. In the computer example three probabilities are needed to describe the behaviour of the system during one time interval:

$\mu_0\delta t$ the probability of a job arriving at the input,
$\mu_1\delta t$ the probability of a job in the loading phase being complete,
$\mu_2\delta t$ the probability of a job in the execution phase being complete.

If $\mu_0\delta t$, $\mu_1\delta t$ and $\mu_2\delta t$ take the values 0.1, 0.2 and 0.3 respectively (as shown in Figure 7.11), the stochastic equation is

$$\begin{bmatrix} x_1^{(k+1)} \\ x_2^{(k+1)} \\ x_3^{(k+1)} \\ x_4^{(k+1)} \\ x_5^{(k+1)} \\ x_6^{(k+1)} \\ x_7^{(k+1)} \end{bmatrix} = \begin{bmatrix} 0.9 & 0.3 & & & & & \\ & 0.6 & 0.2 & & & & \\ 0.1 & & 0.7 & 0.3 & & & \\ & 0.1 & & 0.6 & 0.2 & & \\ & & 0.1 & & 0.7 & 0.3 & \\ & & & 0.1 & & 0.7 & 0.2 \\ & & & & 0.1 & & 0.8 \end{bmatrix} \begin{bmatrix} x_1^{(k)} \\ x_2^{(k)} \\ x_3^{(k)} \\ x_4^{(k)} \\ x_5^{(k)} \\ x_6^{(k)} \\ x_7^{(k)} \end{bmatrix} \tag{7.71}$$

which has the long-run probability vector $(1/121)\{36 \quad 12 \quad 24 \quad 12 \quad 18 \quad 10 \quad 9\}$.

For a job entering a phase at time t_0 and having a probability of completion during interval δt of $\mu\delta t$, then, with $\delta t \to 0$, the chance of this job still being in the phase at time t can be shown to be $e^{-\mu(t-t_0)}$. Hence the occupancy of the phases by jobs must be negative exponential in order that the stochastic matrix is independent of time. The constant μ is the reciprocal of the mean occupancy time of the phase.

In the analysis of more complex systems the number of possible states can be very large, resulting in large but also very sparse stochastic matrices. These matrices will almost certainly be unsymmetric.

7.12 SOME EIGENVALUE PROPERTIES OF STOCHASTIC MATRICES

Consider the Gerschgorin disc associated with the j-th column of the stochastic matrix. This disc has centre p_{jj} and, on account of equations (7.64) and (7.65), must pass through the point (1, 0) on the Argand diagram (Figure 7.12). The column with the smallest diagonal element will give rise to the largest disc. This largest disc will contain all of the other column discs and hence also the full set of eigenvalues of the matrix. Since no diagonal element can be negative, the limiting range for the eigenvalues of any stochastic matrix is the unit circle in the Argand diagram. Hence every eigenvalue λ of a stochastic matrix must obey the condition

$$|\lambda| \leqslant 1 \tag{7.72}$$

Because the column sums of the stochastic matrix are unity it follows that

$$[1 \quad 1 \quad \ldots \quad 1] \begin{bmatrix} & & \\ & \mathbf{P} & \\ & & \end{bmatrix} = [1 \quad 1 \quad \ldots \quad 1] \tag{7.73}$$

Hence there must always be an eigenvalue of $\mathbf{P}$ such that $\lambda_1 = 1$ having a left eigenvector $\{1 \quad 1 \quad \ldots \quad 1\}$.

The Markov process of premultiplying a vector by the stochastic matrix corresponds to a power method iteration to obtain the dominant eigenvalue and eigenvector (see section 10.1) in which the scaling factors are unnecessary. The same theory can be used to establish that convergence is to the right eigenvector corresponding to the eigenvalue of largest modulus. Hence if a stochastic matrix

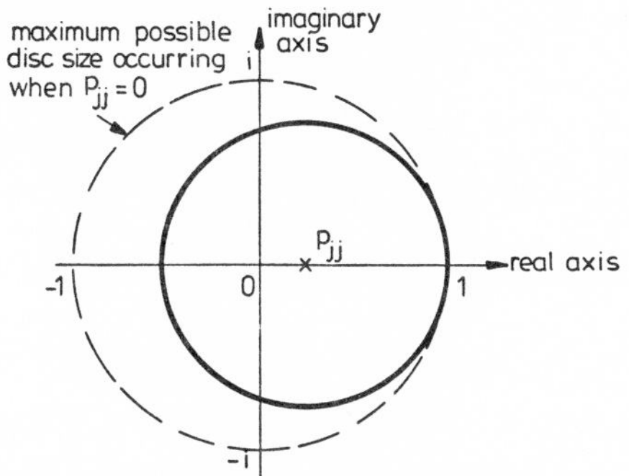

Figure 7.12 Gerschgorin disc for column j of a stochastic matrix

Table 7.5 Eigensolution of stochastic matrix (7.66)

Left eigenvector	Eigenvalue	Right eigenvector
{1 1 1}	1	{0.3125 0.1875 0.5}
{−9 7 3}	0.2	{1 −1 0}
{1 1 −1}	−0.2	{−1 −3 4}

has only one eigenvalue of unit modulus, this must be $\lambda_1 = 1$ and its right eigenvector will be the long-run probability vector, whatever is the initial probability vector. An example is the stochastic matrix (7.66) whose eigensolution is given in Table 7.5.

Two circumstances may complicate the convergence condition of Markov chains, both of which can be identified in terms of the eigenvalues of the stochastic matrix.

(a) When eigenvalues occur on the unit circle other than at $\lambda = 1$

An example is the matrix shown in Table 7.6. If Markov chains are generated with matrices of this type they will not normally converge to a steady long-run probability vector, but instead will converge to a cyclic behaviour. This type of behaviour can only occur when at least one of the diagonal elements is zero, for it is only then that Gerschgorin discs touch the unit circle at other than $\lambda = 1$.

Table 7.6 A stochastic matrix having a cyclic Markov process

Matrix	Eigenvalue	Right eigenvector
$\begin{bmatrix} 0 & 0 & 1 \\ 1 & 0 & 0 \\ 0 & 1 & 0 \end{bmatrix}$	1	{1 1 1}
	$-\tfrac{1}{2}(1 - \sqrt{3}i)$	$\{1 \;\; -\tfrac{1}{2}(1 + \sqrt{3}i) \;\; -\tfrac{1}{2}(1 - \sqrt{3}i)\}$
	$-\tfrac{1}{2}(1 + \sqrt{3}i)$	$\{1 \;\; -\tfrac{1}{2}(1 - \sqrt{3}i) \;\; -\tfrac{1}{2}(1 + \sqrt{3}i)\}$

(b) When multiple eigenvalues occur at $\lambda = 1$

An example is the *reducible* matrix

$$\mathbf{P} = \begin{bmatrix} 0.8 & 0.6 & 0.1 & & & \\ & & 0.4 & & & \\ & & & & & \\ 0.2 & 0.3 & & 1 & & \\ & & 0.4 & & 0.4 & 1 \\ & 0.1 & 0.1 & & 0.6 & \end{bmatrix} \tag{7.74}$$

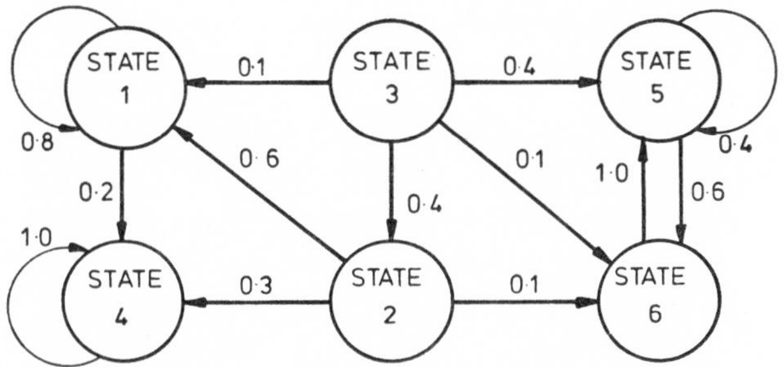

Figure 7.13 Probabilities for a system with two recurrent chains (stochastic matrix 7.74)

which has right eigenvectors $\{0 \quad 0 \quad 0 \quad 1 \quad 0 \quad 0\}$ and $\{0 \quad 0 \quad 0 \quad 0 \quad 0.625 \quad 0.375\}$, both corresponding to eigenvalues $\lambda = 1$. Examining the directed graph of this stochastic matrix, shown in Figure 7.13, it is evident that there are two *recurrent chains*, i.e. sequences of events from which there is no exit, one being at state 4 and the other comprising states 5 and 6 together. The eigenvectors corresponding to the multiple eigenvalues $\lambda = 1$ describe probability vectors associated with both of these recurrent chains. Any Markov chain for this matrix converges to a steady long-run probability vector which is a linear combination of these two eigenvectors and is therefore not unique. The particular result depends on the initial probability vector, thus for

$$\mathbf{x}^{(0)} = \{1 \quad 0 \quad 0 \quad 0 \quad 0 \quad 0\}, \quad \mathbf{x}^{(k)} \to \{0 \quad 0 \quad 0 \quad 1 \quad 0 \quad 0\}$$

for

$$\mathbf{x}^{(0)} = \{0 \quad 0 \quad 0 \quad 0 \quad 1 \quad 0\}, \quad \mathbf{x}^{(k)} \to \{0 \quad 0 \quad 0 \quad 0 \quad 0.625 \quad 0.375\}$$

and for

$$\mathbf{x}^{(0)} = \{0 \quad 0 \quad 1 \quad 0 \quad 0 \quad 0\}, \quad \mathbf{x}^{(k)} \to \{0 \quad 0 \quad 0 \quad 0.46 \quad 0.3375 \quad 0.2025\}$$

From the biorthogonality condition between left and right eigenvectors established in section 8.8 it follows that all right eigenvectors corresponding to subdominant eigenvalues, $\lambda < 1$, of a stochastic matrix must be orthogonal to the left eigenvector $\{1 \quad 1 \quad \ldots \quad 1\}$. Hence the sum of the elements in each of these eigenvectors will be zero and they cannot constitute valid probability vectors. By continuing the parallel between the Markov process and the power method of obtaining the dominant eigenvalue of a matrix (section 10.1) it follows that the Markov chain converges to the long-run probability vector (or its cyclic equivalent) at a rate which is primarily governed by the magnitude of the subdominant eigenvalues each have a modulus of 0.2 and hence the convergence is rapid. Conversely, equations (7.71) yield a slow convergence rate. Starting with $\mathbf{x}^{(0)} = \{1 \quad 0 \quad 0 \quad 0 \quad 0 \quad 0 \quad 0\}$ it takes ninety-three transitions before all of the probabilities agree with the long-run probabilities to three decimal figures. In the same problem, if a smaller value of δt were to be chosen, then the convergence rate would be correspondingly slower.

One method of determining the long-run probability vector is by converting equation (7.70) to the homogeneous simultaneous equations

$$(\mathbf{P} - \mathbf{I})\mathbf{q} = \mathbf{0} \tag{7.75}$$

Since one of these equations is a linear combination of the others, it can be replaced by the condition $\Sigma q_i = 1$. The modified set of simultaneous equations may then be solved to determine $\mathbf{q}$. Modifying the first equation for the stochastic matrix (7.66) gives

$$\left.\begin{aligned} q_1 + \quad q_2 + \quad q_3 &= 1 \\ -0.8q_2 + 0.3q_3 &= 0 \\ 0.6q_1 + 0.6q_2 - 0.6q_3 &= 0 \end{aligned}\right\} \tag{7.76}$$

from which the solution $\mathbf{q} = \{0.3125 \quad 0.1875 \quad 0.5\}$ may be derived. A disadvantage of adopting this procedure is that the presence of recurrent chains or cyclic behaviour patterns is not diagnosed, and also no knowledge is gained of the convergence rate of the Markov process. It may therefore be more satisfactory to obtain the eigenvalues of largest modulus and their associated eigenvectors. The knowledge that at least one eigenvalue $\lambda_1 = 1$ should be present may be used as a partial check on the analysis. If the largest subdominant eigenvalue is also obtained the convergence rate of the Markov chain will be established. For large sparse stochastic matrices a full eigensolution is unnecessary since only a few of the eigenvalues and associated eigenvectors of largest modulus are of any significance.

BIBLIOGRAPHY

Bailey, N. T. J. (1964). *The Elements of Stochastic Processes: with Applications to the Natural Sciences*, Wiley, New York.

Bishop, R. E. D., Gladwell, G. M. L., and Michaelson, S. (1964). *The Matrix Analysis of Vibration*, Cambridge University Press, Cambridge.

Clough, R. W. and Penzien, J. (1975). *Dynamics of Structures*, McGraw-Hill, New York.

Craddock, J. M., and Flood, C. R. (1969). 'Eigenvectors for representing the 500 mb geopotential surface over the Northern Hemisphere'. *Quarterly J. of the Royal Meteorological Soc.*, 95, 576–593.

Frazer, R. A., Duncan, W. J., and Collar, A. R. (1938). *Elementary Matrices and Some Applications to Dynamics and Differential Equations*, Cambridge University Press, Cambridge.

Fry, T. C. (1965). *Probability and its Engineering Uses*, 2nd ed. Van Nostrand Reinhold, New York. (Chapter on 'Matrix Methods and Markov Processes'.)

Jennings, A. (1963). 'The elastic stability of rigidly jointed frames'. *Int. J. Mech. Science*, 5, 99–113.

Morrison, D. F. (1967). *Multivariate Statistical Methods*, McGraw-Hill, New York. (Chapter on 'Principal Components'.)

Scheerer, A. E. (1969). *Probability on Discrete Sample Spaces, with applications*, International Textbook Co., Scranton, Pennsylvania. (Discusses Markov chains.)

Seal, H. L. (1964). *Multivariate Statistical Analysis for Biologists*, Methuen, London. (Chapter on 'Principal Components'.)

Wilkinson, J. H., and Reinsch, C. (1971). *Handbook for Automatic Computation, Vol. II, Linear Algebra*, Springer-Verlag, Berlin. (An algorithm by R. S. Martin and J. H. Wilkinson on reduction of the symmetric eigenproblem $\mathbf{Ax} = \lambda\mathbf{Bx}$ and related problems to standard form.)

Williams, D. (1960). *An Introduction to the Theory of Aircraft Structures*, Edward Arnold, London. (Considers finite difference buckling analysis on page 108.)

Chapter 8
Transformation Methods for Eigenvalue Problems

8.1 ORTHOGONAL TRANSFORMATION OF A MATRIX

The characteristic equation method of determining eigenvalues given in Chapter 1 is not very suitable for computer implementation. In contrast transformation methods are easily automated. Of the methods given in this chapter, only Jacobi's method was available before 1954. The other methods together constitute what is possibly the most significant recent development in numerical analysis.

A transformation method transforms the matrix under investigation into another with the same eigenvalues. Usually many such transformations are carried out until either the eigenvalues can be obtained by inspection or the matrix is in a form which can be easily analysed by alternative procedures. The most general transformation which retains the eigenvalues of a matrix is a *similarity transformation*

$$\bar{\mathbf{A}} = \mathbf{N}^{-1}\mathbf{A}\mathbf{N} \tag{8.1}$$

where $\mathbf{N}$ may be any non-singular matrix of the same order as $\mathbf{A}$. However, in the first eight sections of this chapter, eigensolution methods for symmetric matrices will be discussed and for this purpose only orthogonal transformation matrices will be required. If $\mathbf{N}$ is orthogonal then, since $\mathbf{N}^{-1} = \mathbf{N}^T$, equation (8.1) becomes

$$\bar{\mathbf{A}} = \mathbf{N}^T\mathbf{A}\mathbf{N} \tag{8.2}$$

From section 1.14 it follows that, if $\mathbf{A}$ is symmetric, $\bar{\mathbf{A}}$ will also be symmetric. Therefore the use of orthogonal transformations ensures that symmetry is preserved.

If a symmetric matrix $\mathbf{A}$ has eigenvalues expressed in the diagonal matrix form $\boldsymbol{\Lambda} = \lceil \lambda_1 \;\; \lambda_2 \;\; \ldots \;\; \lambda_n \rfloor$ and corresponding right eigenvectors expressed as a matrix $\mathbf{Q} = [\mathbf{q}_1 \;\; \mathbf{q}_2 \;\; \ldots \;\; \mathbf{q}_n]$ such that $\mathbf{A}\mathbf{Q} = \mathbf{Q}\boldsymbol{\Lambda}$ (equation 1.112), then

$$\bar{\mathbf{A}}(\mathbf{N}^T\mathbf{Q}) = \mathbf{N}^T\mathbf{A}\mathbf{Q} = (\mathbf{N}^T\mathbf{Q})\boldsymbol{\Lambda} \tag{8.3}$$

signifying that the eigenvalues of $\bar{\mathbf{A}}$ are the same as those of $\mathbf{A}$, and the matrix of right eigenvectors of $\bar{\mathbf{A}}$ is

$$\bar{\mathbf{Q}} = \mathbf{N}^T\mathbf{Q} \tag{8.4}$$

If a symmetric matrix can be transformed until all of its off-diagonal elements are zero, then the full set of eigenvalues are, by inspection, the diagonal elements. Hence the object of the transformation methods will be to eliminate off-diagonal elements so that the matrix becomes more nearly diagonal.

8.2 JACOBI DIAGONALIZATION

Each Jacobi transformation eliminates one pair of off-diagonal elements in a symmetric matrix. In order to eliminate the pair of equal elements a_{pq} and a_{qp} an orthogonal transformation matrix

$$\mathbf{N} = \begin{array}{r} \\ \\ \text{row } p \\ \\ \text{row } q \\ \\ \end{array} \left[\begin{array}{cccccc} 1 & & \text{col } p & & \text{col } q & \\ & 1 & & & & \\ & & \cos\alpha & & -\sin\alpha & \\ & & & 1 & & \\ & & \sin\alpha & & \cos\alpha_1 & \\ & & & & & 1 \end{array}\right] \tag{8.5}$$

is employed. The multiplication $\bar{\mathbf{A}} = \mathbf{N}^T\mathbf{A}\mathbf{N}$ only affects rows p and q and columns p and q as shown in Figure 8.1 and can be considered to be a rotation in the plane of the p-th and q-th variables. The choice of α in the transformation matrix must be such that elements $\bar{a}_{pq}$ $(= \bar{a}_{qp}) = 0$. Constructing $\bar{a}_{pq}$ from the transformation gives

$$\bar{a}_{pq} = (-a_{pp} + a_{qq})\cos\alpha\sin\alpha + a_{pq}(\cos^2\alpha - \sin^2\alpha) = 0 \tag{8.6}$$

Hence

$$\tan 2\alpha = \frac{2a_{pq}}{a_{pp} - a_{qq}} \tag{8.7}$$

An alternative manipulation of equation (8.6) gives

$$\cos^2\alpha = ½ + \frac{a_{pp} - a_{qq}}{2r} \tag{8.8a}$$

$$\sin^2\alpha = ½ - \frac{a_{pp} - a_{qq}}{2r} \tag{8.8b}$$

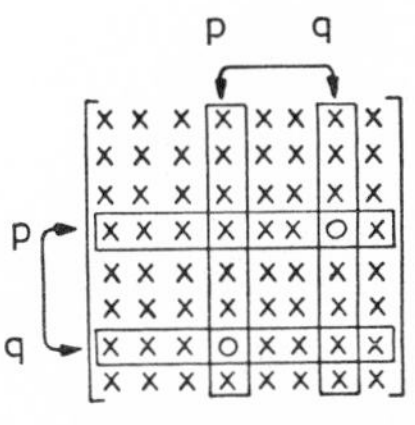

Figure 8.1 A single Jacobi transformation to eliminate elements a_{pq} and a_{qp} (boxes indicate modified rows and columns)

and

$$\sin\alpha\cos\alpha = \frac{a_{pq}}{r} \tag{8.8c}$$

where

$$r^2 = (a_{pp} - a_{qq})^2 + 4a_{pq}^2 \tag{8.8d}$$

Because α is not required explicitly, it is better to use these equations to derive $\sin\alpha$ and $\cos\alpha$ than to use equation (8.7). The signs of α and $\sin\alpha$ may both be chosen as positive, in which case $\cos\alpha$ must be given the same sign as a_{pq}. If $a_{pp} > a_{qq}$ equation (8.8a) should be used to find $\cos\alpha$ and then equation (8.8c) to find $\sin\alpha$. Alternatively, if $a_{pp} < a_{qq}$ equation (8.8b) should be used to find $\sin\alpha$ and then equation (8.8c) to find $\cos\alpha$. By proceeding in this way, not only is one of the square root computations avoided but also there is no significant loss in accuracy with floating-point arithmetic when a_{pq}^2 is much smaller than $(a_{pp} - a_{qq})^2$. The two diagonal elements that are modified by the transformation become

$$\left.\begin{aligned} \bar{a}_{pp} &= a_{pp}\cos^2\alpha + a_{qq}\sin^2\alpha + 2a_{pq}\sin\alpha\cos\alpha \\ &\text{and} \\ \bar{a}_{qq} &= a_{pp}\sin^2\alpha + a_{qq}\cos^2\alpha - 2a_{pq}\sin\alpha\cos\alpha \end{aligned}\right\} \tag{8.9}$$

which may either be computed from the above formulae or from the following:

$$\left.\begin{aligned} \bar{a}_{pp} &= \tfrac{1}{2}(a_{pp} + a_{qq} + r) \\ &\text{and} \\ \bar{a}_{qq} &= \tfrac{1}{2}(a_{pp} + a_{qq} - r) = \frac{a_{pp}a_{qq} - a_{pq}^2}{\bar{a}_{pp}} \end{aligned}\right\} \tag{8.10}$$

The other elements affected by the transformation are modified according to

$$\left.\begin{aligned} \bar{a}_{ip} &= \bar{a}_{pi} = a_{ip}\cos\alpha + a_{iq}\sin\alpha \\ \bar{a}_{iq} &= \bar{a}_{qi} = -a_{ip}\sin\alpha + a_{iq}\cos\alpha \end{aligned}\right\} \tag{8.11}$$

The normal procedure is to perform a series of transformations of the type described above, with each transformation eliminating the off-diagonal element, having the largest modulus, present in the matrix at that stage. Unfortunately elements which have been eliminated do not necessarily stay zero, and hence the method is iterative in character. If the matrix after $k - 1$ transformations is designated $\mathbf{A}^{(k)}$ then the k-th transformation may be written as

$$\mathbf{A}^{(k+1)} = \mathbf{N}_k^T\mathbf{A}^{(k)}\mathbf{N}_k \tag{8.12}$$

and the eigenvector matrices of $\mathbf{A}^{(k)}$ and $\mathbf{A}^{(k+1)}$ are related by

$$\mathbf{Q}^{(k+1)} = \mathbf{N}_k^T\mathbf{Q}^{(k)} \tag{8.13}$$

If a total of s transformations are necessary to diagonalize the matrix

$$\mathbf{Q}^{(s+1)} = \mathbf{N}_s^T \ldots \mathbf{N}_2^T \mathbf{N}_1^T \mathbf{Q}^{(1)} \tag{8.14}$$

However, since $\mathbf{Q}^{(s+1)}$ is the matrix of eigenvectors of a diagonal matrix,

$$\mathbf{Q}^{(s+1)} = \mathbf{I} \tag{8.15}$$

Hence the eigenvectors of the original matrix $\mathbf{A}^{(1)}$ appear as columns of

$$\mathbf{Q}^{(1)} = \mathbf{N}_1 \mathbf{N}_2 \ldots \mathbf{N}_s \tag{8.16}$$

Table 8.1 shows the sequence of matrices $\mathbf{A}^{(k)}$ in the Jacobi diagonalization of a 3 x 3 matrix, together with the values of p, q, r, $\cos\alpha$ and $\sin\alpha$ for each transformation. The eigenvalues of the matrix are the diagonal elements of $\mathbf{A}^{(7)}$ to four-figure accuracy. Because $\bar{a}_{pp} > \bar{a}_{qq}$ in equations (8.10), and p and q have been chosen so that $p < q$ for every transformation, the eigenvalues appear in descending order in the diagonal matrix. If, instead, $p > q$ had been adopted for every transformation then the eigenvalues would have appeared in ascending order.

Table 8.1 Jacobi diagonalization of a 3 x 3 matrix

k	Matrix $\mathbf{A}^{(k)}$	Transformation k			
		p, q	r	$\cos\alpha$	$\sin\alpha$
1	$\begin{bmatrix} 3.5 & -6 & 5 \\ -6 & 8.5 & -9 \\ 5 & -9 & 8.5 \end{bmatrix}$	2,3	18	−0.7071	0.7071
2	$\begin{bmatrix} 3.5 & -7.7782 & -0.7071 \\ -7.7782 & 17.5 & 0 \\ -0.7071 & 0 & -0.5 \end{bmatrix}$	1,2	20.9285	−0.4069	0.9135
3	$\begin{bmatrix} 20.9643 & 0 & 0.2877 \\ 0 & 0.0358 & -0.6459 \\ 0.2877 & -0.6459 & -0.5 \end{bmatrix}$	2,3	1.3985	−0.8316	0.5554
4	$\begin{bmatrix} 20.9643 & -0.1598 & 0.2392 \\ -0.1598 & 0.4672 & 0 \\ 0.2392 & 0 & -0.9314 \end{bmatrix}$	1,3	21.9009	0.9999	0.0109
5	$\begin{bmatrix} 20.9669 & -0.1598 & 0 \\ -0.1598 & 0.4672 & -0.0017 \\ 0 & -0.0017 & -0.9340 \end{bmatrix}$	1,2	20.5022	−1.0000	0.0078
6	$\begin{bmatrix} 20.9681 & 0 & 0.0000 \\ 0 & 0.4659 & 0.0017 \\ 0.0000 & 0.0017 & -0.9340 \end{bmatrix}$	2,3	1.3999	1.0000	0.0012
7	$\begin{bmatrix} 20.9681 & 0.0000 & 0.0000 \\ 0.0000 & 0.4659 & 0 \\ 0.0000 & 0 & -0.9340 \end{bmatrix}$				

It can be established that one transformation increases the sum of the squares of the diagonal elements by $2a_{pq}^2$ and, at the same time, decreases the sum of the squares of the off-diagonal elements by the same amount. If at each transformation a_{pq} is chosen to be the off-diagonal element of largest modulus, the transformation must reduce the sum of the squares of the off-diagonal elements by at least a factor $[1 - \{2/n(n-1)\}]$. Hence, to reduce the Euclidean norm of the matrix of off-diagonal elements by three decimal places requires no more than s transformations where

$$\left(1 - \frac{2}{n(n-1)}\right)^{s/2} = 10^{-3} \tag{8.17}$$

which gives approximately

$$s < 6.9n(n-1) \tag{8.18}$$

Since the number of multiplications per transformation is approximately $4n$, no more than about $28n^3$ multiplications are required to reduce the Euclidean norm of the matrix of off-diagonal elements by three decimal places. For a sparse matrix the initial convergence rate must be faster than this, but the matrix soon becomes fully populated. Furthermore, this rule is unduly pessimistic for all matrices because the convergence rate increases as iteration proceeds till in the limit the convergence rate measured over $n(n-1)/2$ transformations becomes quadratic. A figure of about $8n^3$ multiplications for a full diagonalization might be more realistic, but this will depend on how many large off-diagonal elements there are in the original matrix.

8.3 COMPUTER IMPLEMENTATION OF JACOBI DIAGONALIZATION

Using equations (8.8), (8.11) and (8.9) or (8.10) to perform a transformation it is not necessary to form or store the individual transformation matrices. The matrix $\mathbf{A}^{(k+1)}$ can overwrite $\mathbf{A}^{(k)}$ using only a small amount of temporary store. The main problem associated with computer implementation is the searching time required to locate the largest off-diagonal element before each transformation can take place. In hand computation this time is likely to be insignificant, but a computer may take at least as long to perform a search as to perform a transformation. The ratio of searching time/arithmetic time will increase proportionally with n. The following are three methods of reducing the searching time.

(a) Monitored searching

Initially each row of the matrix is searched and the modulus and column number of the largest off-diagonal element are recorded. A $(1 \times n)$ real array and a $(1 \times n)$ integer array will be required to store this information for the whole matrix. The off-diagonal element of largest modulus can then be located by searching the real array for its largest element and then reading the column number from the appropriate entry in the integer array. During a transformation these arrays can be updated so that they can continue to be used as a short-cut to the searching process. If a transformation eliminates elements a_{pq} and a_{qp}, then elements p and q of the arrays will have to be revised; however, other elements in the arrays will only need to be revised if either of the two new elements in the row have a modulus greater than the previous maximum, or if the previous maximum off-diagonal element in the row occurred at either column p or q.

(b) Serial Jacobi

The serial Jacobi method eliminates elements a_{pq} in a systematic way, e.g. $(p, q) = (1, 2), (1, 3), \ldots, (1, n)$ then $(2, 3), (2, 4)$, etc. When all of the elements have been eliminated once, the process is repeated as many times as necessary. The searching process has therefore been completely dispensed with, but at the expense of performing a larger number of transformations. Convergence is still assured.

(c) Threshold Jacobi

The threshold Jacobi method is a compromise between the classical Jacobi and the serial Jacobi methods. Here the elements are eliminated in a serial order, except that elements having a modulus below a given threshold value are left unaltered. When all of the elements have a modulus below the threshold value, the threshold value is reduced and the process continued. Iteration will be complete when a threshold value has been satisfied for all the off-diagonal elements which corresponds to a suitable tolerance.

In order to determine the eigenvectors by equation (8.16) it is necessary to allocate an $n \times n$ array store at the start of the eigensolution into which the unit matrix is entered. As each transformation takes place the matrix held in this array store is postmultiplied by $\mathbf{N}_k$ (where k is the current transformation number). This only involves modifying columns p and q of the array. At the end of the iteration this array will contain the full set of eigenvectors stored by columns. The amount of computation required to find the full set of eigenvectors in this way is of the same order as the amount of computation performed in transforming the matrix. If only a few eigenvectors are required it may be more advantageous to obtain these separately by inverse iteration (section 10.3) once the eigenvalues have been determined.

Jacobi's method is useful as a simple procedure for the complete eigensolution of small symmetric matrices. It may also be competitive for larger matrices if the off-diagonal elements are initially very small. However, for most computational requirements involving complete eigensolution it has been superceded by tridiagonalization methods.

8.4 GIVENS' TRIDIAGONALIZATION

Givens' (1954) method adopts the Jacobi transformation to produce a tridiagonal matrix having the same eigenvalues. The process is non-iterative and more efficient

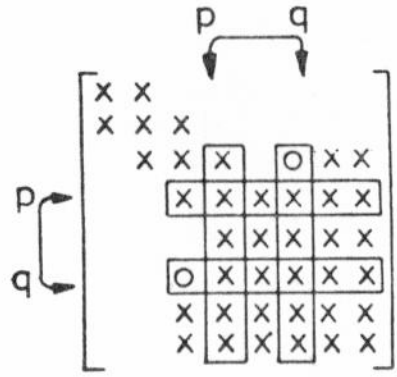

Figure 8.2 A single Givens transformation to eliminate elements a_{pq} and a_{qp}

than Jacobi diagonalization, although it does require the resulting tridiagonal matrix to be analysed separately. The procedure is to use a series of transformations, representing rotations in planes (p, q), in the order $(2, 3), (2, 4), \ldots, (2, n)$ followed by $(3, 4), (3, 5), \ldots, (3, n)$, etc. Any particular rotation (p, q) is used to eliminate the two elements in positions $(p-1, q)$ and $(q, p-1)$. Figure 8.2 shows the position of non-zero elements in an 8×8 matrix after transformation (4, 6). At this stage the first two rows and columns have been reduced to the required form and the reduction of the third row and column is under way. It can be verified that, once a pair of elements have been eliminated, they remain zero and so do not need to be operated on again.

If $\mathrm{A}^{(k)}$ is the matrix before transformation (p, q), equation (8.11) gives

$$a_{p-1,q}^{(k+1)} = -a_{p-1,p}^{(k)} \sin\alpha + a_{p-1,q}^{(k)} \cos\alpha = 0 \tag{8.19}$$

Hence

$$\tan\alpha = \frac{a_{p-1,q}^{(k)}}{a_{p-1,p}^{(k)}} \tag{8.20}$$

It is not necessary to find α explicitly. Instead, r can be computed from

$$r^2 = (a_{p-1,p}^{(k)})^2 + (a_{p-1,q}^{(k)})^2 \tag{8.21}$$

(taking the positive square root), whereupon

$$\left.\begin{aligned} \cos\alpha &= \frac{a_{p-1,p}^{(k)}}{r} \\ \text{and} \qquad & \\ \sin\alpha &= \frac{a_{p-1,q}^{(k)}}{r} \end{aligned}\right\} \tag{8.22}$$

The values of $\cos\alpha$ and $\sin\alpha$ may then be used to perform the transformation, noting that

$$a_{p-1,p}^{(k+1)} = a_{p,p-1}^{(k+1)} = r \tag{8.23}$$

The total number of multiplications required to reduce an $n \times n$ symmetric matrix to tridiagonal form by Givens' method is approximately $4n^3/3$. In the computer implementation, as each pair of elements is eliminated, the storage space they occupied can be used to hold the values of $\sin \alpha$ and $\cos \alpha$ which characterize the transformation. In this way details of the transformations can be retained in case this information is later required to generate the eigenvectors. It is also possible to perform a Givens' reduction in a triangular store. In this case only $\sin \alpha$ or $\cos \alpha$ can be retained, and the other must be regenerated if it is required later.

Eigenvectors can only be determined after the eigensolution of the tridiagonal matrix has been accomplished. If $\mathbf{p}_i$ is an eigenvector of the tridiagonal matrix then the corresponding eigenvector of the original matrix is

$$\mathbf{q}_i = \mathbf{N}_1\mathbf{N}_2 \ldots \mathbf{N}_s\mathbf{p}_i \tag{8.24}$$

where s signifies the last Givens' transformation, i.e. $s = (n-1)(n-2)/2$. Hence vector $\mathbf{q}_i$ can be determined by a series of operations on the eigenvector $\mathbf{p}_i$ which apply the rotations in reverse order.

Eigensolution of tridiagonal matrices may be carried out by LR, QR or Sturm sequence method, all of which are described later.

Recently Gentleman and Golub have proposed modifications to the procedure for implementing Givens' transformations which reduce the amount of computation required and hence make it competitive with Householder's method given below. The fast Givens' transformation and a more economical storage scheme for the plane rotations are discussed by Hammarling (1974) and Stewart (1976) respectively.

8.5 HOUSEHOLDER'S TRANSFORMATION

Householder's method also reduces a symmetric matrix to tridiagonal form but is computationally more efficient than the basic Givens' method. The appropriate elements of an entire column are eliminated in one transformation. The transformation matrix is of the form

$$\mathbf{P} = \mathbf{I} - 2\mathbf{w}\mathbf{w}^T \tag{8.25}$$

where $\mathbf{w}$ is a column vector whose Euclidean norm is unity, i.e.

$$\mathbf{w}^T\mathbf{w} = 1 \tag{8.26}$$

The transformation matrix is orthogonal since

$$\mathbf{P}^T\mathbf{P} = (\mathbf{I} - 2\mathbf{w}\mathbf{w}^T)(\mathbf{I} - 2\mathbf{w}\mathbf{w}^T) = \mathbf{I} \tag{8.27}$$

and, because it is also symmetric,

$$\mathbf{P} = \mathbf{P}^T = \mathbf{P}^{-1} \tag{8.28}$$

Consider the first stage of the tridiagonalization of an $n \times n$ matrix $\mathbf{A}$. The transformation may be expressed by the product

$$\mathbf{A}^{(2)} = \mathbf{P}_1\mathbf{A}\mathbf{P}_1 \tag{8.29}$$

The vector $\mathbf{w}$ is chosen to have the form $\mathbf{w} = \{0 \quad w_2 \quad w_3 \quad \ldots \quad w_n\}$ and, in consequence,

$$\mathbf{P}_1 = \begin{bmatrix} 1 & & & & \\ & 1-2w_2^2 & -2w_2w_3 & \cdots & -2w_2w_n \\ & -2w_2w_3 & 1-2w_3^2 & \cdots & -2w_3w_n \\ & \cdot & \cdot & \cdot & \cdot \\ & -2w_2w_n & -2w_3w_n & \cdots & 1-2w_n^2 \end{bmatrix} \tag{8.30}$$

The off-diagonal elements in the first row of $\mathbf{A}^{(2)}$ must all be zero except for the superdiagonal element $a_{12}^{(2)}$. Expanding the relevant elemental equations of matrix equation (8.29) gives

$$\left.\begin{aligned} a_{12}^{(2)} &= a_{12} - 2w_2b = r \text{ (say)} \\ a_{13}^{(2)} &= a_{13} - 2w_3b = 0 \\ &\cdots \\ a_{1n}^{(2)} &= a_{1n} - 2w_nb = 0 \end{aligned}\right\} \tag{8.31}$$

where

$$b = a_{12}w_2 + a_{13}w_3 + \cdots + a_{1n}w_n \tag{8.32}$$

But from equation (8.26)

$$w_2^2 + w_3^2 + \cdots + w_n^2 = 1 \tag{8.33}$$

These equations are sufficient to determine the vector $\mathbf{w}$ and also r. The following equations may be derived:

(a) by squaring equations (8.31) and adding

$$r^2 = a_{12}^2 + a_{13}^2 + \cdots + a_{1n}^2 \tag{8.34}$$

(b) by scaling equations (8.31) by $a_{12}, a_{13}, \ldots, a_{1n}$ respectively and adding

$$2b^2 = r^2 - a_{12}r \tag{8.35}$$

The transformation parameters can be computed by forming r (equation 8.34), b (equation 8.35) and then using equations (8.31) to determine the vector $\mathbf{w}$. The sign of r may be chosen to avoid cancellation in the equation for b, i.e. r should be of opposite sign to a_{12}. However, the second square root evaluation for b may be avoided if, instead of $\mathbf{w}$, the vector

$$\mathbf{v} = 2b\mathbf{w} = \{0, a_{12} - r, a_{13}, \ldots, a_{1n}\} \tag{8.36}$$

is obtained. Then the transformation matrix $\mathbf{P}$ can be written as

$$\mathbf{P} = \mathbf{I} - \frac{1}{2b^2}\mathbf{v}\mathbf{v}^T \tag{8.37}$$

For the elimination of the appropriate elements in row k of the matrix, the computation of the transformation parameters is according to

$$\left.\begin{aligned} &r = -(\operatorname{sign} a_{k,k+1})\sqrt{(a_{k,k+1}^2 + a_{k,k+2}^2 + \cdots + a_{k,n}^2)} \\ &2h^2 = r^2 - ra_{k,k+1} \\ &\text{and} \\ &\mathbf{v} = \{0, \ldots, 0, a_{k,k+1} - r, a_{k,k+2}, \ldots, a_{k,n}\} \end{aligned}\right\} \tag{8.38}$$

As with Givens' method, there is little loss of accuracy involved in computing the tridiagonal form. Wilkinson (1965) has shown that the tridiagonal matrices obtained by both methods are the same apart from sign changes within corresponding rows and columns.

8.6 IMPLEMENTATION OF HOUSEHOLDER'S TRIDIAGONALIZATION (Wilkinson, 1960)

It would be grossly inefficient both in computing time and storage space to form each transformation matrix explicitly, and then perform each transformation by means of matrix multiplication operations. Hence an important feature of Householder's method is the way in which it may be efficiently implemented.

The transformation for the k-th row of the matrix may be written as

$$\mathbf{A}^{(k+1)} = \mathbf{P}\mathbf{A}^{(k)}\mathbf{P} = \left(\mathbf{I} - \frac{1}{2h^2}\mathbf{v}\mathbf{v}^T\right)\mathbf{A}^{(k)}\left(\mathbf{I} - \frac{1}{2h^2}\mathbf{v}\mathbf{v}^T\right) \tag{8.39}$$

where it is assumed that $\mathbf{P}$, $\mathbf{v}$ and h relate to the k-th transformation. This expands to

$$\left.\begin{aligned} &\mathbf{A}^{(k+1)} = \mathbf{A}^{(k)} - \frac{1}{2h^2}\mathbf{v}\mathbf{u}^T - \frac{1}{2h^2}\mathbf{u}\mathbf{v}^T + \frac{\gamma}{4h^4}\mathbf{v}\mathbf{v}^T \\ &\text{where} \\ &\mathbf{u} = \mathbf{A}^{(k)}\mathbf{v} \quad \text{and the scalar} \quad \gamma = \mathbf{v}^T\mathbf{u} \end{aligned}\right\} \tag{8.40}$$

If

$$\left.\begin{aligned} &\mathbf{M} = \mathbf{z}\mathbf{v}^T \\ &\text{where} \\ &\mathbf{z} = \frac{1}{2h^2}\mathbf{u} - \frac{\gamma}{8h^4}\mathbf{v} \end{aligned}\right\} \tag{8.41}$$

then

$$\mathbf{A}^{(k+1)} = \mathbf{A}^{(k)} - \mathbf{M} - \mathbf{M}^T \tag{8.42}$$

For the k-th transformation of a symmetric matrix, the elements modified by $\mathbf{M}$ and $\mathbf{M}^T$ are shown in Figure 8.3. If the transformations are performed through $\mathbf{v}$, $\mathbf{u}$, γ, $\mathbf{z}$ and $\mathbf{M}$, it can be shown that, for large n, a complete transformation of a

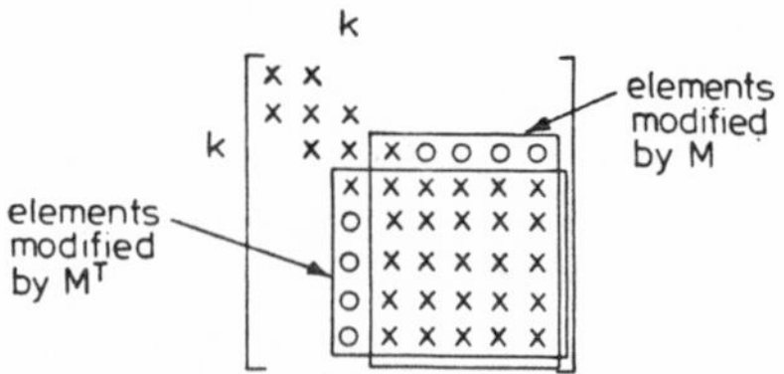

Figure 8.3 The kth Householder transformation of a symmetric matrix

Table 8.2 Householder's tridiagonalization of a symmetric matrix

k	Matrix $\mathbf{A}^{(k)}$				r	$2b^2$	$\mathbf{v}$	$\mathbf{u}$	γ	$\mathbf{z}$
1	1	−3	−2	1	3.7417	25.2250	0	25.2250	313.6996	1
	−3	10	−3	6			−6.7417	−55.4166		−0.535[illegible]
	−2	−3	3	−2			−2	12.2250		0.977[illegible]
	1	6	−2	1			1	−35.4499		−1.651[illegible]
2	1	3.7417	0	0	−5.2465	40.7510	0	0	957.0121	0
	3.7417	2.7857	2.5208	−4.6012			0	40.7510		1
	0	2.5208	6.9106	−6.2814			7.7673	82.5782		−0.211[illegible]
	0	−4.6012	−6.2814	4.3037			−4.6012	−68.5914		−0.357[illegible]
3	1	3.7417	0	0						
	3.7417	2.7857	−5.2465	0						
	0	−5.2465	10.1993	−4.4796						
	0	0	−4.4796	1.0150						

fully populated symmetric matrix to tridiagonal form can be implemented in approximately $2n^3/3$ multiplications and is therefore more efficient than the basic Givens' method. Table 8.2 shows the stages in the tridiagonalization of a 4 × 4 symmetric matrix.

Any eigenvector of the matrix **A** can be derived from the corresponding eigenvector of the tridiagonal matrix by a series of transformations of the form

$$\mathbf{q}^{(k)} = \mathbf{P}_k\mathbf{q}^{(k+1)} = \mathbf{q}^{(k+1)} - \alpha_k\mathbf{v}_k \tag{8.43}$$

where $\mathbf{v}_k$ is the vector $\mathbf{v}$ for the k-th transformation and

$$\alpha_k = \frac{1}{2b_k^2}\mathbf{v}_k^T\mathbf{q}^{(k+1)} \tag{8.44}$$

where b_k is the value of b for the k-th transformation. Table 8.3 shows the stages in the transformation of an eigenvector $\mathbf{q}^{(3)}$ of the tridiagonal matrix of Table 8.2 to obtain the corresponding eigenvector of the original matrix. Generation of all the

Table 8.3 Transformation of an eigenvector $\mathbf{q}^{(3)}$ of the tridiagonal matrix (Table 8.2) to obtain the corresponding eigenvector $\mathbf{q}$ of the original matrix

$\mathbf{q}^{(3)}$	α_2	$\mathbf{q}^{(2)}$	α_1	$\mathbf{q}^{(1)}$	$\mathbf{q}$ (normalized)
$\begin{bmatrix} -0.1403 \\ -0.5000 \\ 1 \\ -0.3364 \end{bmatrix}$	0.2286	$\begin{bmatrix} -0.1403 \\ -0.5000 \\ -0.7755 \\ 0.7154 \end{bmatrix}$	0.2235	$\begin{bmatrix} -0.1403 \\ 1.0066 \\ -0.3286 \\ 0.4919 \end{bmatrix}$	$\begin{bmatrix} -0.1394 \\ 1 \\ -0.3264 \\ 0.4886 \end{bmatrix}$

eigenvectors of A from the eigenvectors of the tridiagonal matrix takes approximately n^3 multiplications. However, a proportionate amount of this computation can be avoided if only a few eigenvectors are required.

On a digital computer the only array storage required for Householder tridiagonalization is an array which initially stores the upper triangle of the original matrix, together with an array to store a $1 \times n$ vector. The transformations are performed within the triangular store. The $1 \times n$ vector store is required for the current vector $\mathbf{u}$ which may be overwritten by the required part of the current vector $\mathbf{z}$ when it is formed. Furthermore, if the elements $a_{ki}^{(k)}$ $(i > k + 1)$ are left in the triangular store after the transformation for row k has taken place, they can be used as elements of the vector $\mathbf{v}_k$ if needed later for eigenvector transformations. Additional array storage, to hold information for the eigenvector transformations until after the eigensolution of the tridiagonal matrix, is required only for the first non-zero element in each vector $\mathbf{v}_k$. (The factors $2b_k^2$ may be kept in the unrequired part of the storage for $\mathbf{u}$.)

8.7 TRANSFORMATION OF BAND SYMMETRIC MATRICES (Rutishauser, 1963)

If either the standard Givens or Householder tridiagonalization procedure is applied to sparse matrices, nearly all of the zero elements will become non-zero after very few transformations have taken place. However, it is possible to transform a symmetric band matrix to tridiagonal form within a band store by using the Givens or Householder transformations in a non-standard way. This is done by eliminating outer elements of the band using transformations involving neighbouring rows and columns only.

Consider a 12×12 symmetric matrix which has a bandwidth of 7. The first transformation eliminates elements (1, 4) and (4, 1) by modifying rows and columns 3 and 4. Householder's method is no more efficient than Givens' method, where only two rows and columns are involved, and hence a plane rotation of the third and fourth variables is likely to be adopted. Figure 8.4(a) shows the elements modified by this transformation, and it will be noted that new non-zero elements arise at positions (3, 7) and (7, 3). These elements are immediately eliminated by a further plane rotation of the sixth and seventh variables (Figure 8.4(b)) which

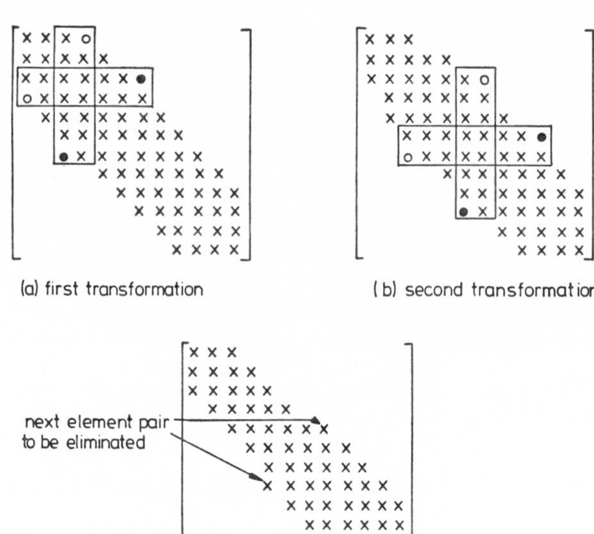

(a) first transformation

(b) second transformation

(c) after four element pairs have been peeled from the band

Figure 8.4 Transformation of a symmetric band matrix to tridiagonal form

introduces new elements in positions (6, 10) and (10, 6). After elimination of elements (6, 10) and (10, 6) the matrix is restored to its original pattern of non-zero elements, but with elements (1, 4) and (4, 1) missing. The next step is to eliminate elements (2, 5) and (5, 2) using a similar method of chasing the loose elements to the bottom of the matrix. The process of stripping the outer elements from the band can proceed layer by layer until just a tridiagonal matrix is left (Figure 8.4(c)).

If $2b - 1$ is the bandwidth of a matrix of order n and $2 \ll b \ll n$, then to eliminate an outer element of the upper triangle in row k and chase out the loose elements requires approximately $8(n - k)$ multiplications. Hence the tridiagonalization of the band matrix involves approximately $4n^2(b - 2)$ multiplications. There is not sufficient storage space vacated by the eliminated elements to retain information about the transformations, and hence inverse iteration is likely to be adopted for obtaining any required eigenvectors (section 10.3).

8.8 EIGENVALUE PROPERTIES OF UNSYMMETRIC MATRICES

Methods of eigensolution of symmetric matrices may be extended to complex matrices which are Hermitian (i.e. $\mathbf{A}^H = \mathbf{A}$). In this case a *unitary* transformation matrix $\mathbf{N}$ obeying the property

$$\mathbf{N}^H\mathbf{N} = \mathbf{I} \tag{8.45}$$

takes the place of the orthogonal transformation matrix. All the eigenvalues of a Hermitian matrix can be shown to be real. Alternatively, a Hermitian eigenvalue problem of order n having an equation $\mathbf{A}\mathbf{q} = \lambda\mathbf{q}$ can be converted into a real symmetric eigenvalue problem of order $2n$, namely

$$\begin{bmatrix} \mathbf{A}_r & -\mathbf{A}_i \\ \mathbf{A}_i & \mathbf{A}_r \end{bmatrix} \begin{bmatrix} \mathbf{q}_r \\ \mathbf{q}_i \end{bmatrix} = \lambda \begin{bmatrix} \mathbf{q}_r \\ \mathbf{q}_i \end{bmatrix} \tag{8.46}$$

where $\mathbf{A}_r$, $\mathbf{A}_i$, $\mathbf{q}_r$ and $\mathbf{q}_i$ contain the real and imaginary components of $\mathbf{A}$ and $\mathbf{q}$. (The supermatrix is symmetric because $\mathbf{A}_r = \mathbf{A}_r^T$ and $\mathbf{A}_i = -\mathbf{A}_i^T$.) However, the transformation methods already discussed do not easily extend to general unsymmetric matrices. Before discussing transformation methods which are suitable for real unsymmetric matrices, some further eigenvalue and eigenvector properties will be presented.

It has been noted (section 1.17) that a real matrix may have complex conjugate eigenvalues. Whenever a real matrix has complex eigenvalues both the left and right corresponding eigenvectors will also be complex. In view of the significance of the Hermitian transpose in relation to eigenvalue problems, when left eigenvectors may be complex they are often defined in Hermitian transpose form, i.e. for left eigenvector $\mathbf{p}_j$,

$$\mathbf{p}_j^H\mathbf{A} = \lambda_j\mathbf{p}_j^H \tag{8.47}$$

If λ_i and λ_j are not coincident eigenvalues, then premultiplying the standard right eigenvector equation for λ_i by $\mathbf{p}_j^H$ gives

$$\mathbf{p}_j^H\mathbf{A}\mathbf{q}_i = \lambda_i\mathbf{p}_j^H\mathbf{q}_i \tag{8.48}$$

However, postmultiplying equation (8.47) by $\mathbf{q}_i$ gives

$$\mathbf{p}_j^H\mathbf{A}\mathbf{q}_i = \lambda_j\mathbf{p}_j^H\mathbf{q}_i \tag{8.49}$$

Since $\lambda_i \neq \lambda_j$ then

$$\mathbf{p}_j^H\mathbf{A}\mathbf{q}_i = \mathbf{p}_j^H\mathbf{q}_i = 0 \tag{8.50}$$

The property $\mathbf{p}_j^H\mathbf{q}_i = 0$ is described as a *biorthogonal relationship* between the left and right eigenvectors. If the eigenvectors are real this relationship becomes $\mathbf{p}_j^T\mathbf{q}_i = 0$, a condition which can easily be verified for all of the appropriate combinations of vectors in Table 1.7.

By compounding the eigenvectors such that $\mathbf{P} = [\mathbf{p}_1 \;\; \mathbf{p}_2 \;\; \cdots \;\; \mathbf{p}_n]$ and $\mathbf{Q} = [\mathbf{q}_1 \;\; \mathbf{q}_2 \;\; \cdots \;\; \mathbf{q}_n]$, the biorthogonality condition gives

$$\mathbf{P}^H\mathbf{Q} = \mathbf{I} \tag{8.51}$$

provided that all the eigenvalues are distinct and that each left or right eigenvector has been scaled so that

$$\mathbf{p}_i^H\mathbf{q}_i = 1 \tag{8.52}$$

Premultiplying equation (1.112) by matrix $\mathbf{P}^H$ gives

$$\mathbf{P}^H\mathbf{A}\mathbf{Q} = \mathbf{P}^H\mathbf{Q}\mathbf{\Lambda} = \mathbf{\Lambda} \tag{8.53}$$

Table 8.4 Full eigensolution of an unsymmetric matrix

Matrix	Eigenvalues	Corresponding left eigenvectors	Corresponding right eigenvectors
$\begin{bmatrix} -1 & -1 & 5 \\ -4 & -2 & 4 \\ 1 & -5 & 3 \end{bmatrix}$	4	$\{1 \;\; -1 \;\; 1\}$	$\{1 \;\; 0 \;\; 1\}$
	$-2+4i$	$\{1 \;\; i \;\; -1\}$	$\{1 \;\; 1+i \;\; i\}$
	$-2-4i$	$\{1 \;\; -i \;\; -1\}$	$\{1 \;\; 1-i \;\; -i\}$

Thus the left and right eigenvectors of a matrix can be used to transform it into a diagonal matrix containing the eigenvalues as elements. Table 8.4 shows the full eigensolution of a real unsymmetric matrix having a pair of complex conjugate eigenvalues. Scaling the left eigenvectors by ½, ¼(1 − i) and ¼(1 + i) respectively, so that the biorthogonality condition is satisfied, and forming matrices **P** and **Q**, gives $\mathbf{P}^H\mathbf{AQ} = \mathbf{\Lambda}$ as

$$\begin{bmatrix} \tfrac{1}{2} & -\tfrac{1}{2} & \tfrac{1}{2} \\ \tfrac{1}{4}+\tfrac{1}{4}i & \tfrac{1}{4}-\tfrac{1}{4}i & -\tfrac{1}{4}-\tfrac{1}{4}i \\ \tfrac{1}{4}-\tfrac{1}{4}i & \tfrac{1}{4}+\tfrac{1}{4}i & -\tfrac{1}{4}+\tfrac{1}{4}i \end{bmatrix}\begin{bmatrix} -1 & -1 & 5 \\ -4 & -2 & 4 \\ 1 & -5 & 3 \end{bmatrix}\begin{bmatrix} 1 & 1 & 1 \\ 0 & 1+i & 1-i \\ 1 & i & -i \end{bmatrix} = \begin{bmatrix} 4 & 0 & 0 \\ 0 & -2+4i & 0 \\ 0 & 0 & -2-4i \end{bmatrix} \tag{8.54}$$

Although it is not possible to guarantee that transformations with *real* biorthogonal matrices will always completely diagonalize a real unsymmetric matrix, it is normally possible, using real arithmetic only, to reach a form in which single elements and/or 2 x 2 submatrices are arranged in diagonal block form. For instance, the matrix (Table 8.4) may be transformed biorthogonally as follows:

$$\begin{bmatrix} \tfrac{1}{2} & -\tfrac{1}{2} & \tfrac{1}{2} \\ \tfrac{1}{2} & \tfrac{1}{2} & -\tfrac{1}{2} \\ 0 & 1 & 0 \end{bmatrix}\begin{bmatrix} -1 & -1 & 5 \\ -4 & -2 & 4 \\ 1 & -5 & 3 \end{bmatrix}\begin{bmatrix} 1 & 1 & 0 \\ 0 & 0 & 1 \\ 1 & -1 & 1 \end{bmatrix} = \begin{bmatrix} 4 & 0 & 0 \\ 0 & -6 & 4 \\ 0 & -8 & 2 \end{bmatrix} \tag{8.55}$$

revealing the real eigenvalue immediately and the complex conjugate pair by eigensolution of the 2 x 2 matrix $\begin{bmatrix} -6 & 4 \\ -8 & 2 \end{bmatrix}$.

A further complication in the eigensolution of unsymmetric matrices is the possibility that the matrix is *defective*. A defective matrix has two or more equal eigenvalues for which there is only one distinct corresponding left or right eigenvector. A simple example of such a matrix is $\begin{bmatrix} 8 & 1 \\ 0 & 8 \end{bmatrix}$ which, according to the the determinantal equation, has two equal eigenvalues $\lambda = 8$, yet it is only possible to find one right eigenvector $\{1 \;\; 0\}$. If a slight adjustment is made to a defective matrix two almost parallel eigenvectors usually appear, e.g. the matrix $\begin{bmatrix} 8+\epsilon & 1 \\ 0 & 8 \end{bmatrix}$ has right eigenvectors $\{1 \;\; 0\}$ and $\{1 \;\; -\epsilon\}$. Due to the presence of rounding errors, matrices will rarely appear as defective in numerical computation. However, there is a distinct possibility that almost parallel eigenvectors can occur, and this must be allowed for if eigensolution methods are to be comprehensive and reliable.

8.9 SIMILARITY TRANSFORMATIONS

For unsymmetric matrices any similarity transformation may be adopted in the eigensolution process. If **N** is any non-singular matrix of order $n \times n$ the transformed matrix $\bar{\mathbf{A}} = \mathbf{N}^{-1}\mathbf{A}\mathbf{N}$ (equation 8.1) is such that

$$\bar{\mathbf{A}}(\mathbf{N}^{-1}\mathbf{Q}) = \mathbf{N}^{-1}\mathbf{A}\mathbf{Q} = (\mathbf{N}^{-1}\mathbf{Q})\mathbf{\Lambda} \tag{8.56}$$

Hence it must have the same eigenvalues as **A**, the matrix of right eigenvectors being

$$\bar{\mathbf{Q}} = \mathbf{N}^{-1}\mathbf{Q} \tag{8.57}$$

Since the biorthogonal condition of equation (8.51) implies that $\mathbf{P}^H$ is the inverse of **Q**, a similarity transformation using the matrix of right eigenvectors produces a diagonal matrix whose elements are the eigenvalues.

Unsymmetric eigensolution methods use a sequence of transformations in order to eliminate off-diagonal elements of a matrix. However, three reasons mitigate against aiming at a full diagonalization of the matrix, namely:

(a) The eigenvalues of a triangular matrix are equal to the diagonal elements (section 1.17), and hence transformation to triangular form is sufficient.

(b) If the matrix is defective or nearly defective it may still be reduced to triangular form by similarity transformations, whereas a full diagonalization would either not be possible or would run into problems involving accuracy.

(c) Where the original matrix is real it is advantageous to keep the similarity transformations entirely real.

8.10 REDUCTION TO UPPER HESSENBERG FORM

It is possible to transform any real matrix to Hessenberg form by stable similarity transformations in real arithmetic by a non-iterative process. The upper Hessenberg form consists of an upper triangle together with the subdiagonal, and therefore has the following pattern of non-zero elements:

$$\mathbf{M} = \begin{bmatrix} \times & \times & \times & \times & \times & \times \\ \times & \times & \times & \times & \times & \times \\ & \times & \times & \times & \times & \times \\ & & \times & \times & \times & \times \\ & & & \times & \times & \times \\ & & & & \times & \times \end{bmatrix} \tag{8.58}$$

This reduction may be performed by the orthogonal transformations of Givens and Householder in approximately $10n^3/3$ and $5n^3/3$ multiplications respectively (if the original matrix is fully populated). However, it is more efficient to use elementary stabilized transformations which require only about $5n^3/6$ multiplications.

Reduction by elementary stabilized transformations is accomplished one column at a time so that $n - 2$ transformations of the form

$$\mathbf{A}^{(k+1)} = \mathbf{N}_k^{-1}\mathbf{A}^{(k)}\mathbf{N}_k \tag{8.59}$$

are required. The transformation matrix $\mathbf{N}_k$ used to eliminate the appropriate elements in column k is a unit matrix in which subdiagonal elements have been added in column $k + 1$.

Thus

$$\left.\begin{aligned}
\mathbf{N} &= \begin{bmatrix} 1 & & & & & \\ & \ddots & & & & \\ & & 1 & & & \\ & & & 1 & & \\ & & & n_{k+2,k+1} & 1 & \\ & & & \vdots & & \ddots & \\ & & & n_{n,k+1} & & & 1 \end{bmatrix} \\
&\text{and} \\
\mathbf{N}^{-1} &= \begin{bmatrix} 1 & & & & & \\ & \ddots & & & & \\ & & 1 & & & \\ & & & 1 & & \\ & & & -n_{k+2,k+1} & 1 & \\ & & & \vdots & & \ddots & \\ & & & -n_{n,k+1} & & & 1 \end{bmatrix}
\end{aligned}\right\} \tag{8.60}$$

Figure 8.5 shows the row and column operations within the matrix due to the k-th transformation.

The only divisor in the transformation is the element $a_{k+1,k}^{(k)}$, which can be considered as a pivot. The transformation is numerically stable if this pivot is the element of largest modulus below the diagonal in column k. To achieve this the element of largest modulus of the set $a_{k+1,k}, a_{k+2,k}, \ldots, a_{n,k}$ is located, and then corresponding rows and columns are interchanged to place it in the pivot position (see section 1.17, property (h)).

If the eigenvectors of the original matrix are to be determined from the eigenvectors of the upper Hessenberg matrix when these are later evaluated, details of the transformations and also the interchanges have to be stored. The interchanges

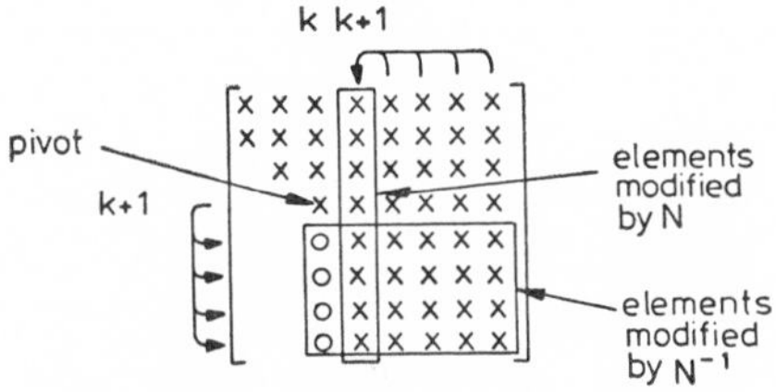

Figure 8.5 Elementary transformation to eliminate elements on column k

can be recorded in a one-dimensional integer array acting as a permutation vector. If the matrix $\mathbf{A}^{(1)}$ is the original matrix with rows and columns permuted into their final order, the upper Hessenberg matrix can be written as

$$\mathbf{H} = \mathbf{A}^{(n-1)} = \mathbf{N}_{n-2}^{-1} \ldots \mathbf{N}_2^{-1}\mathbf{N}_1^{-1}\mathbf{A}^{(1)}\mathbf{N}_1\mathbf{N}_2 \ldots \mathbf{N}_{n-2} \tag{8.61}$$

But it can be shown that

$$\mathbf{N}_1\mathbf{N}_2 \ldots \mathbf{N}_{n-2} = \begin{bmatrix} 1 & & & & & \\ & 1 & & & & \\ & n_{32} & 1 & & & \\ & \cdot & \cdot & \cdot & & \\ & n_{n-1,2} & n_{n-1,3} & \cdots & 1 & \\ & n_{n2} & n_{n3} & \cdots & n_{n,n-1} & 1 \end{bmatrix} = \mathbf{N} \tag{8.62}$$

Hence the whole reduction to upper Hessenberg form can be represented by

$$\mathbf{A}^{(1)}\mathbf{N} = \mathbf{N}\mathbf{H} \tag{8.63}$$

If $\bar{\mathbf{q}}_i$ is an eigenvector of $\mathbf{H}$

$$\mathbf{H}\bar{\mathbf{q}}_i = \lambda_i\bar{\mathbf{q}}_i \tag{8.64}$$

and

$$\mathbf{A}^{(1)}\mathbf{N}\bar{\mathbf{q}}_i = \lambda_i\mathbf{N}\bar{\mathbf{q}}_i \tag{8.65}$$

Thus the corresponding eigenvector of $\mathbf{A}^{(1)}$ is given by

$$\mathbf{q}_i^{(1)} = \mathbf{N}\bar{\mathbf{q}}_i \tag{8.66}$$

the corresponding eigenvector of the original matrix being recovered from $q_i^{(1)}$ by reference to the permutation vector. In the computer implementation of this method the elements n_{ij} can be stored in the locations vacated by elements $a_{i,j-1}$ as they are eliminated.

Table 8.5 Reduction to upper Hessenberg form of matrix (8.67). (The boxed elements in the bottom left are the transformation parameters which contribute to the lower triangle of N)

Transformation	Permutation vector	Permuted matrix					Transformed matrix			
First	1	−2	2	−1	−2		−2	1.625	−1	−2
	4	−8	3	−14	−8	→	−8	0.25	−14	−8
	3	−1	0	2	0		0.125	0.2188	3.75	1
	2	−1	−2	−13	−3		0.125	−4.0313	−11.25	−2
Second	1	−2	1.625	−2	−1		−2	1.625	−1.9457	−1
	4	−8	0.25	−8	−14	→	−8	0.25	−7.2403	−14
	2	0.125	−4.0313	−2	−11.25		0.125	−4.0313	−1.3895	−11.25
	3	0.125	0.2188	1	3.75		0.125	−0.0543	0.7211	3.1395

Table 8.5 shows the operations in the reduction of the matrix

$$\mathbf{A} = \begin{bmatrix} -2 & -2 & -1 & 2 \\ -1 & -3 & -13 & -2 \\ -1 & 0 & 2 & 0 \\ -8 & -8 & -14 & 3 \end{bmatrix} \tag{8.67}$$

to upper Hessenberg form, which is achieved by just two transformations. The lower triangular elements of **N** are shown shifted one space to the right. These elements may be used for eigenvector transformations. For instance, an eigenvector of **H** corresponding to the eigenvalue $\lambda = 1$ is $\mathbf{q}^{(3)} = \{-0.3478 \quad 0.3478 \quad 1 \quad -0.3370\}$. The permuted eigenvector of the original matrix is therefore given by

$$\mathbf{q}^{(1)} = \begin{bmatrix} 1 & & & \\ & 1 & & \\ & 0.125 & 1 & \\ & 0.125 & -0.0543 & 1 \end{bmatrix} \begin{bmatrix} -0.3478 \\ 0.3478 \\ 1 \\ -0.3370 \end{bmatrix} = \begin{bmatrix} -0.3478 \\ 0.3478 \\ 1.0435 \\ -0.3478 \end{bmatrix} \tag{8.68}$$

Hence the corresponding normalized eigenvector of matrix (8.67) is $\{-0.3333 \quad 1 \quad -0.3333 \quad 0.3333\}$.

8.11 THE LR TRANSFORMATION

Rutishauser's LR method uses a similarity transformation involving the triangular factors of a matrix to produce another matrix with greater diagonal dominance. If this transformation is applied iteratively, the diagonal elements normally converge to the eigenvalues of the original matrix. It is not an efficient method for fully populated matrices. However, when the matrix has certain patterns of zero elements, these are retained by the transformation, so improving the efficiency. In particular, band and Hessenberg matrices may be transformed within their own store.

If $\mathbf{A}^{(k)}$ is the matrix obtained from the $(k-1)$-th transformation, its triangular decomposition may be written as

$$\mathbf{A}^{(k)} = \mathbf{L}_k \mathbf{R}_k \tag{8.69}$$

where, in the simple form, $\mathbf{L}_k$ is a lower triangular matrix with unit diagonal elements and $\mathbf{R}_k$ is an upper triangular matrix. Multiplying the factors in reverse order completes the transformation, i.e.

$$\mathbf{A}^{(k+1)} = \mathbf{R}_k \mathbf{L}_k \tag{8.70}$$

Since $\mathbf{L}_k$ is non-singular

$$\mathbf{A}^{(k+1)} = \mathbf{L}_k^{-1}\mathbf{A}^{(k)}\mathbf{L}_k \tag{8.71}$$

which shows that $\mathbf{A}^{(k+1)}$ is obtained from $\mathbf{A}^{(k)}$ by a similarity transformation and hence has the same eigenvalues.

Consider the transformation of a 3 x 3 tridiagonal matrix. If

$$\mathbf{A}^{(k)} = \begin{bmatrix} a_{11} & a_{12} & \\ a_{21} & a_{22} & a_{23} \\ & a_{32} & a_{33} \end{bmatrix} = \begin{bmatrix} 1 & & \\ l_{21} & 1 & \\ & l_{32} & 1 \end{bmatrix} \begin{bmatrix} r_{11} & r_{12} & \\ & r_{22} & r_{23} \\ & & r_{33} \end{bmatrix} \tag{8.72}$$

then

$$\left.\begin{aligned} &r_{11} = a_{11}, \quad r_{12} = a_{12}, \quad l_{21} = a_{21}/r_{11}, \quad r_{22} = a_{22} - l_{21}r_{12} \\ &r_{23} = a_{23}, \quad l_{32} = a_{32}/r_{22}, \quad r_{33} = a_{33} - l_{32}r_{23} \end{aligned}\right\} \tag{8.73}$$

and

$$\mathbf{A}^{(k+1)} = \begin{bmatrix} a_{11}^* & a_{12}^* & \\ a_{21}^* & a_{22}^* & a_{23}^* \\ & a_{32}^* & a_{33}^* \end{bmatrix} = \begin{bmatrix} r_{11} & r_{12} & \\ & r_{22} & r_{23} \\ & & r_{33} \end{bmatrix} \begin{bmatrix} 1 & & \\ l_{21} & 1 & \\ & l_{32} & 1 \end{bmatrix} \tag{8.74}$$

giving

$$\left.\begin{aligned} &a_{11}^* = r_{11} + r_{12}l_{21}, \quad a_{12}^* = r_{12}(=a_{12}), \quad a_{21}^* = r_{22}l_{21} \\ &a_{22}^* = r_{22} + r_{23}l_{32}, \quad a_{23}^* = r_{23}(=a_{23}), \quad a_{32}^* = r_{33}l_{32}, \quad a_{33}^* = r_{33} \end{aligned}\right\} \tag{8.75}$$

From equations (8.73) and (8.75) it is apparent that, in computer implementation, the elements of $\mathbf{L}_k$ and $\mathbf{R}_k$ can overwrite the corresponding element of $\mathbf{A}^{(k)}$, which can themselves be overwritten by the elements of $\mathbf{A}^{(k+1)}$, provided that they are formed in the order specified. The operations involved for a 3 x 3 tridiagonal matrix are shown in Table 8.6. The computed eigenvalues are 3.2469, 1.5550 and 0.1981. After $\mathbf{A}^{(6)}$ has been determined the last row and column remain unchanged (to an accuracy of four decimal figures), and so they may be omitted from subsequent transformations by deflating the matrix to order 2 x 2.

A possible variation of the LR method for symmetric positive definite matrices is to use Choleski decomposition for the triangular factorization. This has the advantage that symmetry is maintained in the sequence of transformed matrices, but is likely to involve more computing time when matrices having a small bandwidth are analysed if the computation of square roots is not very rapid.

8.12 CONVERGENCE OF THE LR METHOD

It can be proved that the eigenvalues of lowest modulus tend to converge first and that the convergence rate for λ_i, when $\lambda_{i+1}, \ldots, \lambda_n$ have already been obtained, depends on the ratio $|\lambda_i|/|\lambda_{i-1}|$. Thus in Table 8.6 elements $a_{32}^{(k)}$ reduces by a

Table 8.6 LR transformation of a 3 × 3 tridiagonal matrix

k	$\mathbf{A}^{(k)}$	$\mathbf{L}_k\backslash\mathbf{R}_k$
1	$\begin{bmatrix} 2 & -1 & \\ -1 & 2 & -1 \\ & -1 & 1 \end{bmatrix}$	$\begin{bmatrix} 2 & -1 & \\ -0.5 & 1.5 & -1 \\ & -0.6667 & 0.3333 \end{bmatrix}$
2	$\begin{bmatrix} 2.5 & -1 & \\ -0.75 & 2.1667 & -1 \\ & -0.2222 & 0.3333 \end{bmatrix}$	$\begin{bmatrix} 2.5 & -1 & \\ -0.3 & 1.8667 & -1 \\ & -0.1190 & 0.2143 \end{bmatrix}$
3	$\begin{bmatrix} 2.8 & -1 & \\ -0.56 & 1.9857 & -1 \\ & -0.0255 & 0.2143 \end{bmatrix}$	$\begin{bmatrix} 2.8 & -1 & \\ -0.2 & 1.7857 & -1 \\ & -0.0143 & 0.2000 \end{bmatrix}$
4	$\begin{bmatrix} 3 & -1 & \\ -0.3571 & 1.8000 & -1 \\ & -0.0028 & 0.2000 \end{bmatrix}$	$\begin{bmatrix} 3 & -1 & \\ -0.1190 & 1.6810 & -1 \\ & -0.0017 & 0.1983 \end{bmatrix}$
5	$\begin{bmatrix} 3.1190 & -1 & \\ -0.2001 & 1.6827 & -1 \\ & -0.0003 & 0.1983 \end{bmatrix}$	$\begin{bmatrix} 3.1190 & -1 & \\ -0.0642 & 1.6185 & -1 \\ & -0.0002 & 0.1981 \end{bmatrix}$
6	$\begin{bmatrix} 3.1832 & -1 & \\ -0.1038 & 1.6187 & -1 \\ & -0.0000 & 0.1981 \end{bmatrix}$	$\begin{bmatrix} 3.1832 & -1 & \\ -0.0326 & 1.5861 & -1 \\ & -0.0000 & 0.1981 \end{bmatrix}$
	-- -- --	-- -- --
15	$\begin{bmatrix} 3.2469 & -1 & \\ -0.0001 & 1.5550 & -1 \\ & 0.0000 & 0.1981 \end{bmatrix}$	$\begin{bmatrix} 3.2469 & -1 & \\ 0.0000 & 1.5550 & -1 \\ & 0.0000 & 0.1981 \end{bmatrix}$

factor of approximately 0.198/1.555 at each iteration and in later iterations element $a_{21}^{(k)}$ reduces by a factor of approximately 1.5550/3.2469. By adopting a shift of origin towards the lowest eigenvalue the convergence rate can be improved. Thus, by modifying the procedure of Table 8.6 so that

$$\bar{\mathbf{A}}^{(7)} = \mathbf{A}^{(7)} - 1.5344\mathbf{I} \tag{8.76}$$

the subsequent iterations with the deflated matrix become

$$\left.\begin{aligned} &\bar{\mathbf{A}}^{(7)} = \begin{bmatrix} 1.6814 & -1 \\ -0.0517 & 0.0517 \end{bmatrix}, \quad \bar{\mathbf{A}}^{(8)} = \begin{bmatrix} 1.7122 & -1 \\ -0.0006 & 0.0209 \end{bmatrix}, \\ &\bar{\mathbf{A}}^{(9)} = \begin{bmatrix} 1.7126 & -1 \\ 0.0000 & 0.0206 \end{bmatrix} \end{aligned}\right\} \tag{8.77}$$

and the first two eigenvalues of the original matrix are obtained by restoring the shift, i.e. $\lambda_1 = 1.7126 + 1.5344$ and $\lambda_2 = 0.0206 + 1.5344$.

However, the process will break down if a zero pivot is encountered in the triangular decomposition, and the accuracy of the results will be seriously impaired if a small pivot is used. For symmetric matrices strong pivots can be assured by ensuring that the shift places the origin outside (and below) the set of Gerschgorin discs formed from the rows of the matrix. (The sequence of transformed matrices obtained by the Choleski LR technique will then all be positive definite, and those obtained by the standard LR technique will be similar, except for scaling factors applied to the rows and columns.) The shift adopted in equation (8.76) is the maximum possible which satisfies this criterion. The shift may be adjusted at each transformation so that the maximum convergence rate is obtained, consistent with the condition that the origin lies below the range of possible eigenvalues. If this is done with the matrix shown in Table 8.6, the number of transformations for the eigensolution of the matrix is reduced from fifteen to six.

An alternative method for ensuring that pivots are strong, which can be employed for unsymmetric as well as symmetric matrices, is to adopt a row interchange scheme in the triangular decomposition. However, the alternative QR method is favoured because of its greater reliability.

8.13 THE QR TRANSFORMATION

The QR transformation (Francis, 1961) is similar to the LR transformation except that the lower triangular matrix is replaced by an orthogonal matrix. Thus if

$$\mathbf{A}^{(k)} = \mathbf{Q}_k \mathbf{R}_k \tag{8.78}$$

where $\mathbf{Q}_k$ is an orthogonal matrix and $\mathbf{R}_k$ is an upper triangular matrix,

$$\mathbf{A}^{(k+1)} = \mathbf{R}_k \mathbf{Q}_k \tag{8.79}$$

It follows that, since

$$\mathbf{A}^{(k+1)} = \mathbf{Q}_k^{-1} \mathbf{A}^{(k)} \mathbf{Q}_k \tag{8.80}$$

this is again a similarity transformation and the eigenvalues are preserved. In practice equation (8.78) is converted to the form

$$\mathbf{Q}_k^T \mathbf{A}^{(k)} = \mathbf{R}_k \tag{8.81}$$

where $\mathbf{Q}_k^T$ is the product of orthogonal matrices of Givens or Householder type. For the case of either a tridiagonal or an upper Hessenberg matrix, there will be $n-1$ orthogonal matrices $\mathbf{N}_i$ each eliminating one subdiagonal element $a_{i+1,i}$, giving

$$\left.\begin{aligned} &\mathbf{N}_{n-1}^T \ldots \mathbf{N}_2^T \mathbf{N}_1^T \mathbf{A}^{(k)} = \mathbf{R}_k \\ &\text{and} \\ &\mathbf{A}^{(k+1)} = \mathbf{R}_k \mathbf{N}_1 \mathbf{N}_2 \ldots \mathbf{N}_{n-1} \end{aligned}\right\} \tag{8.82}$$

as the operative equations for one iteration. For this particular use Householder's transformation technique is no more efficient than Givens' transformation technique.

The operations to perform the first QR iteration for the 3 x 3 tridiagonal matrix (Table 8.6) using Givens' transformations are as follows:

$$\underset{\mathbf{N}_1^T}{\begin{bmatrix} 0.8944 & -0.4472 & \\ 0.4472 & 0.8944 & \\ & & 1 \end{bmatrix}} \underset{\mathbf{A}^{(1)}}{\begin{bmatrix} 2 & -1 & \\ -1 & 2 & -1 \\ & -1 & 1 \end{bmatrix}} = \underset{\text{Intermediate matrix}}{\begin{bmatrix} 2.2361 & -1.7889 & 0.4472 \\ & 1.3416 & -0.8944 \\ & -1 & 1 \end{bmatrix}}$$

$$\underset{\mathbf{N}_2^T}{\begin{bmatrix} 1 & & \\ & 0.8018 & -0.5976 \\ & 0.5976 & 0.8018 \end{bmatrix}} \underset{\text{Intermediate matrix}}{\begin{bmatrix} 2.2361 & -1.7889 & 0.4472 \\ & 1.3416 & -0.8944 \\ & -1 & 1 \end{bmatrix}}$$

$$= \underset{\mathbf{R}_1}{\begin{bmatrix} 2.2361 & -1.7889 & 0.4472 \\ & 1.6733 & -1.3148 \\ & & 0.2673 \end{bmatrix}}$$

$$\underset{\mathbf{R}_1}{\begin{bmatrix} 2.2361 & -1.7889 & 0.4472 \\ & 1.6733 & -1.3148 \\ & & 0.2673 \end{bmatrix}} \underset{\mathbf{N}_1}{\begin{bmatrix} 0.8944 & 0.4472 & \\ -0.4472 & 0.8944 & \\ & & 1 \end{bmatrix}} \underset{\mathbf{N}_2}{\begin{bmatrix} 1 & & \\ & 0.8018 & 0.5976 \\ & -0.5976 & 0.8018 \end{bmatrix}}$$

$$= \underset{\mathbf{A}^{(2)}}{\begin{bmatrix} 2.8000 & -0.7483 & 0.0000 \\ -0.7483 & 1.9857 & -0.1597 \\ & -0.1597 & 0.2143 \end{bmatrix}} \qquad (8.83)$$

8.14 ORIGIN SHIFT WITH THE QR METHOD

The QR method is similar to the LR method insofar as convergence is to the eigenvalue of least modulus, and also the rate of convergence for λ_i (assuming that $\lambda_{i+1}, \ldots, \lambda_n$ have already been obtained) depends on the ratio $|\lambda_i|/|\lambda_{i-1}|$). However, the amount of computation per iteration tends to be greater. For a full upper Hessenberg matrix approximately $4n^2$ multiplications are required per iteration as opposed to approximately n^2 multiplications for an LR iteration. The main advantage of the QR method is that any shift strategy can be adopted without incurring either the need to interchange rows or the likelihood of loss of accuracy due to small pivots. Accuracy is important for iterative transformation methods because of possible accumulating effects due to errors being carried forward from one transformation to the next.

For upper Hessenberg and tridiagonal matrices, a shift strategy may be adopted which leads to rapid convergence. In both of these types of matrix, the 2 x 2 submatrix formed by extracting rows and columns $n-1$ and n may be analysed and its eigenvalue of lowest modulus taken as an estimate to the next eigenvalue of the matrix. If μ_k is this estimate for a matrix $\mathbf{A}^{(k)}$, then a full QR iteration with shift and restoration is described by

$$\left.\begin{aligned} \mathbf{Q}_k^T(\mathbf{A}^{(k)} - \mu_k\mathbf{I}) &= \mathbf{R}_k \\ \mathbf{A}^{(k+1)} &= \mathbf{R}_k\mathbf{Q}_k + \mu_k\mathbf{I} \end{aligned}\right\} \tag{8.84}$$

Considering the upper Hessenberg matrix derived in Table 8.5, successive QR iterations with shift and restoration are

$$\left.\begin{aligned}
\mathbf{A}^{(1)} &= \begin{bmatrix} -2 & 1.625 & -1.9457 & -1 \\ -8 & 0.25 & -7.2403 & -14 \\ & -4.0313 & -1.3895 & -11.25 \\ & & 0.7211 & 3.1395 \end{bmatrix} \text{(no shift)} \\
\mathbf{A}^{(2)} &= \begin{bmatrix} -1.3824 & -4.1463 & 3.8908 & -16.7771 \\ -4.1782 & -1.5417 & 9.0249 & -6.9591 \\ & -0.8854 & 2.4424 & -2.5105 \\ & & -0.5673 & 0.4816 \end{bmatrix}, \mu_2 = 0.0824 \\
\mathbf{A}^{(3)} &= \begin{bmatrix} -4.1232 & -0.6246 & -13.0936 & -9.0262 \\ -3.2933 & 1.5194 & -4.4614 & -13.0121 \\ & -0.5541 & 1.8971 & -1.3854 \\ & & 0.1679 & 0.7067 \end{bmatrix}, \mu_3 = 0.9532 \\
\mathbf{A}^{(4)} & \begin{bmatrix} -4.2406 & -10.1146 & 10.2397 & -14.0418 \\ -0.5364 & 0.3621 & 2.0738 & 4.3242 \\ & 1.5113 & 2.8890 & 4.6577 \\ & & -0.0023 & 0.9895 \end{bmatrix}, \mu_4 = 0.9952 \\
\mathbf{A}^{(5)} &= \begin{bmatrix} -5.2726 & 12.4960 & 6.5306 & -13.5382 \\ -0.1593 & 2.9106 & 1.1057 & 3.0299 \\ & -1.4219 & 1.3620 & -6.7360 \\ & & 0.0000 & 1.0000 \end{bmatrix}, \mu_5 = 1.0001 \\
\mathbf{A}^{(6)} &= \begin{bmatrix} -4.9545 & -5.1034 & -13.3464 & -13.4570 \\ -0.0542 & 2.2859 & 1.7645 & -7.0021 \\ & -0.5966 & 1.6682 & 2.7783 \\ & & 0.0000 & 1.0000 \end{bmatrix}, \mu_6 = 1.0000
\end{aligned}\right\} \tag{8.85}$$

At this stage the eigenvalue $\lambda = 1$ has been predicted to at least four decimal places. Once the subdiagonal element in the last row has become zero the last row and column can be ignored and hence the matrix is deflated.

8.15 DISCUSSION OF THE QR METHOD

Complex conjugate eigenvalues

The technique as described is sufficient for the eigensolution of real matrices having real eigenvalues. However, for the case of a general real matrix which may have complex conjugate eigenvalues, effective acceleration by shifting is not possible using entirely real arithmetic. If iteration is continued in the example, equation (8.85), using $\mu = 1$, then

$$\left.\begin{aligned}
\mathbf{A}^{(7)} &= \begin{bmatrix} -5.0005 & -0.8500 & -14.2243 \\ -0.0133 & 1.7405 & -1.9183 \\ & 0.5642 & 2.2600 \end{bmatrix}, \\
\mathbf{A}^{(8)} &= \begin{bmatrix} -5.0024 & 9.2740 & -10.8232 \\ -0.0021 & 1.2947 & 1.1532 \\ & -1.2977 & 2.7077 \end{bmatrix}, \\
\mathbf{A}^{(9)} &= \begin{bmatrix} -4.9992 & -12.5929 & 6.6782 \\ -0.0005 & 2.6697 & -0.9881 \\ & 1.4665 & 0.3295 \end{bmatrix}, \\
\mathbf{A}^{(10)} &= \begin{bmatrix} -5.0001 & 5.0580 & 13.3259 \\ -0.0002 & 2.3239 & 1.8616 \\ & -0.5936 & 1.6762 \end{bmatrix}
\end{aligned}\right\} \tag{8.86}$$

Although element a_{32} is not stabilizing at all, element a_{21} is tending to zero, and the eigenvalues of the 2 x 2 submatrix from rows and columns 2 and 3 are tending to the complex conjugate eigenvalues of the matrix, namely $2 \pm i$. It is possible to accelerate the convergence by resorting to complex arithmetic when the eigenvalue predictions become complex. However, the use of the double QR step is a much more satisfactory procedure to adopt.

Double QR transformation

This is an extension of the QR method which performs the equivalent of two basic QR transformations in one step. This not only has the effect of allowing acceleration of complex conjugate pairs using real arithmetic only, but also produces some economy in computation. The double QR step using a Householder-type transformation can be accomplished in approximately $5n^2$ multiplications (as opposed to $8n^2$ multiplications for a double application of the basic QR iteration).

Equal eigenvalues

The QR method is able to cope satisfactorily with matrices having eigenvalues of equal modulus, equal eigenvalues and non-linear elementary divisors. However, in the last two cases, attempts to compute the corresponding eigenvectors must take into account their special properties.

Symmetric band matrices

When applied to symmetric band matrices, the QR transformation maintains symmetry. (This may be verified by replacing $\mathbf{Q}_k^{-1}$ with $\mathbf{Q}_k^T$ in equation 8.80.) Since the bandwidth of a symmetric matrix is retained, the QR method provides a method for the eigensolution of tridiagonal and other band symmetric matrices. The example, equations (8.83), illustrates this. By evaluating just the lower triangle of the sequence of matrices computation is simplified, and, since the eigenvalues will be real, the double-step procedure is not necessary.

It has been proved that, for a symmetric matrix, the QR transformation produces, in one iteration, the same result that the Choleski LR transformation produces in two iterations (assuming that the same shift is employed).

Calculation of eigenvectors

On a computer it is not very practical to retain the transformation parameters for later use. It is usual to determine any required eigenvectors by inverse iterations. (This is relatively efficient for upper Hessenberg and tridiagonal matrices.)

Computational requirements

The average number of iterations per eigenvalue is about two or three. The reduction in size of the active part of the matrix, as eigenvalues are predicted, should be allowed for in assessing the computational requirement. Hence, to evaluate all the eigenvalues of an $n \times n$ unsymmetric matrix by first reducing it to upper Hessenberg form and then adopting a QR iteration requires approximately $4n^3$ multiplications, of which only about one-fifth are required for the reduction to upper Hessenberg form. For a band symmetric matrix approximately $2nb^2$ multiplications are required per iteration, giving approximately $2.5n^2b^2$ multiplications for a full eigensolution by deflation. From this result it is seen to be numerically more efficient to first reduce a band symmetric matrix to tridiagonal form by the technique of section 8.7 rather than to apply QR transformations directly if a full eigensolution is required. However, this does not apply in the case where only a few of the lowest modulus eigenvalues are required.

8.16 THE APPLICATION OF TRANSFORMATION METHODS

Transformation methods are particularly suitable for obtaining all the eigenvalues of fully populated matrices. Useful general procedures are as follows:

(a) *for symmetric matrices:* Householder reduction to tridiagonal form, then QR iteration;
(b) *for unsymmetric matrices:* elementary stabilized transformations to upper Hessenberg form, then double QR iteration.

In both of these cases the amount of computation is a function of n^3, with a significant amount of further computation being necessary to evaluate any required eigenvectors. These methods are very powerful for the eigensolution of matrices whose order does not exceed 100. However, for larger problems, the computing time and storage requirements rise rapidly and, in the region of $n = 1{,}000$, start to become prohibitive even on the largest computers.

For symmetric matrices advantage may be taken of sparseness if the equations can be arranged in a narrow-band form. This may be done either by using the Rutishauser reduction to tridiagonal form or by direct application of an LR or QR algorithm. However, it is not possible to take full advantage of the variable bandwidth format as may be done in the solution of simultaneous equations which are symmetric and positive definite. For unsymmetric matrices it is possible to take advantage of the sparseness if they can be arranged so that the non-zero elements in the lower triangle all lie close to the leading diagonal. This may then be reduced to upper Hessenberg form, using stable similarity transformations, by peeling away the lower bands in a similar way to that in Rutishauser's method where symmetric band matrices are reduced to tridiagonal form.

Moler and Stewart (1973) have developed the QZ algorithm, as an extension of the QR algorithm, for the solution of the linearized eigenvalue problem $\mathbf{Ax} = \lambda\mathbf{Bx}$. They advocate its use when the matrix $\mathbf{B}$ is singular or near singular since it does not require either the explicit or implicit inversion of $\mathbf{B}$. It is less efficient than reducing the problem to standard eigenvalue form by means of equation (7.26) and then proceeding as in (b) above, which is possible if $\mathbf{B}$ is well-conditioned. However, even if $\mathbf{B}$ is singular or near singular, it is usually possible to perform a transformation to standard eigenvalue form by means of equation (7.35) thus providing a more efficient solution procedure than is possible using the QZ algorithm. Large order linearized eigenvalue problems are discussed in the next two chapters.

BIBLIOGRAPHY

Fox, L. (1964). *An Introduction to Numerical Linear Algebra*, Clarendon Press, Oxford.

Francis, J. G. F. (1961, 1962). 'The QR transformation, Parts I and II'. *Computer J.*, 4, 265–271 and 332–345.

Froberg, C. E. (1969). *Introduction to Numerical Linear Algebra*, 2nd ed. Addison-Wesley, Reading, Massachusetts.

Givens, J. W. (1954). 'Numerical computation of the characteristic values of a real symmetric matrix'. *Oak Ridge National Laboratory Report* ORNL-1574.

Gourlay, A. R., and Watson, G. A. (1973). *Computational Methods for Matrix Eigenproblems*, Wiley, London.

Hammarling, S. (1974). 'A note on modifications to the Givens plane rotation'. *J. Inst. Maths. Applics.*, 13, 215–218.

Moler, C. B., and Stewart, G. W. (1973). 'An algorithm for generalized matrix eigenvalue problems'. *SIAM J. Numer. Anal.*, 10, 241–256. (The QZ algorithm.)

Rutishauser, H. (1958). 'Solution of eigenvalue problems with the LR transformation'. *Nat. Bur. Standards. Appl. Math. Ser.*, 49, 47–81.

Rutishauser, H. (1963). 'On Jacobi rotation patterns'. *Proc. AMS Symp. in Appl. Math.*, 15, 219–239. (Reduction of band symmetric matrices to tridiagonal form.)

Schwarz, H. R., Rutishauser, H., and Stiefel, E. (1973). *Numerical Analysis of Symmetric Matrices*, English translation: Prentice-Hall, Englewood Cliffs, New Jersey. (Includes some ALGOL procedures.)

Stewart, G. W. (1973). *Introduction to Matrix Computations*, Academic Press, New York.

Stewart, G. W. (1976). 'The economical storage of plane rotations'. *Numer. Math.*, 25, 137–138.

Wilkinson, J. H. (1960). 'Householder's method for the solution of the algebraic eigenproblem'. *Computer J.*, 3, 23–27.

Wilkinson, J. H. (1965). *The Algebraic Eigenvalue Problem*, Clarendon Press, Oxford. (The most comprehensive general reference on transformation methods.)

Wilkinson, J. H., and Reinsch, C. (1971). *Handbook for Automatic Computation, Vol. II Linear Algebra*, Springer-Verlag, Berlin. (Contains algorithms for most transformation methods discussed in this chapter.)

Chapter 9
Sturm Sequence Methods

9.1 THE CHARACTERISTIC EQUATION

The characteristic equation of an $n \times n$ matrix A may be derived by expansion of the determinantal equation

$$|\mathrm{A} - \lambda \mathrm{I}| = 0 \tag{9.1}$$

into an n-th order polynomial in λ (section 1.16). If this polynomial equation can be determined, it can in turn be solved to give the eigenvalues of the matrix. However, it is difficult to organize a storage scheme in which it is possible to obtain this polynomial by direct expansion for a general $n \times n$ matrix. Also the computational requirement is excessive for all but very small values of n.

An alternative method is to determine

$$f_i = |\mathrm{A} - \mu_i \mathrm{I}| \tag{9.2}$$

for $n + 1$ values of μ_i, and then to find the polynomial

$$\mu^n + c_{n-1}\mu^{n-1} + \cdots + c_1\mu + c_0 = f \tag{9.3}$$

which passes through the points (μ_i, f_i). Equating the left-hand side of equation (9.3) to zero gives the characteristic equation. Although the storage problem is now simplified the computational requirement is still excessively large for fully populated matrices. (To obtain the $n + 1$ values of f_i for a fully populated real unsymmetric matrix requires approximately $n^4/3$ multiplications.) Although the computational requirement is not excessive for band and Hessenberg matrices the eigenvalues computed by this method are very sensitive to errors arising both in the determination of the values f_i and also in the subsequent solution of the polynomial equation, particularly if this is derived explicitly. Hence this method is not normally recommended for computer implementation.

9.2 THE STURM SEQUENCE PROPERTY

For the case of symmetric matrices the *Sturm sequence* property of the principal minors of equation (9.1) greatly enhances the characteristic equation formulation.

In order to establish this property it is necessary to obtain a relationship between the characteristic equations of two symmetric matrices $\mathbf{A}_k$ and $\mathbf{A}_{k-1}$ where $\mathbf{A}_{k-1}$, of order $k-1$, is formed from $\mathbf{A}_k$, of order k, by omitting the last row and column; thus

$$\mathbf{A}_k = \left[\begin{array}{cccc|c} & & & & a_{k1} \\ & & & & a_{k2} \\ & \mathbf{A}_{k-1} & & & \cdot \\ & & & & a_{k,k-1} \\ \hline a_{k1} & a_{k2} & \cdots & a_{k,k-1} & a_{k,k} \end{array}\right] \tag{9.4}$$

Let $\bar{\lambda}_1, \bar{\lambda}_2, \ldots, \bar{\lambda}_{k-1}$ be the eigenvalues of $\mathbf{A}_{k-1}$ such that $\bar{\lambda}_1 \geqslant \bar{\lambda}_2 \geqslant \cdots \geqslant \bar{\lambda}_{k-1}$. If $\bar{\mathbf{Q}} = [\bar{\mathbf{q}}_1 \quad \bar{\mathbf{q}}_2 \quad \ldots \quad \bar{\mathbf{q}}_{k-1}]$ is the matrix of corresponding eigenvectors, then since

$$\bar{\mathbf{Q}}^T \mathbf{A}_{k-1} \bar{\mathbf{Q}} = \begin{bmatrix} \bar{\lambda}_1 & & \\ & \bar{\lambda}_2 & \\ & & \ddots \\ & & & \bar{\lambda}_{k-1} \end{bmatrix} \tag{9.5}$$

the matrix $\mathbf{A}_k$ may be transformed to give

$$\mathbf{G}_k = \left[\begin{array}{ccc|c} & & & 0 \\ & \bar{\mathbf{Q}}^T & & 0 \\ & & & \cdot \\ & & & 0 \\ \hline 0 & 0 \ldots & 0 & 1 \end{array}\right] \left[\begin{array}{c} \\ \mathbf{A}_k \\ \\ \end{array}\right] \left[\begin{array}{ccc|c} & & & 0 \\ & \bar{\mathbf{Q}} & & 0 \\ & & & \cdot \\ & & & 0 \\ \hline 0 & 0 \ldots & 0 & 1 \end{array}\right]$$

$$= \left[\begin{array}{cccc|c} \bar{\lambda}_1 & & & & c_1 \\ & \bar{\lambda}_2 & & & c_2 \\ & & \ddots & & \cdot \\ & & & \bar{\lambda}_{k-1} & c_{k-1} \\ \hline c_1 & c_2 & \cdots & c_{k-1} & a_{kk} \end{array}\right] \tag{9.6}$$

where c_i is the inner product of $\bar{\mathbf{q}}_i$ and the off-diagonal elements in the k-th row of $\mathbf{A}_k$. Since the transformation is orthogonal, $\mathbf{G}_k$ and $\mathbf{A}_k$ must have the same eigenvalues and hence the same characteristic polynomial, namely

$$f_k(\mu) = (\lambda_1 - \mu)(\lambda_2 - \mu) \ldots (\lambda_k - \mu) = |\mathbf{A}_k - \mu\mathbf{I}| = |\mathbf{G}_k - \mu\mathbf{I}| \tag{9.7}$$

It may be seen from equation (9.6) that $\mathbf{G}_k - \mu\mathbf{I}$ has a zero diagonal element when μ takes any of the $k-1$ values $\bar{\lambda}_i$, and hence

$$f_k(\bar{\lambda}_i) = -(\lambda_1 - \bar{\lambda}_i)(\lambda_2 - \bar{\lambda}_i) \ldots (\lambda_{i-1} - \bar{\lambda}_i)(\lambda_{i+1} - \bar{\lambda}_i) \ldots (\lambda_{k-1} - \bar{\lambda}_i) c_i^2 \tag{9.8}$$

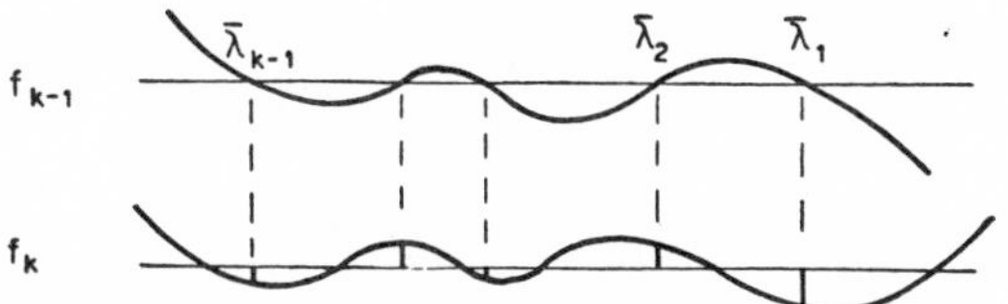

Figure 9.1 Characteristic functions for $\mathbf{A}_{k-1}$ and $\mathbf{A}_k$

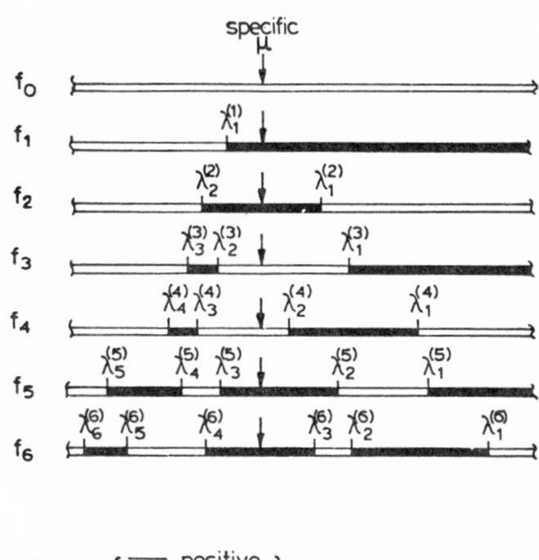

Figure 9.2 The eigenvalue pyramid

Provided that $f_k(\bar{\lambda}_i) \neq 0$ for all values of i, it follows that the sequence $f_k(-\infty)$, $f_k(\bar{\lambda}_{k-1}), f_k(\bar{\lambda}_{k-2}), \ldots, f_k(\bar{\lambda}_1), f(\infty)$ alternates in sign. Since there are just k roots of the equation $f_k(\mu) = 0$, one and only one of these must lie in each of the k intervals of the sequence. But the roots of $f_k(\mu) = 0$ are the eigenvalues λ_i of the matrix $\mathbf{A}_k$. Hence the eigenvalues of $\mathbf{A}_{k-1}$ must *interlace* the eigenvalues of $\mathbf{A}_k$, the characteristic equations for $k = 6$ being as shown in Figure 9.1.

If the matrix $\mathbf{A}_k$ has eigenvalues $\lambda_i^{(k)}$, then the complete set of eigenvalues for $k = 1, 2, \ldots, n$ form a pyramid structure, illustrated for $n = 6$ in Figure 9.2. The sign of the characteristic function in each region is also given. Since

$$f_k(\mu) = |\mathbf{A}_k - \mu\mathbf{I}| \tag{9.9}$$

is the leading principal minor of order k of $\mathbf{A} - \mu\mathbf{I}$, it can be deduced that the number of changes of sign in the sequence of leading principal minors of $\mathbf{A} - \mu\mathbf{I}$ is equal to the number of eigenvalues of $\mathbf{A}$ which are less than μ. This is illustrated in Figure 9.2 by examining the sequence $f_0(\mu), f_1(\mu), \ldots, f_6(\mu)$, where $f_0(\mu)$ is taken to be positive.

The importance of this Sturm sequence property lies in the fact that leading principal minors of $\mathbf{A}$ may be computed during the evaluation of $|\mathbf{A} - \mu\mathbf{I}|$. The information they give can be used either to enhance techniques based on the characteristic equation or to isolate the eigenvalues by bisection.

9.3 BISECTION FOR TRIDIAGONAL MATRICES

For a symmetric tridiagonal matrix having leading diagonal elements α_i ($i = 1, 2, \ldots, n$) and subdiagonal elements β_i ($i = 1, 2, \ldots, n-1$)

$$f_k(\mu) = |\mathbf{A}_k - \mu\mathbf{I}| = \begin{vmatrix} \alpha_1 - \mu & \beta_1 & & & \\ \beta_1 & \alpha_2 - \mu & \beta_2 & & \\ & \cdot & \cdot & \cdot & \\ & & \cdots & \cdots & \beta_{k-1} \\ & & & \beta_{k-1} & \alpha_k - \mu \end{vmatrix} \tag{9.10}$$

It may be shown that

$$f_k(\mu) = (\alpha_k - \mu) f_{k-1}(\mu) - \beta_{k-1}^2 f_{k-2}(\mu) \tag{9.11}$$

where $f_0(\mu) = 1$ and $f_1(\mu) = \alpha_1 - \mu$, which is a convenient recursive formula for obtaining the Sturm sequence for specific values of μ.

Equation (9.11) can also be used to show that, if all the coefficients β_i are non-zero, f_{k-1} and f_k cannot have coincident zeros, and hence $\lambda_i^{(k)} \neq \lambda_{i+1}^{(k)}$. In the case where a particular β_i coefficient is zero, the analysis can be simplified by separating the matrix into two tridiagonal submatrices of orders i and $n - i$ and carrying out an eigensolution of each.

In the method of bisection the Sturm sequence property is used to restrict the interval in which a particular eigenvalue must lie, until the eigenvalue is predicted to sufficient accuracy. For example, consider the isolation of the largest eigenvalue of

Table 9.1 Bisection for the largest eigenvalue of the 3 × 3 matrix (Table 8.6)

Range	μ	f_0	f_1	f_2	f_3	No. of sign changes
0 – 4	2	1	0	−1	1	2
2 – 4	3	1	−1	0	1	2
3 – 4	3.5	1	−1.5	1.25	−1.625	3
3 –3.5	3.25	1	−1.25	0.563	−0.016	3
3 –3.25	3.125	1	−1.125	0.266	0.561	2
3.125 –3.25	3.1875	1	−1.188	0.410	0.290	2
3.1875 –3.25	3.21875	1	−1.219	0.485	0.142	2
3.21875 –3.25	3.234375	1	−1.234	0.524	0.064	2
3.234375 –3.25	3.2421875	1	−1.242	0.543	0.025	2
3.2421875–3.25	3.2460937	1	−1.246	0.553	0.005	2
3.2460937–3.25	3.2480468	1	−1.248	0.557	−0.006	3
3.2460937–3.2480468	3.2470702	1	−1.247	0.555	−0.001	3
3.246 –3.247 approx						

Table 9.2 Bisection for the second largest eigenvalue of the 3 x 3 matrix (Table 8.6)

Range	μ	f_0	f_1	f_2	f_3	No. of sign changes
0 – 4			(as Table 9.1)			2
0 – 2	1	1	1	0	−1	1
1 – 2	1.5	1	0.5	−0.75	−0.125	1
1.5– 2	1.75	1	0.25	−0.938	0.453	2
1.5–1.75	1.625	1	0.375	−0.859	0.162	2
1.5–1.625						
	etc.					

the matrix shown in Table 8.6. By applying Gerschgorin's theorem it can be ascertained that all of the eigenvalues are in the range $0 \leqslant \lambda \leqslant 4$. The interval in which the largest eigenvalue must lie may be halved by obtaining the Sturm sequence with $\mu = 2$, as shown in the first row of Table 9.1. Since there are two changes of sign in this Sturm sequence, λ_1 must lie in the range $2 < \lambda_1 \leqslant 4$. Subsequent bisections of this range are shown in Table 9.1 until the stage is reached at which λ_1 has been isolated to approximately four-figure accuracy. The information regarding the successive ranges and the number of changes of sign within the Sturm sequence for the bisection (i.e. the first and last columns in Table 9.1) are useful in starting the determination of neighbouring eigenvalues. For Table 9.1 only the first row is valid for the determination λ_2, the isolation of which is shown in Table 9.2. In the isolation of λ_3 the first and second rows of Table 9.2 are relevant.

9.4 DISCUSSION OF BISECTION FOR TRIDIAGONAL MATRICES

Convergence

If the interval for a particular eigenvalue is halved at each step it follows that the number of steps, s, required to determine a particular eigenvalue to p decimal places is such that $2^s \simeq 10^p$, i.e. $s \simeq 3.3p$. Since it is reliable and easy to automate it is a useful alternative to the LR and QR methods given in Chapter 8. One useful feature of the bisection method is that the eigenvalues within a specific interval can be obtained easily without having to determine any of the eigenvalues outside the interval.

Preventing overflow or underflow

If equation (9.11) is used without modification for constructing the principal minors, it is possible for the magnitude of these to grow or diminish rapidly as k increases until a number generated either overflows or underflows the storage. For instance, if the elements of a 50 x 50 tridiagonal matrix are of order 10^3, then

$|f_{50}(\mu)|$ could be of the order of 10^{150}. This problem may be overcome by modifying the process so that the sequence

$$q_k(\mu) = \frac{f_k(\mu)}{f_{k-1}(\mu)} \tag{9.12}$$

is determined by using the recurrence relation

$$q_k(\mu) = \alpha_k - \mu - \frac{\beta_{k-1}^2}{q_{k-1}(\mu)} \tag{9.13}$$

If the computed value of any $q_k(\mu)$ is zero, then its value should be replaced by a small quantity so that a division failure will not be encountered in subsequent computation.

Interpolation

It is possible to improve upon the method of bisection in the later stages of convergence by using the knowledge that the characteristic polynomial is a smooth function. For instance, after five bisections in Table 9.1 the interval has been reduced to $3.125 < \lambda_1 < 3.25$ and it is known that the nearest neighbouring value is well away from this interval ($\mu_2 < 2$). A linear interpolation of $f_3(\mu)$ between $\mu = 3.125$ and $\mu = 3.25$ provides a prediction of 3.2466 for the required eigenvalue. Using $\mu = 3.2466$ instead of the bisection $\mu = 3.125$ leads to the reduced interval $3.2466 < \lambda_1 < 3.25$. A linear prediction over this interval yields $\lambda_1 = 3.24697$ which has an error of 0.00001. The length of the interval for λ can no longer be used to decide when a satisfactory convergence has been obtained because it will not, in general, tend to zero. If bisection is replaced by linear interpolation at too early a stage, the successive predictions may be only weakly convergent, as illustrated by Figure 9.3. Here the initial interval for λ is comparable in magnitude with the spacing between the neighbouring eigenvalues, and after four linear interpolations the estimate for λ lies well outside the interval for λ obtained if a corresponding

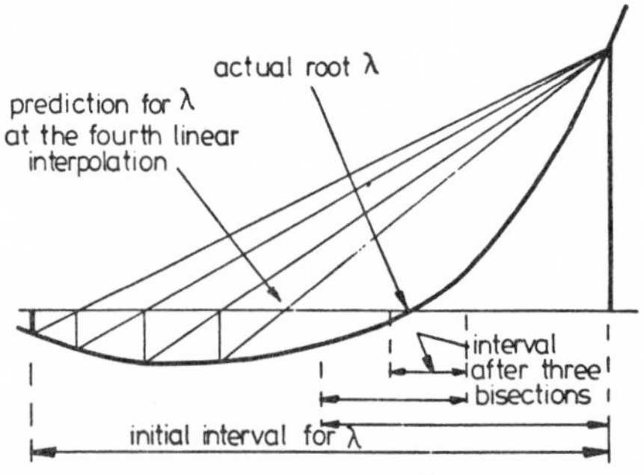

Figure 9.3 Predicting a root of the characteristic function $f_n(\lambda)$ – a case where linear interpolation has poor convergence

amount of computation had been devoted to continuing the bisection process (i.e. three more bisections).

More effective prediction methods may be developed by judicious use of quadratic or higher order interpolation formulae based on three or more values of μ and $f_n(\mu)$.

9.5 BISECTION FOR GENERAL SYMMETRIC MATRICES

From section 9.2 it may be seen that the Sturm sequence property can be applied to symmetric matrices which are not tridiagonal. Evans (1975) has developed a recurrence relationship for evaluating the Sturm sequence of a quindiagonal matrix. However, recurrence relationships are not available for more general symmetric matrices, and hence elimination techniques have to be employed.

If Gaussian elimination without interchanges is applied to a symmetric matrix, it is possible to show that the principal minor of order k is equal to the product of the first k pivots. Hence the number of sign changes of the Sturm sequence is equal to the number of negative pivots. The difficulty in performing such a reduction is that $\mathbf{A} - \mu\mathbf{I}$ will not normally be positive definite, and hence a zero or small pivot could be encountered making the process breakdown. If a large number of significant figures are being retained in the computation and the eigenvalues are not required to high precision it will be rare for a pivot to be sufficiently small to affect the accuracy of the process disastrously. It is thus possible to risk a breakdown of the elimination procedure, through zero pivot or an occasional wrong prediction from the Sturm sequence, provided that:

(a) If a zero pivot is encountered the elimination is restarted with a slightly modified μ value.

(b) The bisection or interpolation procedure is constructed so that it can recover if one single μ value gives erroneous results.

Although the symmetric pivoting technique (section 4.14) allows strong pivots to be chosen, and can be used to obtain the sequence of leading principal minors, it is not very effective when applied to sparse or band matrices.

A reliable alternative advocated by Wilkinson (1965) is to adopt a row interchange reduction technique which makes no attempt to retain the symmetry of the original matrix. Consider the matrix

$$\mathbf{A} - \mu\mathbf{I} = \begin{bmatrix} 1 & 5 & 10 \\ 5 & 5 & -10 \\ 10 & -10 & 20 \end{bmatrix} \tag{9.14}$$

whose first leading principal minor is 1. In evaluating the second principal minor rows 1 and 2 are interchanged so that the larger value 5 may be used as pivot. The interchange and elimination is shown as the first stage in Table 9.3. Because the sign of a determinant is reversed by interchanging two rows, the second principal minor of $\mathbf{A} - \mu\mathbf{I}$ can be evaluated as $-(5 \times 4) = -20$. In determining the third

Table 9.3 Unsymmetric reduction of matrix (9.14) allowing for Sturm sequence evaluation

Row order	Matrix after row interchange		Matrix after elimination
2, 1, 3	$\begin{bmatrix} 5 & 5 & -10 \\ 1 & 5 & 10 \\ 10 & -10 & 20 \end{bmatrix}$	$\longrightarrow$	$\begin{bmatrix} 5 & 5 & -10 \\ 0 & 4 & 12 \\ 10 & -10 & 20 \end{bmatrix}$
3, 1, 2	$\begin{bmatrix} 10 & -10 & 20 \\ 0 & 4 & 12 \\ 5 & 5 & -10 \end{bmatrix}$	$\longrightarrow$	$\begin{bmatrix} 10 & -10 & 20 \\ 0 & 4 & 12 \\ 0 & 10 & -20 \end{bmatrix}$
3, 2, 1	$\begin{bmatrix} 10 & -10 & 20 \\ 0 & 10 & -20 \\ 0 & 4 & 12 \end{bmatrix}$	$\longrightarrow$	$\begin{bmatrix} 10 & -10 & 20 \\ 0 & 10 & -20 \\ 0 & 0 & 20 \end{bmatrix}$

leading principal minor it may be necessary to interchange row 3 with either or both of the previous rows in order to maintain strong pivots. In Table 9.3 two more interchanges have been performed giving $|\mathbf{A} - \mu\mathbf{I}|$ as $-(10 \times 10 \times 20) = -200$.

9.6 BISECTION FOR BAND MATRICES

If the unsymmetric reduction technique is adopted for a symmetric matrix having a bandwidth of $2b - 1$, the row interchanges may result in elements to the right of the band becoming non-zero. However, there is still a band structure involved in the

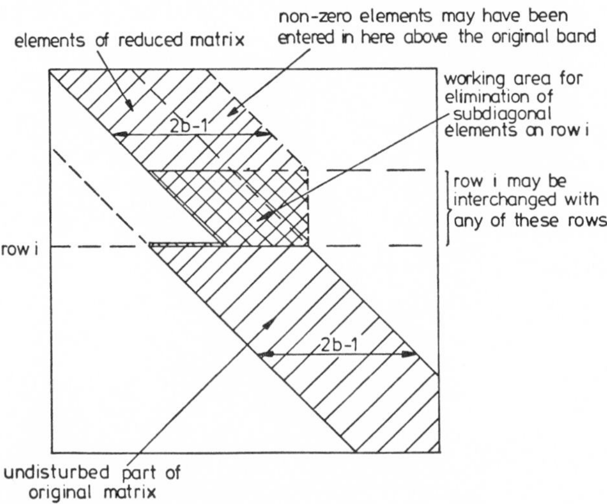

Figure 9.4 Pattern of non-zero elements when eliminating elements on row i during unsymmetric reduction of a band matrix, allowing for Sturm sequence evaluation

reduction, as illustrated by Figure 9.4. The precise number of multiplications required to investigate a single value of μ will depend on the amount of interchanging of rows but must lie between nb^2 and $3nb^2/2$ (as opposed to $nb^2/2$ if row interchanges are not carried out). These values are of the same order of magnitude as the computational requirements of the LR and QR methods and hence bisections will have a comparable efficiency if only a few bisections are required for each eigenvalue.

Four comments may be made regarding the use of bisection for band matrices:

(a) Economy in use of μ values

It is more important than in the case of tridiagonal matrices to use as few μ values as possible, and hence effective use should be made of interpolation methods. Alternatively, inverse iteration should be adopted to predict the eigenvalues more precisely once they have been separated from their neighbours. Inverse iteration is, in any case, the most suitable technique for determining the corresponding eigenvectors, if these are required.

(b) Partial eigensolution

If only a few eigenvalues of a band matrix are required, it is economical to apply the Sturm sequence method directly to the band matrix rather than to apply it to the tridiagonal matrix obtained by Rutishauser's transformations.

(c) Extension to the linearized eigenvalue problem

When applied to symmetric matrices of the form $\mathbf{A} - \mu\mathbf{B}$ the Sturm sequence count gives the number of eigenvalues less than μ such that $\mathbf{Ax} = \lambda\mathbf{Bx}$. This property is useful when $\mathbf{A}$ and $\mathbf{B}$ are large band matrices, for then $\mathbf{A} - \mu\mathbf{B}$ will also be of band form. In comparison, transformation methods need to operate on the full matrix $\mathbf{L}^{-1}\mathbf{A}\mathbf{L}^{-T}$ (where $\mathbf{B} = \mathbf{L}\mathbf{L}^T$), which may be much less efficient.

9.7 NON-LINEAR SYMMETRIC EIGENVALUE PROBLEMS

The non-linear eigenvalue problem may be described as one in which non-trivial solutions are required of the equations

$$\mathbf{Mx} = \mathbf{0} \tag{9.15}$$

where the elements of $\mathbf{M}$ are functions of a parameter λ, some functions being non-linear. Williams and Wittrick (1970) have shown that the Sturm sequence property can be used for the solution of certain non-linear symmetric eigenvalue problems in which the elements of $\mathbf{M}$ are transcendental functions of λ. These non-linear eigenvalue problems may be obtained from the analysis of stability and vibration of certain classes of structure.

BIBLIOGRAPHY

Evans, D. J. (1975). 'A recursive algorithm for determining the eigenvalues of a quindiagonal matrix'. *Computer J.*, 18, 70–73.

Gupta, K. K. (1973). 'Eigenproblem solution by a combined Sturm sequence and inverse iteration technique'. *Int. J. Num. Mech. Engng.*, 7, 17–42.

Peters, G., and Wilkinson, J. H. (1969). 'Eigenvalues of $\mathbf{Ax} = \lambda\mathbf{Bx}$ with band symmetric A and B'. *Computer J.*, 12, 398–404.

Wilkinson, J. H. (1965). *The Algebraic Eigenvalue Problem*, Clarendon Press, Oxford.

Wilkinson, J. H., and Reinsch, C. (1971). *Handbook for Automatic Computation, Vol. II Linear Algebra*, Springer-Verlag, Berlin. (An algorithm by W. Barth, R. S. Martin and J. H. Wilkinson on calculation of the eigenvalues of a symmetric tridiagonal matrix by the method of bisection.)

Williams, F. W., and Wittrick, W. H. (1970). 'An automatic computational procedure for calculating natural frequencies of skeletal structures'. *Int. J. Mech. Sci.*, 12, 781–791.

Chapter 10
Vector Iterative Methods for Eigenvalues

10.1 THE POWER METHOD

The basic power method may be used to compute the eigenvalue of largest modulus and the corresponding eigenvector of a general matrix. It is sometimes attributed to von Mises (e.g. Bodewig, 1956) who obtained it as a development of the Stodola method of determining the lowest natural frequency of a vibrating system.

Let the eigenvalues of a matrix **A** be ordered in such a way that $|\lambda_1| \geq |\lambda_2| \geq \cdots \geq |\lambda_n|$. In section 1.20 it was shown that, if **A** is symmetric, any arbitrary vector $\mathbf{u}^{(0)}$ may be expressed as a linear combination of the eigenvectors of the matrix. Thus

$$\mathbf{u}^{(0)} = c_1\mathbf{q}_1 + c_2\mathbf{q}_2 + \cdots + c_n\mathbf{q}_n \tag{10.1}$$

Such an expression for $\mathbf{u}^{(0)}$ is also valid when **A** is unsymmetric, provided that the matrix of eigenvectors $\mathbf{Q} = [\mathbf{q}_1 \quad \mathbf{q}_2 \quad \ldots \quad \mathbf{q}_n]$ is non-singular (in other words, when **A** is not defective). If the arbitrary vector is premultiplied by **A**, a vector $\mathbf{u}^{(1)}$ is obtained such that

$$\mathbf{u}^{(1)} = \mathbf{A}\mathbf{u}^{(0)} = \sum_{i=1}^{n} c_i\mathbf{A}\mathbf{q}_i = \sum_{i=1}^{n} \lambda_i c_i \mathbf{q}_i \tag{10.2}$$

If premultiplication by **A** is performed k times, a vector $\mathbf{u}^{(k)}$ is obtained which satisfies

$$\mathbf{u}^{(k)} = \mathbf{A}^k\mathbf{u}^{(0)} = \lambda_1^k c_1\mathbf{q}_1 + \lambda_2^k c_2\mathbf{q}_2 + \cdots + \lambda_n^k c_n\mathbf{q}_n \tag{10.3}$$

Provided that $|\lambda_1| > |\lambda_2|$, it follows that when k is large and c_1 is non-zero

$$\mathbf{u}^{(k)} \simeq \lambda_1^k c_1\mathbf{q}_1 \tag{10.4}$$

Hence $\mathbf{u}^{(k)}$ tends to become proportional to the dominant eigenvector $\mathbf{q}_1$. Since an eigenvector can be arbitrarily scaled it is convenient to normalize the trial vector after each premultiplication. An iterative algorithm to determine $\mathbf{q}_1$ may therefore be expressed by the two equations

$$\left.\begin{aligned} &\mathbf{v}^{(k)} = \mathbf{A}\mathbf{u}^{(k)} \\ \text{and} \qquad & \\ &\mathbf{u}^{(k+1)} = \frac{1}{\alpha}\mathbf{v}^{(k)} \end{aligned}\right\} \qquad (10.5)$$

where

$$\alpha = \pm \| \mathbf{v}^{(k)} \|$$

To demonstrate the use of the method consider its application to the matrix **FM** obtained from the cantilever vibration problem (equation 7.15). Choosing the initial trial vector as $\mathbf{u}^{(0)} = \{1 \quad 1 \quad 1\}$ and adopting the maximum element norm as the criterion for vector normalization yields:

Iteration 1:

$$\begin{bmatrix} 528.2 & 547.6 & 156.4 \\ 273.8 & 312.8 & 98.0 \\ 78.2 & 98.0 & 39.1 \end{bmatrix} \begin{bmatrix} 1 \\ 1 \\ 1 \end{bmatrix} = \begin{bmatrix} 1232.2 \\ 684.6 \\ 215.3 \end{bmatrix} = 1232.2 \begin{bmatrix} 1 \\ 0.5556 \\ 0.1747 \end{bmatrix}$$

Iteration 2:

$$\begin{bmatrix} & & \\ & \text{as above} & \\ & & \end{bmatrix} \begin{bmatrix} 1 \\ 0.5556 \\ 0.1747 \end{bmatrix} = \begin{bmatrix} 859.8 \\ 464.7 \\ 139.5 \end{bmatrix} = 859.8 \begin{bmatrix} 1 \\ 0.5405 \\ 0.1622 \end{bmatrix}$$

Iteration 3:

$$\begin{bmatrix} & & \\ & \text{as above} & \\ & & \end{bmatrix} \begin{bmatrix} 1 \\ 0.5405 \\ 0.1622 \end{bmatrix} = \begin{bmatrix} 849.5 \\ 458.8 \\ 137.5 \end{bmatrix} = 849.5 \begin{bmatrix} 1 \\ 0.5400 \\ 0.1619 \end{bmatrix}$$

Iteration 4:

$$\begin{bmatrix} & & \\ & \text{as above} & \\ & & \end{bmatrix} \begin{bmatrix} 1 \\ 0.5400 \\ 0.1619 \end{bmatrix} = \begin{bmatrix} 849.2 \\ 458.6 \\ 137.5 \end{bmatrix} = 849.2 \begin{bmatrix} 1 \\ 0.5400 \\ 0.1619 \end{bmatrix}$$

(10.6)

The dominant eigenvector of the matrix (correct to four decimal figures) is therefore $\{1 \quad 0.5400 \quad 0.1619\}$. The dominant eigenvalue is the final value of the normalizing factor, i.e. $\lambda_1 = 849.2$.

When a maximum element norm is used the sign of α should be made to correspond with that of the maximum element. If this is done α will converge to λ_1

even where it is negative, and the sequence of vectors will converge without alternation in sign.

Where an approximation to the dominant eigenvector is available, this may be used as an initial trial vector. In the absence of such information it is necessary to choose a trial vector in a more arbitrary way. It is important to minimize the risk of choosing $\mathbf{u}^{(0)}$ in such a way that the coefficient c_1 is either zero or very small compared with the other coefficients. Whereas specific trial vectors will be quite satisfactory for most matrices, there will always be some for which they are unsuitable. For instance, if the trial vector $\{1 \quad 1 \quad 1 \quad 1\}$ is used for power method iteration with the matrix

$$\mathbf{A} = \begin{bmatrix} 2 & -1 & & \\ -1 & 2 & -1 & \\ & -1 & 2 & -1 \\ & & -1 & 2 \end{bmatrix} \tag{10.7}$$

the coefficient c_1 is zero. Hence, theoretically, convergence is to the second eigenvector rather than the first. In practice rounding errors will normally introduce small components of $\mathbf{q}_1$ into the trial vectors and these components will be magnified by subsequent iteration. Hence convergence is still likely to be to the first eigenvector, although a larger number of iterations would be required than if a more suitable trial vector were to be chosen. For computer implementations of the power method it is useful to have an automatic procedure for constructing initial trial vectors which is unlikely to result in an unsuitable choice. The most reliable procedure appears to be to generate the elements of the trial vector from a set of random numbers within a fixed interval, say $-1 < u_i^{(0)} \leqslant 1$. If this is done, the statistical possibility that c_1 is zero or extremely small will be very remote.

10.2 CONVERGENCE CHARACTERISTICS OF THE POWER METHOD

Rate of convergence

If $\mathbf{u}^{(k)}$ is expressed as the sum of $\mathbf{q}_1$ and an error vector, i.e.

$$\mathbf{u}^{(k)} = \mathbf{q}_1 + \mathbf{e}^{(k)} \tag{10.8}$$

then, since $\mathbf{u}^{(k)}$ may be arbitrarily scaled, equation (10.3) gives

$$\mathbf{e}^{(k)} = \left(\frac{\lambda_2}{\lambda_1}\right)^k \frac{c_2}{c_1}\mathbf{q}_2 + \left(\frac{\lambda_3}{\lambda_1}\right)^k \frac{c_3}{c_1}\mathbf{q}_3 + \cdots + \left(\frac{\lambda_n}{\lambda_1}\right)^k \frac{c_n}{c_1}\mathbf{q}_n \tag{10.9}$$

The component of $\mathbf{q}_2$ will be the slowest to attenuate. Provided that $|c_2| \simeq |c_1|$, s decimal places of accuracy will be achieved when

$$\left(\frac{\lambda_2}{\lambda_1}\right)^k \simeq 10^{-s} \tag{10.10}$$

i.e. when

$$k \simeq \frac{s}{\log_{10}(\lambda_1/\lambda_2)} \tag{10.11}$$

The example given in (10.6) exhibits rapid convergence because the dominant eigenvalue of the matrix is well separated from the others ($\lambda_1/\lambda_2 \simeq 31.65$). This fortuitous circumstance does not always arise for vibration problems. If λ_2/λ_1 is close to unity, the basic power method iteration will converge very slowly.

Acceleration of convergence

If convergence is slow and the eigenvalues of the matrix are real, it is possible to accelerate the convergence rate by adopting an Aitken type method (section 6.12). This will be most effective if λ_2 is close to λ_1, but is well separated from λ_3. However, if several eigenvalues are close to λ_1, it is unlikely that a very substantial improvement in the convergence rate will be obtained.

Eigenvalue predictions

When the matrix is unsymmetric the choice of normalizing procedure may not be important. However, if the matrix is symmetric, more accurate eigenvalue predictions can be obtained by choosing Euclidean normalization rather than maximum element normalization (see Table 10.1). The process of determining the Euclidean norm does not yield a simple criterion for choosing the sign of α. Hence it is necessary to base the choice on whether or not individual elements of the vector have the same sign as the corresponding elements of the previous vector.

Another procedure for eigenvalue prediction is to use the *Rayleigh quotient* (also included in Table 10.1) given by

$$\bar{\lambda} = \frac{[\mathbf{u}^{(k)}]^T \mathbf{A} \mathbf{u}^{(k)}}{[\mathbf{u}^{(k)}]^T \mathbf{u}^{(k)}} = \frac{[\mathbf{u}^{(k)}]^T \mathbf{v}^{(k)}}{[\mathbf{u}^{(k)}]^T \mathbf{u}^{(k)}} \tag{10.12}$$

This prediction has similar accuracy to the prediction based on the Euclidean normalization, but is unambiguous regarding the sign.

Suppose that convergence to the correct solution is almost complete, i.e.

$$\mathbf{u}^{(k)} \simeq c_1 \mathbf{q}_1 + e_2 \mathbf{q}_2 \tag{10.13}$$

Table 10.1 Errors of eigenvalue predictions by the power method for matrix (7.7) with largest eigenvalue $\lambda_1 = 3.7320$ and initial trial vector = $\{1\ 1\ 1\ 1\ 1\}$

Iteration no.	4	6	8	10
Maximum element norm	–0.7320	0.1966	–0.0862	–0.0239
Euclidean norm	–0.3460	–0.0353	–0.0029	–0.0002
Rayleigh quotient	–0.4297	–0.0458	–0.0038	–0.0003

where e_2 is small compared with c_1 and where $q_1^T q_1 = q_2^T q_2 = 1$. Since the matrix is symmetric the eigenvectors q_1 and q_2 are orthogonal ($q_1^T q_2 = 0$). Hence

$$[\mathbf{u}^{(k)}]^T \mathbf{u}^{(k)} \simeq c_1^2 + e_2^2 \tag{10.14}$$

In addition,

$$[\mathbf{u}^{(k)}]^T \mathbf{A}\mathbf{u}^{(k)} \simeq c_1^2 \lambda_1 + e_2^2 \lambda_2 \tag{10.15}$$

Therefore the Rayleigh quotient has a value such that

$$\bar{\lambda} \simeq \lambda_1 \left\{ 1 - \frac{e_2^2}{c_1^2} \left(\frac{\lambda_1 - \lambda_2}{\lambda_1} \right) \right\} \tag{10.16}$$

which shows that the error of the Rayleigh quotient prediction is of order $(e_2/c_1)^2$. If the matrix is unsymmetric, or if the eigenvalue prediction is based on the maximum element norm, the error is of order e_2/c_1.

The corresponding eigenvalue prediction $\bar{\bar{\lambda}}$ using the Euclidean norm is such that

$$\bar{\bar{\lambda}}^2 = \frac{[\mathbf{v}^{(k)}]^T \mathbf{v}^{(k)}}{[\mathbf{u}^{(k)}]^T \mathbf{u}^{(k)}} = \frac{[\mathbf{u}^{(k)}]^T \mathbf{A}^T \mathbf{A}\mathbf{u}^{(k)}}{[\mathbf{u}^{(k)}]^T \mathbf{u}^{(k)}} \tag{10.17}$$

If $\mathbf{A}$ is symmetric this gives

$$\bar{\bar{\lambda}}^2 \simeq \lambda_1^2 \left\{ 1 - \frac{e_2^2}{c_1^2} \left(\frac{\lambda_1^2 - \lambda_2^2}{\lambda_1^2} \right) \right\} \tag{10.18}$$

which indicates a similar order of error to that of the Rayleigh quotient. It is interesting to note that the Rayleigh quotient is the solution of the overdetermined set of equations

$$\left[\mathbf{u}^{(k)} \right] [\bar{\lambda}] = \left[\mathbf{v}^{(k+1)} \right] + \left[\bar{\mathbf{e}}^{(k)} \right] \tag{10.19}$$

such that $\Sigma(\bar{e}_i^{(k)})^2$ is minimized.

It may also be noted that, where $\mathbf{A}$ is symmetric, the Rayleigh quotient and Euclidean norm eigenvalue predictions both give underestimates of the magnitude of the actual eigenvalues (see section 10.8).

Eigenvalues of equal modulus

If the matrix is real and unsymmetric and the largest eigenvalues are a complex conjugate pair, neither the vectors nor the eigenvalue predictions will converge. However, it is possible to modify the iterative procedure in such a way that two dominant eigenvalues are predicted. To achieve this, two iterations are performed without an intermediate normalization, namely

$$\mathbf{v}^{(k)} = \mathbf{A}\mathbf{u}^{(k)}, \qquad \mathbf{v}^{(k+1)} = \mathbf{A}\mathbf{v}^{(k)} \tag{10.20}$$

If sufficient iterations have been performed to wash out all but the components of the eigenvectors $\mathbf{q}_1$ and $\mathbf{q}_2$, it can be shown that

$$\mathbf{v}^{(k+1)} + \beta \mathbf{v}^{(k)} + \gamma \mathbf{u}^{(k)} = \mathbf{0} \tag{10.21}$$

where β and γ are coefficients of the quadratic equation

$$\lambda^2 + \beta\lambda + \gamma = 0 \tag{10.22}$$

having roots λ_1 and λ_2 (see Fox, 1964, p. 219). A suitable method of determining β and γ is to find the least squares solution of

$$\begin{bmatrix} \mathbf{u}^{(k)} & \mathbf{v}^{(k)} \end{bmatrix} \begin{bmatrix} \gamma \\ \beta \end{bmatrix} = -\begin{bmatrix} \mathbf{v}^{(k+1)} \end{bmatrix} + \begin{bmatrix} \bar{\mathbf{e}}^{(k+1)} \end{bmatrix} \tag{10.23}$$

where $\bar{\mathbf{e}}^{(k+1)}$ is the error vector. Having found β and γ, the error vector may be computed in order to detect convergence, and estimates for λ_1 and λ_2 may be obtained by solving the quadratic equation (10.22). The eigenvectors can be evaluated from

$$\left.\begin{aligned} \mathbf{q}_1 &\propto \mathbf{v}^{(k+1)} - \lambda_2 \mathbf{v}^{(k)} \\ &\text{and} \\ \mathbf{q}_2 &\propto \mathbf{v}^{(k+1)} - \lambda_1 \mathbf{v}^{(k)} \end{aligned}\right\} \tag{10.24}$$

As an example consider the sequence of vectors $\mathbf{u}^{(k)} = \{0.6 \quad 1 \quad 0.4\}$, $\mathbf{v}^{(k)} = \{0.4 \quad -2.8 \quad -3.2\}$ and $\mathbf{v}^{(k+1)} = \{-13.6 \quad -8.8 \quad 4.8\}$ which could arise when applying a power method iteration to the matrix shown in Table 8.4. Substituting in equation (10.23) gives

$$\begin{bmatrix} 0.6 & 0.4 \\ 1 & -2.8 \\ 0.4 & -3.2 \end{bmatrix} \begin{bmatrix} \gamma \\ \beta \end{bmatrix} = \begin{bmatrix} 13.6 \\ 8.8 \\ -4.8 \end{bmatrix} + \begin{bmatrix} \bar{\mathbf{e}}^{(k+1)} \end{bmatrix} \tag{10.25}$$

Applying the procedure given in section 2.5 for solving overdetermined equations using unit weighting factors gives

$$\begin{bmatrix} 1.52 & -3.84 \\ -3.84 & 18.24 \end{bmatrix} \begin{bmatrix} \gamma \\ \beta \end{bmatrix} = \begin{bmatrix} 15.04 \\ -3.84 \end{bmatrix} \tag{10.26}$$

Hence $\beta = 4$ and $\gamma = 20$. It follows that $\bar{\mathbf{e}}^{(k+1)} = \mathbf{0}$ and the eigenvalues are $2 \pm 4i$. The corresponding eigenvectors are therefore

$$\mathbf{q} = \begin{bmatrix} -13.6 \\ -8.8 \\ 4.8 \end{bmatrix} + (2 \pm 4i) \begin{bmatrix} 0.4 \\ -2.8 \\ -3.2 \end{bmatrix} = \begin{bmatrix} -12.8 \pm 1.6i \\ -14.4 \mp 11.2i \\ -1.6 \mp 12.8i \end{bmatrix} \tag{10.27}$$

which can be scaled to yield $\{1 \quad 1 \pm i \quad \pm i\}$.

The above technique for predicting two eigenvalues can also be employed when the matrix has two dominant eigenvalues which are equal but opposite in sign ($\lambda_1 = -\lambda_2$, $|\lambda_1| > |\lambda_3|$). In this case the sequences of vectors, $\mathbf{u}^{(k)}$, $\mathbf{u}^{(k+2)}$, $\mathbf{u}^{(k+4)}, \ldots,$ and $\mathbf{u}^{(k+1)}$, $\mathbf{u}^{(k+3)}, \ldots,$ will converge to separate limits, and the coefficients β and γ will converge to 0 and $-\lambda_1^2$ respectively.

If the matrix has two coincident eigenvalues such that $\lambda_1 = \lambda_2$ and $|\lambda_1| > |\lambda_3|$ then, as k becomes large,

$$\mathbf{u}^{(k)} \to \lambda_1^k(c_1\mathbf{q}_1 + c_2\mathbf{q}_2) \tag{10.28}$$

In this case the normal power method iteration converges to give λ_1 as the dominant eigenvalue, and $c_1\mathbf{q}_1 + c_2\mathbf{q}_2$ as the corresponding eigenvector. It may be noted that this is a valid but not unique eigenvector. The presence of coincident eigenvalues can only be detected from the power method iteration by repeating the process with a different initial trial vector.

There are some real unsymmetric matrices for which the method will not converge by any of the above techniques, namely when there are more than two dominant but distinct eigenvalues which have equal modulus (e.g. the matrix shown in Table 7.6).

10.3 EIGENVALUE SHIFT AND INVERSE ITERATION

The power method can be modified in several useful ways. One such modification is to replace $\mathbf{A}$ by $\bar{\mathbf{A}} = \mathbf{A} - \mu\mathbf{I}$. Since

$$(\mathbf{A} - \mu\mathbf{I})\mathbf{q}_i = (\lambda_i - \mu)\mathbf{q}_i \tag{10.29}$$

it follows that the eigenvalues of $\bar{\mathbf{A}}$ are the same as those of $\mathbf{A}$ except that they have all been shifted by an amount μ. The eigenvectors remain unaffected by the shift. Consider a case in which the power method with shift is applied to a matrix all of whose eigenvalues are real, with the extreme eigenvalues being -4 and 6. If $\mu < 1$ then convergence will be to $\lambda = 6$ as would be the case if the normal power method were applied. However, if $\mu > 1$ then convergence will be to the lower extreme eigenvalue $\lambda = -4$.

In the case of the column buckling problem (section 7.1) the smallest eigenvalue satisfying equation (7.7) will be obtained by using a shifted power method iteration with $\mu > 2$, the most rapid convergence rate being achieved for $\mu = 2.366$. However, the detailed knowledge of the eigenvalue spectrum necessary to evaluate the optimum value of μ will not generally be available. Consequently, it may be necessary to use eigenvalue properties to determine a value of μ which is sufficiently large to ensure convergence to the lower extreme eigenvalue. In the column buckling problem the Gerschgorin discs of the stiffness matrix of equation (7.7) restrict the eigenvalues to the interval $0 \leqslant \lambda \leqslant 4$ and also (from the trace property of section 1.17) the average of all of the eigenvalues must be 2. Hence, with the choice $\mu = 3$, convergence will be to the smallest eigenvalue.

Another modification of the power method consists of replacing $\mathbf{A}$ by its inverse. In this case the reciprocal eigenvalue $(1/\lambda_i)$ of largest modulus is obtained. It is not necessary to evaluate $\mathbf{A}^{-1}$ explicitly since, if the triangular decomposition $\mathbf{A} = \mathbf{LU}$ is performed, the two operations:

$$\left.\begin{array}{ll} \text{Solve by forward-substitution} & \mathbf{L}\mathbf{x}^{(k)} = \mathbf{u}^{(k)} \\ \text{and} & \\ \text{Solve by backsubstitution} & \mathbf{U}\mathbf{v}^{(k)} = \mathbf{x}^{(k)} \end{array}\right\} \tag{10.30}$$

are equivalent to the premultiplication $\mathbf{v}^{(k)} = \mathbf{A}^{-1}\mathbf{u}^{(k)}$.

Consider again the column buckling problem. A triangular decomposition of the stiffness matrix is

$$\begin{bmatrix} 2 & -1 & & & \\ -1 & 2 & -1 & & \\ & -1 & 2 & -1 & \\ & & -1 & 2 & -1 \\ & & & -1 & 2 \end{bmatrix}$$

$$= \begin{bmatrix} 1 & & & & \\ -0.5 & 1 & & & \\ & -0.6667 & 1 & & \\ & & -0.75 & 1 & \\ & & & -0.8 & 1 \end{bmatrix} \begin{bmatrix} 2 & -1 & & & \\ & 1.5 & -1 & & \\ & & 1.3333 & -1 & \\ & & & 1.25 & -1 \\ & & & & 1.2 \end{bmatrix} \tag{10.31}$$

If inverse iteration is performed, starting with a trial vector proportional to {1 1 1 1 1}, convergence is achieved in the manner shown in Table 10.2. The lowest eigenvalue, which is the most important one from the engineering viewpoint, can therefore be obtained either by an eigenvalue shift technique or by an inverse iteration technique. It is useful to know which is the more expedient of the two methods to use. Clearly the shift technique is easier to apply since it does not require the triangular decomposition of $\bar{A}$. However, it yields a slower convergence rate. The eigenvalues of $\bar{A}$ are −2.732, −2, −1, 0.732 and 0. Hence, from equation (10.11), it can be shown that approximately twenty-three iterations are required to achieve three-figure accuracy in the predicted value of q_1. On the other hand, the eigenvalues of A^{-1} are 3.7320, 1, 0.5, 0.3333 and 0.2679, indicating that a similar accuracy can be achieved with inverse iteration after only six iterations. (The convergence rate shown in Table 10.2 is even more rapid than this, since the initial trial vector contains no component of the subdominant eigenvector.)

Table 10.2 Inverse iteration (without shift) for matrix (10.31)

$u^{(0)}$	$v^{(0)}$	$u^{(1)}$	$v^{(1)}$	$u^{(2)}$	$v^{(2)}$	$u^{(3)}$	$v^{(3)}$	$u^{(4)}$
0.4472	1.1180	0.3107	1.0874	0.2916	1.0788	0.2891	1.0775	0.2887
0.4472	1.7889	0.4971	1.8641	0.4998	1.8660	0.5000	1.8660	0.5000
0.4472	2.0125	0.5592	2.1437	0.5748	2.1533	0.5770	2.1545	0.5773
0.4472	1.7889	0.4971	1.8641	0.4998	1.8660	0.5000	1.8660	0.5000
0.4472	1.1180	0.3107	1.0874	0.2916	1.0788	0.2891	1.0775	0.2887
Euclidean norm	3.5986		3.7296		3.7320		3.7320	
Eigenvalue prediction	0.2779		0.2681		0.267		0.2679	

Suppose that the column buckling problem is to be analysed by using fifty-nine finite difference variables instead of just the five chosen previously. The eigenvalues of the revised A are 0.00274, 0.0110, 0.0247, . . . , 3.997. With $\mu = 3$, the dominant set of eigenvalues of $\bar{A}$ is -2.99726, -2.9890 and -2.9753, giving a very poor convergence rate for iteration with shift (approximately 2,500 iterations to achieve three-figure accuracy in the eigenvector). On the other hand, the dominant set of eigenvalues of A^{-1} is 364.7, 91.2 and 40.5, giving a good convergence rate for inverse iteration (approximately five iterations to achieve three-figure accuracy in the eigenvector). In general it may be stated that if the smallest eigenvalue of a large symmetric and positive definite matrix is required, the power method with shift may result in a very poor convergence rate compared with that for inverse iteration.

The scope of inverse iteration can be greatly increased by performing the iteration with the inverse of the shifted matrix $\bar{A}$. In this case a typical eigenvalue ϕ_i of $\bar{A}^{-1}$ is related to an eigenvalue λ_i of A according to

$$\phi_i = \frac{1}{\lambda_i - \mu} \tag{10.32}$$

Convergence is to the eigenvalue λ_i which is closest to μ, and, if this eigenvalue is extremely close to μ, the rate of convergence will be very rapid. Inverse iteration therefore provides a means of determining an eigenvector of a matrix for which the corresponding eigenvalue has already been determined to moderate accuracy by an alternative method (e.g. the LR method, the QR method or the Sturm sequence method). For instance, if an eigenvalue of the column buckling matrix has been computed correct to five figures as 0.99999, then, since the eigenvalues of $(A - 0.99999I)^{-1}$ are 100,000, -1.3660, 1.0000, 0.5000 and 0.3660, approximately five figures of accuracy will be gained at every round of the shifted inverse iteration.

When inverse iteration is used to determine eigenvectors corresponding to known eigenvalues, the matrix to be inverted, even if symmetric, will not normally be positive definite, and if unsymmetric will not normally be diagonally dominant. Hence accuracy may be lost unless pivot selection is employed in the decomposition. Theoretically the method breaks down if the chosen value of μ is exactly equal to an eigenvalue, because in this case $(A - \mu I)$ is singular. In practice, rounding errors will usually prevent any of the pivots from becoming identically zero. If this happens inverse iteration will behave normally. However, since there is risk of encountering a zero pivot even when pivot selection procedures are employed, an implementation of the inverse iteration method should incorporate a procedure for coping with such a failure. It is sufficient to replace the zero pivot by a small number whose magnitude is that of a normal rounding error and then re-start the iterative cycle at the place in which the failure occurred.

The computation of an eigenvector corresponding to a complex conjugate eigenvalue by inverse iteration is more difficult than for a real eigenvalue. The computation has to be carried out either in complex arithmetic or else by one of the methods given by Wilkinson (1965).

10.4 SUBDOMINANT EIGENVALUES BY PURIFICATION (Aitken, 1937)

When the dominant eigenvalue and corresponding eigenvector of a matrix have been computed it is possible to remove this eigenvalue from the matrix (i.e. *purify* it) so that a subsequent application of the power method converges onto the eigenvalue immediately subdominant to it.

Using the notation **P** and **Q** for compounded left and right eigenvectors respectively, the biorthogonality condition (8.51) can be written in the form

$$\mathbf{Q}\mathbf{P}^H = \mathbf{I} \tag{10.33}$$

Hence from equation (8.53)

$$\mathbf{A} = \mathbf{Q}\boldsymbol{\Lambda}\mathbf{P}^H = \sum_{i=1}^{n} \lambda_i \mathbf{q}_i \mathbf{p}_i^H \tag{10.34}$$

It follows that the matrix

$$\mathbf{A}^{(2)} = \mathbf{A} - \lambda_1 \mathbf{q}_1 \mathbf{p}_1^H \tag{10.35}$$

has the same eigenvalues and eigenvectors as **A**, with the exception that λ_1 is replaced by zero. When **A** is unsymmetric, $\mathbf{A}^{(2)}$ can only be determined if both the left and right eigenvectors corresponding to λ_1 have been obtained. However, for a symmetric matrix, equation (10.35) may be simplified to

$$\mathbf{A}^{(2)} = \mathbf{A} - \lambda_1 \mathbf{q}_1 \mathbf{q}_1^T \tag{10.36}$$

For instance, if the dominant eigenvalue and eigenvector of the covariance matrix (7.53) have been obtained by power method iteration, then the purified matrix $\mathbf{A}^{(2)}$ determined according to equation (10.36) is

$$\mathbf{A}^{(2)} = \begin{bmatrix} 121.64 & & & \text{symmetric} & & \\ -56.61 & 53.60 & & & & \\ -43.24 & -4.21 & 80.22 & & & \\ -11.21 & -5.84 & 6.13 & 50.60 & & \\ -31.62 & 16.31 & 6.96 & -20.44 & 32.88 & \\ -17.72 & 2.20 & -14.37 & -9.38 & 1.88 & 43.20 \end{bmatrix} \tag{10.37}$$

If the power method is now applied to this matrix, the subdominant eigenvalue $\lambda_2 = 178.35$ and the corresponding eigenvector are determined. The process can be continued to find other subdominant eigenvalues, e.g. in order to obtain the third eigenvalue iteration would need to be performed with

$$\mathbf{A}^{(3)} = \mathbf{A} - \lambda_1 \mathbf{q}_1 \mathbf{q}_1^T - \lambda_2 \mathbf{q}_2 \mathbf{q}_2^T \tag{10.38}$$

If **A** is a sparse matrix, the purification process destroys the sparseness. However, it is possible to modify the power method iteration in such a way that $\mathbf{A}^{(2)}$, $\mathbf{A}^{(3)}$, etc., are not determined explicitly, with the result that the sparseness of **A** is preserved. The premultiplication $\mathbf{v}^{(k)} = \mathbf{A}^{(3)}\mathbf{u}^{(k)}$ for symmetric **A** can be computed as

$$\mathbf{v}^{(k)} = \mathbf{A}\mathbf{u}^{(k)} - \lambda_1 \mathbf{q}_1 \mathbf{q}_1^T \mathbf{u}^{(k)} - \lambda_2 \mathbf{q}_2 \mathbf{q}_2^T \mathbf{u}^{(k)} \tag{10.39}$$

However, since $\lambda_i \mathbf{q}_i = \mathbf{A}\mathbf{q}_i$ this corresponds to

$$\mathbf{v}^{(k)} = \mathbf{A}(\mathbf{u}^{(k)} - \alpha_{13}^{(k)} \mathbf{q}_1 - \alpha_{23}^{(k)} \mathbf{q}_2)$$

where

$$\alpha_{13}^{(k)} = \mathbf{q}_1^T \mathbf{u}^{(k)} \tag{10.40}$$

and

$$\alpha_{23}^{(k)} = \mathbf{q}_2^T \mathbf{u}^{(k)} = \mathbf{q}_2^T (\mathbf{u}^{(k)} - \alpha_{13}^{(k)} \mathbf{q}_1)$$

The vector $\mathbf{u}^{(k)} - \alpha_{13}^{(k)} \mathbf{q}_1 - \alpha_{23}^{(k)} \mathbf{q}_2$ is that obtained from $\mathbf{u}^{(k)}$ by orthogonalizing it with respect to $\mathbf{q}_1$ and $\mathbf{q}_2$ (see section 4.18). In general it is possible to obtain subdominant eigenvalues and corresponding eigenvectors by power method iteration in which the trial vectors are orthogonalized at each iteration with respect to the eigenvectors already determined. Strictly speaking, if the larger eigenvalues have been determined accurately, it may be unnecessary to perform the orthogonalization at every iteration. However, if the orthogonalization process is omitted for too many iterations the error components of the eigenvectors already determined will increase in magnitude so much that they interfere with the convergence.

10.5 SUBDOMINANT EIGENVALUES BY DEFLATION (Duncan and Collar, 1935)

If $\mathbf{a}_1^T$ is the first row of $\mathbf{A}$ and $\mathbf{q}_1$ has been normalized such that its first element is unity, then the first row of the matrix

$$\mathbf{B}^{(2)} = \mathbf{A} - \mathbf{q}_1 \mathbf{a}_1^T \tag{10.41}$$

is null. Furthermore, if $\mathbf{q}_i$ is the i-th right eigenvector of $\mathbf{A}$, normalized in such a way that the first element is unity, then

$$\mathbf{a}_1^T \mathbf{q}_i = \lambda_i \tag{10.42}$$

It follows that for $i \neq 1$

$$\mathbf{B}^{(2)}(\mathbf{q}_i - \mathbf{q}_1) = \lambda_i(\mathbf{q}_i - \mathbf{q}_1) \tag{10.43}$$

Hence it can be shown that if $\mathbf{A}^{(2)}$ is obtained from $\mathbf{B}^{(2)}$ by omitting its first row and column its eigenvalues are $\lambda_2, \lambda_3, \ldots, \lambda_n$, and a typical right eigenvector $\mathbf{q}_i^{(2)}$ is related to the corresponding right eigenvector of $\mathbf{A}$ according to

$$\begin{bmatrix} 0 \\ \mathbf{q}_i^{(2)} \end{bmatrix} = \mathbf{q}_i - \mathbf{q}_1 \tag{10.44}$$

This means that $\mathbf{A}^{(2)}$ is a deflation which may be used to obtain the subdominant eigenvalue λ_2 by power method iteration. Further deflations are possible to obtain

λ_3, λ_4, etc. Once a particular eigenvalue has been obtained, it is possible to compute the corresponding eigenvector of A. For instance, $\mathbf{q}_2$ may be obtained from the formula

$$\left.\begin{aligned} &\mathbf{q}_2 \propto \alpha \mathbf{q}_1 - (\lambda_1 - \lambda_2)\begin{bmatrix} 0 \\ \mathbf{q}_2^{(2)} \end{bmatrix} \\ &\text{where} \\ &\alpha = \mathbf{a}_1^T \begin{bmatrix} 0 \\ \mathbf{q}_2^{(2)} \end{bmatrix} \end{aligned}\right\} \quad (10.45)$$

If the dominant eigenvalue and eigenvector of the covariance matrix (7.53) have been obtained by power method iteration, the deflated matrix may be computed as

$$\mathbf{A}^{(2)} = \begin{bmatrix} 105.42 & 35.37 & 4.42 & 45.26 & 18.42 \\ 29.13 & 105.69 & 12.74 & 25.59 & -3.93 \\ 33.47 & 36.16 & 58.38 & 1.52 & 2.92 \\ 60.15 & 40.46 & -11.76 & 57.37 & 15.60 \\ 40.51 & 14.89 & -1.79 & 23.28 & 55.19 \end{bmatrix} \quad (10.46)$$

If power method iteration is performed on this matrix, convergence is onto the subdominant eigenvalue $\lambda_2 = 178.35$ with the corresponding eigenvector $\{1 \;\; 0.7310 \;\; 0.5227 \;\; 0.7630 \;\; 0.5541\}$. Hence from equation (10.45) the eigenvector of A corresponding to λ_2 is

$$\mathbf{q}_2 \propto 445.80 \begin{bmatrix} 1 \\ 0.9155 \\ 0.5891 \\ 0.6945 \\ 0.7747 \\ 0.6767 \end{bmatrix} - 620.80 \begin{bmatrix} 0 \\ 1 \\ 0.7310 \\ 0.5227 \\ 0.7630 \\ 0.5541 \end{bmatrix} = \begin{bmatrix} 445.8 \\ -212.7 \\ -191.2 \\ -14.9 \\ -128.3 \\ -42.3 \end{bmatrix} \quad (10.47)$$

Some notes on deflation

(a) *Symmetric* A

If A is symmetric, the symmetry will not be preserved through the deflation process.

(b) *Unsymmetric* A

If A is unsymmetric, only the right eigenvectors need to be computed to carry out the deflation.

(c) *Sparse* A

Sparseness in A will not be preserved through the deflation process.

(d) *Need for interchanges*

After the determination of an eigenvector $\mathbf{q}_i^{(i)}$, it is advisable, before deflation, to interchange corresponding rows and columns of the matrix $\mathbf{A}^{(i)}$ in order that the largest element in the vector $\mathbf{q}_i^{(i)}$ appears in the first row. An interchange is clearly necessary if the first element in $\mathbf{q}_i^{(i)}$ is zero.

(e) *Reliability*

Errors in the computed values of the eigenvectors induce errors in the sequence of deflated matrices. In some cases this can cause significant loss of accuracy in the computation of subdominant eigenvalues and eigenvectors.

10.6 A SIMULTANEOUS ITERATION METHOD

Simultaneous iteration (SI) methods are extensions of the power method in which several trial vectors are processed simultaneously in such a way that they converge onto the dominant set of eigenvectors of the matrix A. If $\mathbf{U}^{(k)} = [\mathbf{u}_1^{(k)}\mathbf{u}_2^{(k)} \ldots \mathbf{u}_m^{(k)}]$ is a matrix of m different trial vectors then, at the start of iteration $k + 1$, every SI method has as its basic operation the simultaneous premultiplication of these vectors by A, giving

$$\mathbf{V}^{(k)} = \mathbf{A}\mathbf{U}^{(k)} \tag{10.48}$$

Obviously it is necessary to prevent all of the trial vectors converging onto the dominant eigenvector. In the symmetric form of Bauer's (1958) bi-iteration method this is achieved by obtaining a new set of trial vectors $\mathbf{U}^{(k+1)}$ from $\mathbf{V}^{(k)}$ by an orthonormalization process. This may be considered as a simultaneous application of the vector purification method (section 10.4), and it may be concluded that the convergence rate for an individual eigenvector cannot be any better than the corresponding rate for vector purification or deflation. However, several SI methods have been developed recently which are substantially better than those obtained from the purification or deflation techniques (e.g. Jennings, 1967; Rutishauser, 1969; G. W. Stewart, 1976). The method described in this section, named *lop-sided iteration*, was developed by Jennings and W. J. Stewart (1975). It may be applied to real symmetric or unsymmetric matrices, and uses only one set of trial vectors which converge onto the dominant set of right eigenvectors.

The set of trial vectors $\mathbf{U}^{(k)}$ may be expressed as the product of the full matrix of right eigenvectors of A and an $n \times m$ coefficient matrix as follows:

$$\begin{bmatrix} \mathbf{u}_1^{(k)} & \mathbf{u}_2^{(k)} & \cdots & \mathbf{u}_m^{(k)} \end{bmatrix} = \left[\begin{array}{ccc|ccc} \mathbf{q}_1 \mathbf{q}_2 & \cdots & \mathbf{q}_m & \mathbf{q}_{m+1} & \cdots & \mathbf{q}_n \end{array}\right] \left[\begin{array}{cccc} c_{11}^{(k)} & c_{12}^{(k)} & \cdots & c_{1m}^{(k)} \\ c_{21}^{(k)} & c_{22}^{(k)} & \cdots & c_{2m}^{(k)} \\ \cdot & \cdot & \cdot & \cdot \\ c_{m1}^{(k)} & c_{m2}^{(k)} & \cdots & c_{mm}^{(k)} \\ \hline c_{m+1,1}^{(k)} & c_{m+1,2}^{(k)} & \cdots & c_{m+1,m}^{(k)} \\ \cdot & \cdot & \cdot & \cdot \\ c_{n1}^{(k)} & c_{n2}^{(k)} & \cdots & c_{nm}^{(k)} \end{array}\right] \tag{10.49}$$

The matrix of eigenvectors may be partitioned as shown in equation (10.49) and the submatrices $[\mathbf{q}_1\mathbf{q}_2 \ \cdots \ \mathbf{q}_m]$ and $[\mathbf{q}_{m+1} \ \cdots \ \mathbf{q}_n]$ designated $\mathbf{Q}_A$ and $\mathbf{Q}_B$ respectively. A corresponding row partitioning of the coefficient matrix is also shown. If $\mathbf{C}_A^{(k)}$ and $\mathbf{C}_B^{(k)}$ are submatrices which contain the first m rows and the remaining $n - m$ rows respectively of the coefficient matrix, then equation (10.49) may be re-expressed as

$$\mathbf{U}^{(k)} = \mathbf{Q}_A \mathbf{C}_A^{(k)} + \mathbf{Q}_B \mathbf{C}_B^{(k)} \tag{10.50}$$

Since $\mathbf{Q}_A$ and $\mathbf{Q}_B$ satisfy $\mathbf{A}\mathbf{Q}_A = \mathbf{Q}_A\mathbf{\Lambda}_A$ and $\mathbf{A}\mathbf{Q}_B = \mathbf{Q}_B\mathbf{\Lambda}_B$, where $\mathbf{\Lambda}_A$ and $\mathbf{\Lambda}_B$ are diagonal matrices of the m larger eigenvalues and the $n - m$ smaller eigenvalues respectively, it follows that

$$\mathbf{V}^{(k)} = \mathbf{Q}_A \mathbf{\Lambda}_A \mathbf{C}_A^{(k)} + \mathbf{Q}_B \mathbf{\Lambda}_B \mathbf{C}_B^{(k)} \tag{10.51}$$

The second step in the iterative process is to compute the *interaction matrix* $\mathbf{B}^{(k)}$ which satisfies

$$\mathbf{G}^{(k)}\mathbf{B}^{(k)} = \mathbf{H}^{(k)} \tag{10.52}$$

where

$$\mathbf{G}^{(k)} = [\mathbf{U}^{(k)}]^T\mathbf{U}^{(k)} \quad \text{and} \quad \mathbf{H}^{(k)} = [\mathbf{U}^{(k)}]^T\mathbf{V}^{(k)} \tag{10.53}$$

On account of the premultiplication (equation 10.48), the coefficients of $\mathbf{C}_B^{(k)}$ will become smaller in relation to the principal coefficients of $\mathbf{C}_A^{(k)}$ as k increases. Hence eventually

$$\left.\begin{aligned}\mathbf{U}^{(k)} &\simeq \mathbf{Q}_A\mathbf{C}_A^{(k)}\\ &\text{and}\\ \mathbf{V}^{(k)} &\simeq \mathbf{Q}_A\boldsymbol{\Lambda}_A\mathbf{C}_A^{(k)}\end{aligned}\right\} \tag{10.54}$$

Substituting equations (10.53) and (10.54) into equation (10.52) gives

$$[\mathbf{U}^{(k)}]^T\mathbf{Q}_A\mathbf{C}_A^{(k)}\mathbf{B}^{(k)} \simeq [\mathbf{U}^{(k)}]^T\mathbf{Q}_A\boldsymbol{\Lambda}_A\mathbf{C}_A^{(k)} \tag{10.55}$$

If this equation is premultiplied by the inverse of $[\mathbf{U}^{(k)}]^T\mathbf{Q}_A\mathbf{C}_A^{(k)}$ and

Table 10.3 Lop-sided iteration for matrix (7.53)

k	Trial vectors $\mathbf{U}^{(k)}$	Interaction matrix $\mathbf{B}^{(k)}$	Eigensolution of $\mathbf{B}^{(k)}$ (eigenvalues above, $\mathbf{P}^{(k)}$ below)
0	$\begin{bmatrix} 1 & 1 & 1\\ 1 & 1 & 0\\ 1 & 1 & -1\\ 1 & -1 & -1\\ 1 & -1 & 0\\ 1 & -1 & 1 \end{bmatrix}$	$\begin{bmatrix} 776.45 & 55.74 & 52.98\\ 55.74 & 40.68 & 7.44\\ 79.47 & 11.17 & 115.51 \end{bmatrix}$	$\begin{bmatrix} 787.01 & 109.39 & 36.24 \end{bmatrix}$ $\begin{bmatrix} 1 & -0.0829 & -0.0703\\ 0.0759 & 0.0411 & 1\\ 0.1196 & 1 & -0.0704 \end{bmatrix}$
1	$\begin{bmatrix} 1 & 1 & 1\\ 0.9265 & -0.2426 & -0.2177\\ 0.6139 & -0.8381 & 0.5680\\ 0.7021 & -0.4480 & -0.3512\\ 0.7874 & -0.0821 & -0.2836\\ 0.6926 & 0.2905 & -0.9196 \end{bmatrix}$	$\begin{bmatrix} 798.89 & 17.93 & 8.05\\ 7.39 & 145.48 & 45.29\\ 0.98 & 26.65 & 87.47 \end{bmatrix}$	$\begin{bmatrix} 799.11 & 161.52 & 71.21 \end{bmatrix}$ $\begin{bmatrix} 1 & -0.3268 & 0.3967\\ 0.1143 & 1 & -0.6102\\ 0.1800 & 0.3595 & 1 \end{bmatrix}$
2	$\begin{bmatrix} 1 & 1 & 0.3862\\ 0.9155 & -0.4067 & -0.4325\\ 0.5900 & -0.5021 & 1\\ 0.6941 & -0.1879 & 0.1641\\ 0.7749 & -0.2096 & -0.1692\\ 0.6772 & -0.0550 & -0.8334 \end{bmatrix}$	$\begin{bmatrix} 799.15 & 0.32 & -0.31\\ 0.22 & 175.07 & 16.25\\ -0.08 & 6.86 & 83.79 \end{bmatrix}$	$\begin{bmatrix} 799.15 & 176.27 & 82.59 \end{bmatrix}$ $\begin{bmatrix} 1 & -0.0005 & 0.0005\\ 0.0003 & 1 & 0.1758\\ -0.0001 & 0.0741 & 1 \end{bmatrix}$
3	$\begin{bmatrix} 1 & 1 & 0.1689\\ 0.9154 & -0.4612 & -0.4464\\ 0.5892 & -0.4340 & 1\\ 0.6944 & -0.0827 & 0.2919\\ 0.7745 & -0.2596 & -0.1625\\ 0.6767 & -0.0939 & -0.6301 \end{bmatrix}$	$\begin{bmatrix} 799.15 & 0.01 & 0.00\\ 0.01 & 178.08 & 2.79\\ 0.00 & 1.43 & 85.85 \end{bmatrix}$	$\begin{bmatrix} 799.15 & 178.12 & 85.81 \end{bmatrix}$ $\begin{bmatrix} 1 & 0.0000 & 0.0000\\ 0.0000 & 1 & -0.0303\\ 0.0000 & 0.0155 & 1 \end{bmatrix}$
4	$\begin{bmatrix} 1 & 1 & 0.1176\\ 0.9154 & -0.4739 & -0.4715\\ 0.5891 & -0.4268 & 1\\ 0.6944 & -0.0490 & 0.3583\\ 0.7744 & -0.2787 & -0.1922\\ 0.6766 & -0.0959 & 0.5544 \end{bmatrix}$	$\begin{bmatrix} 799.15 & 0.00 & 0.00\\ 0.00 & 178.33 & 0.54\\ 0.00 & 0.28 & 86.82 \end{bmatrix}$	$\begin{bmatrix} 799.15 & 178.33 & 86.82 \end{bmatrix}$ $\begin{bmatrix} 1 & 0.0000 & 0.0000\\ 0.0000 & 1 & -0.0059\\ 0.0000 & 0.0031 & 1 \end{bmatrix}$

postmultiplied by the inverse of $C_A^{(k)}$ (assuming that such inverses exist), it simplifies to

$$B^{(k)}[C_A^{(k)}]^{-1} \simeq [C_A^{(k)}]^{-1}\Lambda_A \tag{10.56}$$

Hence the eigenvalues of the interaction matrix are approximations to the dominant set of m eigenvalues of A, and the $m \times m$ matrix of right eigenvectors of the interaction matrix is an approximation to $[C_A^{(k)}]^{-1}$. If $P^{(k)}$ is the matrix of right eigenvectors of $B^{(k)}$, since

$$W^{(k)} = V^{(k)}P^{(k)} \simeq Q_A\Lambda_A \tag{10.57}$$

the vectors contained in $W^{(k)}$ are approximations to the dominant set of m right eigenvectors of A. The subsequent set of trial vectors $U^{(k+1)}$ are therefore taken to be equal to the set $W^{(k)}$, except that a normalization is applied to each vector. Table 10.3 shows an application of lop-sided iteration to the covariance matrix (7.53).

In subsequent discussion of simultaneous iteration the iteration number superscript (k) will be omitted.

10.7 THE CONVERGENCE RATE AND EFFICIENCY OF SIMULTANEOUS ITERATION

Convergence rate for eigenvector predictions

Consider the convergence of $u_i^{(k)}$ to the eigenvector q_i. When the interaction matrix eigensolution is used to reorientate the vectors (as in equation 10.57), then as k increases the error component which is slowest to reduce will be that of q_{m+1}. Therefore the prediction for q_i will have an error component whose magnitude will decrease by a factor of approximately λ_i/λ_{m+1} at each iteration. Consider the above application of SI to the covariance matrix (7.53) in which $m = 3$. An evaluation of the errors for the eigenvector predictions is shown in Table 10.4. These results agree reasonably well with the theoretical estimation of the rates of error reduction using the eigenvalue ratios $\lambda_1/\lambda_4 \simeq 13.6$, $\lambda_2/\lambda_4 \simeq 3.1$ and $\lambda_3/\lambda_4 \simeq 1.48$ obtained from the full eigensolution shown in Table 7.2. In general, since $|\lambda_i|/|\lambda_{m+1}| \geqslant |\lambda_i|/|\lambda_{i+1}|$, the convergence rate for any particular

Table 10.4 Errors for eigenvector estimates of the lop-sided iteration shown in Table 10.3. (The approximate value of the maximum element norm of the error vector is shown)

iteration no.	vector no. (i)		
k	1	2	3
0	0.41	1.48	2.00
1	0.025	0.53	0.92
2	0.00084	0.15	0.34
3	0.000037	0.049	0.21
4	0.0000033	0.016	0.15

eigenvector prediction cannot be slower than it is for purification or deflation. It will often be considerably more rapid.

The use of guard vectors

If the dominant set of r eigenvalues and corresponding eigenvectors of a matrix are required, it is normal practice to make the number of trial vectors m greater than r. The extra $m - r$ vectors may be called *guard vectors*. Poor convergence will only then be obtained when $|\lambda_r| \simeq |\lambda_{r+1}| \simeq \cdots \simeq |\lambda_{m+1}|$, the probability of which decreases rapidly as the number of guard vectors increases.

It is convenient to sort the eigenvectors of **B**, which are recorded as columns of **P**, according to the dominance of their corresponding eigenvalues. If this is done the first r trial vectors will converge onto the required eigenvectors of **A**.

Computational efficiency

Formulae for the approximate number of multiplications required to perform the various operations in one cycle of lop-sided iteration are shown in Table 10.5. The amount of computation required to obtain a simultaneous iteration solution is reduced if:

(a) the matrix is sparse,
(b) initial approximations to the eigenvectors are available (e.g. from the eigensolution of a similar system) and
(c) only low accuracy results are required.

Table 10.5 Iteration cycle for lop-sided iteration (n = order of matrix, m = no. of trial vectors, c = average no. of non-zero elements per row of **A**, superscript (k) has been omitted throughout)

Operation	Equation	Notes	Approximate no. of multiplications
Multiply	$\mathbf{V} = \mathbf{AU}$	Advantage may be taken of any sparseness in **A**	ncm
Multiply	$\mathbf{G} = \mathbf{U}^T\mathbf{U}$	Only one triangle needs to be formed	$nm^2/2$
Multiply	$\mathbf{H} = \mathbf{U}^T\mathbf{V}$	—	nm^2
Solve	$\mathbf{GB} = \mathbf{H}$	Take advantage of symmetry and positive definite property of **G**	$7m^3/6$
Full eigensolution	$\mathbf{BP} = \mathbf{P}\Theta$	See section 10.9 for case where **B** has complex conjugate pairs of eigenvalues	$f(m^3)$
Multiply	$\mathbf{W} = \mathbf{VP}$		nm^2
Normalize	$\mathbf{W} \rightarrow \mathbf{U}^{(k+1)}$	—	—
Tolerance test	—	—	—
			Total $= ncm + \frac{5}{2}nm^2 + (f + \frac{7}{6})m^3$

Use of simultaneous iteration

Many eigenvalue problems may be specified in a form in which only the largest eigenvalue or the largest set of eigenvalues is required. The principal component analysis (section 7.8) and the Markov chain analysis (section 7.10) are cases in point. So is the dynamic stability analysis (section 7.6), provided that the equations are transformed according to equation (7.48). Hence SI (or the power method) should be considered for their solution. Where the matrices are large, and particularly where they are also sparse, SI will almost certainly prove to be more efficient than transformation methods. SI may even be feasible for the partial eigensolution of sparse matrices of order 10,000 or more on large computers.

10.8 SIMULTANEOUS ITERATION FOR SYMMETRIC MATRICES

Symmetric form of interaction matrix

If lop-sided iteration is applied to a symmetric matrix **A**, each matrix $\mathbf{H}^{(k)}$ will also be symmetric. Hence it is possible to reduce the number of multiplications required to compute it from nm^2 to approximately $nm^2/2$. Since, in general, the interaction

Table 10.6 Iteration cycle for a symmetric simultaneous iteration procedure (notation as for Table 10.5)

Operation	Equation	Notes	Approximate no. of multiplications
Multiply	$\mathbf{V} = \mathbf{AU}$	Advantage may be taken of any sparseness in **A**	ncm
Multiply	$\mathbf{G} = \mathbf{U}^T\mathbf{U}$	Only one triangle needs to be formed	$nm^2/2$
Multiply	$\mathbf{H} = \mathbf{U}^T\mathbf{V}$	Only one triangle needs to be formed	$nm^2/2$
Choleski decomposition	$\mathbf{G} = \mathbf{LL}^T$	–	$m^3/6$
Solve	$\mathbf{LX} = \mathbf{H}$	Forward-substitution	$m^3/2$
Transpose	$\mathbf{X} \to \mathbf{X}^T$	–	–
Solve	$\mathbf{LB}_S = \mathbf{X}^T$	Forward-substitution in which only upper triangle needs to be formed	$m^3/6$
Full eigensolution	$\mathbf{B}_S\mathbf{P}_S = \mathbf{P}_S\Theta$	Advantage may be taken of symmetry of $\mathbf{B}_S$	$f_S(m^3)$
Solve	$\mathbf{L}^T\mathbf{P} = \mathbf{P}_S$	Backsubstitution	$m^3/2$
Multiply	$\mathbf{W} = \mathbf{VP}$	–	nm^2
Normalize	$\mathbf{W} \to \mathbf{U}^{(k+1)}$	–	–
Tolerance test	–	–	–
			Total $= ncm + 2nm^2 + (f_S + \frac{4}{3})m^3$

$$\mathbf{U} = \begin{bmatrix} 1 & 1 & 1 \\ 0.9265 & -0.2426 & -0.2177 \\ 0.6139 & -0.8381 & 0.5680 \\ 0.7021 & -0.4480 & -0.3512 \\ 0.7874 & -0.0821 & -0.2836 \\ 0.6926 & 0.2905 & -0.9196 \end{bmatrix} \quad \mathbf{V} = \begin{bmatrix} 807.60 & 196.85 & 149.21 \\ 742.27 & -43.32 & -65.56 \\ 479.10 & -103.54 & 18.57 \\ 562.60 & -22.84 & -2.44 \\ 628.02 & -16.78 & -26.88 \\ 548.45 & 25.79 & -56.15 \end{bmatrix} \quad \mathbf{G} = \begin{bmatrix} 3.8281 & & \text{symmetric} \\ 0.0828 & 2.0531 & \\ 0.0402 & 0.4903 & 2.4194 \end{bmatrix}$$

$$\mathbf{H} = \begin{bmatrix} 3058.85 & & \text{symmetric} \\ 81.76 & 313.24 & \\ 38.10 & 136.53 & 234.14 \end{bmatrix} \quad \mathbf{L} = \begin{bmatrix} 1.9565 & & \\ 0.0423 & 1.4322 & \\ 0.0205 & 0.3417 & 1.5173 \end{bmatrix} \quad \mathbf{B} = \begin{bmatrix} 799.06 & & \text{symmetric} \\ 5.58 & 151.68 & \\ 0.76 & 28.22 & 81.10 \end{bmatrix}$$

$$\boldsymbol{\Theta} = \begin{bmatrix} 799.11 & 161.52 & 71.21 \end{bmatrix}$$

$$\mathbf{P_s} = \begin{bmatrix} 1 & -0.0092 & 0.0016 \\ 0.0087 & 1 & -0.3507 \\ 0.0014 & 0.3508 & 1 \end{bmatrix} \quad \mathbf{P} = \begin{bmatrix} 0.5110 & -0.0210 & 0.0026 \\ 0.0058 & 0.6431 & -0.4021 \\ 0.0009 & 0.2312 & 0.6591 \end{bmatrix} \quad \mathbf{W} = \begin{bmatrix} 413.95 & 144.11 & 21.29 \\ 378.97 & -58.61 & -23.84 \\ 244.22 & -72.36 & 55.13 \\ 287.34 & -27.07 & 9.05 \\ 320.77 & -30.20 & -9.33 \\ 280.34 & -7.92 & -45.94 \end{bmatrix}$$

Figure 10.1 Second iteration for the example shown in Table 10.3 using the symmetric form of the interaction matrix

matrix has real eigenvalues. If

$$\mathbf{G} = \mathbf{L}\mathbf{L}^T \tag{10.58}$$

is a Choleski decomposition of **G**, equation (10.52) gives

$$\mathbf{B} = \mathbf{L}^{-T}(\mathbf{L}^{-1}\mathbf{H}\mathbf{L}^{-T})\mathbf{L}^T \tag{10.59}$$

Hence **B** is similar to the symmetric matrix $\mathbf{L}^{-1}\mathbf{H}\mathbf{L}^{-T}$ and consequently has the same eigenvalues.

There are several possible ways of modifying the SI process so that the symmetric matrix

$$\mathbf{B_s} = \mathbf{L}^{-1}\mathbf{H}\mathbf{L}^{-T} \tag{10.60}$$

is used as an interaction matrix instead of **B**. For instance, $\mathbf{B_s}$ may be obtained from the solution of the equation

$$\mathbf{L}\mathbf{B_s} = \mathbf{X}^T \tag{10.61}$$

where **X** has previously been obtained by solving the equation

$$\mathbf{L}\mathbf{X} = \mathbf{H} \tag{10.62}$$

If $\mathbf{B_s}$ has $\mathbf{P_s}$ as its matrix of right eigenvectors then **P** can be obtained from $\mathbf{P_s}$ by solving the equation

$$\mathbf{L}^T\mathbf{P} = \mathbf{P_s} \tag{10.63}$$

An alternative SI procedure based on these observations is presented in Table 10.6.

The modification of the second iteration cycle of Table 10.3 to this symmetric form of simultaneous iteration is shown in Figure 10.1. The eigenvalue estimates and the revised set of trial vectors are the same as for the basic lop-sided method.

A lower bound property for eigenvalue estimates

Let $\mathbf{Y} = [\mathbf{y}_1 \mathbf{y}_2 \ \ldots \ \mathbf{y}_m]$ be a set of vectors related to $\mathbf{U}$ by the equation

$$\mathbf{U} = \mathbf{Y}\mathbf{L}^T \tag{10.64}$$

where $\mathbf{L}$ is defined in equation (10.58). The symmetric interaction matrix (10.60) may be written as

$$\mathbf{B}_s = \mathbf{L}^{-1}\mathbf{U}^T\mathbf{A}\mathbf{U}\mathbf{L}^{-T} = \mathbf{Y}^T\mathbf{A}\mathbf{Y} \tag{10.65}$$

However, since

$$\mathbf{U}^T\mathbf{U} = \mathbf{G} = \mathbf{L}\mathbf{L}^T \tag{10.66}$$

it follows from equation (10.64) that

$$\mathbf{Y}^T\mathbf{Y} = \mathbf{L}^{-1}\mathbf{U}^T\mathbf{U}\mathbf{L}^{-T} = \mathbf{I} \tag{10.67}$$

It is possible to define (but not uniquely) a matrix $\mathbf{Z}$ of order $n \times (n - m)$ which combines with $\mathbf{Y}$ to form an orthogonal matrix, i.e.

$$\begin{bmatrix} \mathbf{Y}^T \\ \mathbf{Z}^T \end{bmatrix} \begin{bmatrix} \mathbf{Y} & \mathbf{Z} \end{bmatrix} = \mathbf{I} \tag{10.68}$$

The matrix

$$\bar{\mathbf{A}} = \begin{bmatrix} \mathbf{Y}^T \\ \mathbf{Z}^T \end{bmatrix} \begin{bmatrix} \mathbf{A} \end{bmatrix} \begin{bmatrix} \mathbf{Y} & \mathbf{Z} \end{bmatrix} = \begin{bmatrix} \mathbf{B}_s & \mathbf{Y}^T\mathbf{A}\mathbf{Z} \\ \mathbf{Z}^T\mathbf{A}\mathbf{Y} & \mathbf{Z}^T\mathbf{A}\mathbf{Z} \end{bmatrix} \tag{10.69}$$

being an orthogonal transformation of $\mathbf{A}$, must have the same eigenvalues. However, since $\mathbf{B}_s$ is a leading submatrix of $\bar{\mathbf{A}}$, the Sturm sequence property of $\bar{\mathbf{A}}$ may be used to show that each eigenvalue of $\mathbf{B}_s$ has a modulus less than that of the corresponding eigenvalue of $\mathbf{A}$ (see the eigenvalue pyramid, Figure 9.2). Therefore the eigenvalue estimates converge from below and the magnitude of any particular eigenvalue estimate is a lower bound to the magnitude of the eigenvalue to which it is converging. This property is illustrated in Table 10.3 where all of the eigenvalue estimates are less than or equal to the correct values.

Accuracy of the eigenvalue estimates

If the eigenvectors of $\mathbf{A}$ are normalized to satisfy $\mathbf{q}_i^T\mathbf{q}_i = 1$ then, with $\mathbf{A}$ symmetric, the sets of eigenvectors $\mathbf{Q}_A$ and $\mathbf{Q}_B$ must satisfy the following equations:

$$\left.\begin{aligned} \mathbf{Q}_A^T\mathbf{Q}_A &= \mathbf{I}, & \mathbf{Q}_A^T\mathbf{Q}_B &= \mathbf{0} \\ \mathbf{Q}_B^T\mathbf{Q}_A &= \mathbf{0}, & \mathbf{Q}_B^T\mathbf{Q}_B &= \mathbf{I} \end{aligned}\right\} \tag{10.70}$$

Therefore, using the expressions for **U** and **V** given in equations (10.50) and (10.51), it may be shown that

$$\mathbf{G} = \mathbf{U}^T\mathbf{U} = \mathbf{C}_A^T\mathbf{C}_A + \mathbf{C}_B^T\mathbf{C}_B \tag{10.71}$$

and

$$\mathbf{H} = \mathbf{V}^T\mathbf{V} = \mathbf{C}_A^T\mathbf{\Lambda}_A\mathbf{C}_A + \mathbf{C}_B^T\mathbf{\Lambda}_B\mathbf{C}_B \tag{10.72}$$

Since the eigenvalue estimates and also the revised trial vectors will be correct if $\mathbf{C}_B = \mathbf{0}$, the coefficients of $\mathbf{C}_B$ may be considered to be error terms. From the equations above it may be seen that both **G** and **H** contain only second-order error terms. Hence from equation (10.52) it may be shown that the error terms in **B** will be of second order.

However, if **B** is changed to **B** + d**B**, where the elements of d**B** are small, it can be shown that the diagonal matrix of eigenvalues **Θ** changes to **Θ** + d**Θ** where

$$\mathrm{d}\mathbf{\Theta} \simeq \mathbf{P}^T\mathrm{d}\mathbf{B}\mathbf{P} \tag{10.73}$$

Table 10.7 Approximate errors in eigenvalue estimates of the covariance matrix for the solution shown in Table 10.3 (computation carried out to approximately twelve decimal places)

Iteration no. k	Eigenvalue $\lambda_1 \simeq 799$	$\lambda_2 \simeq 178$	$\lambda_3 \simeq 87$
0	12	69	57
1	0.042	17	16
2	0.00016	2.1	4.8
3	0.71×10^{-6}	0.23	1.6
4	0.34×10^{-8}	0.024	0.56

Hence the errors in the eigenvalue estimates will be of second order. Close to convergence the error in the estimated value of λ_i at iteration k will be of order $(\lambda_i/\lambda_{m+1})^{2k}$. That is, the eigenvalue estimates will be correct to approximately twice as many significant figures as the eigenvector estimates. This may be observed by comparing the errors shown in Tables 10.4 and 10.7.

Orthogonal vector methods

If the trial vectors are orthonormal then

$$\mathbf{G} = \mathbf{U}^T\mathbf{U} = \mathbf{I} \tag{10.74}$$

Hence, since

$$\mathbf{B} = \mathbf{B}_s = \mathbf{H} = \mathbf{U}^T\mathbf{V} \tag{10.75}$$

the reorientation process is simplified. However, if an iterative process incorporationg this extra requirement is to be formed, the reoriented set of vectors

W will need to be orthonormalized rather than just normalized to obtain the trial vector set for the next iteration. The Gram–Schmidt process is suitable for carrying out the orthonormalization.

Rutishauser's reorientation

In Rutishauser's (1969) method the trial vectors are assumed to be orthonormal and the interaction matrix

$$\mathbf{B}_{\mathrm{R}} = \mathbf{V}^T\mathbf{V} = \mathbf{U}^T\mathbf{A}^2\mathbf{U} \tag{10.76}$$

is used. If the full eigensolution of this matrix is such that

$$\mathbf{B}_{\mathrm{R}}\mathbf{P}_{\mathrm{R}} = \mathbf{P}_{\mathrm{R}}\mathbf{\Phi} \tag{10.77}$$

the reoriented set of vectors is given by

$$\mathbf{W} = \mathbf{V}\mathbf{P}_{\mathrm{R}} \tag{10.78}$$

However,

$$\mathbf{W}^T\mathbf{W} = \mathbf{P}_{\mathrm{R}}^T\mathbf{V}^T\mathbf{V}\mathbf{P}_{\mathrm{R}} = \mathbf{P}_{\mathrm{R}}^T\mathbf{B}_{\mathrm{R}}\mathbf{P}_{\mathrm{R}} = \mathbf{\Phi} \tag{10.79}$$

This means that, theoretically, the reoriented vectors are orthogonal and only a normalization is required to produce a satisfactory set of trial vectors for the next iteration. However, since rounding errors may disturb the orthogonality, it is not safe to omit the orthogonalization process from many consecutive iterations.

As the iteration process continues, successive values of each eigenvalue ϕ_i of $\mathbf{B}_{\mathrm{R}}$ converge to λ_i^2 (where λ_i is the corresponding eigenvalue from the dominant set $\mathbf{\Lambda}_A$). Hence the magnitude, but not the sign, of the required eigenvalues can be determined from the eigenvalues of $\mathbf{B}_{\mathrm{R}}$. Where necessary, signs may be resolved by comparing corresponding elements of $\mathbf{u}_i$ and $\mathbf{v}_i$. When $m = 1$ Rutishauser's method degenerates to the power method with a Euclidean norm prediction for the dominant eigenvalue. On the other hand, if interaction matrices $\mathbf{B}$ or $\mathbf{B}_{\mathrm{S}}$ are used, the equivalent power method has a Rayleigh quotient prediction for the eigenvalue.

10.9 SIMULTANEOUS ITERATION FOR UNSYMMETRIC MATRICES

If the lop-sided method is used to compute a partial eigensolution of an unsymmetric matrix, then the eigenvalue estimates will not, in general, be any more accurate than the corresponding eigenvector estimates; nor will the lower bound property given in section 10.8 be applicable. However, satisfactory convergence will be obtained subject to the following two constraints.

Complex eigenvalues

If **A** is a real matrix and iteration is commenced with a set of real trial vectors, the sequence of trial vectors will not remain real after an interaction matrix has been encountered which has one or more complex conjugate pairs of eigenvalues. To perform the computation in complex arithmetic requires extra storage space and also incurs a time penalty. However, it is possible to avoid the use of complex trial vectors by modifying the procedure in the following way.

If a real interaction matrix **B** has a complex conjugate pair of eigenvalues with corresponding eigenvectors $\mathbf{r} \pm i\mathbf{s}$, then, instead of including these complex vectors in **P**, their real and imaginary components may be used. If they are to be entered into columns j and $j + 1$ of **P**, then $\mathbf{p}_j = \mathbf{r}$ and $\mathbf{p}_{j+1} = \mathbf{s}$. The multiplication $\mathbf{W} = \mathbf{VP}$ will involve only real arithmetic, and columns j and $j + 1$ of the reoriented set of vectors **W** will contain the real and imaginary components of the prediction for the complex pair of eigenvectors of **A**.

In order to ensure convergence it is also necessary to modify the normalization process. Suppose that $\mathbf{w}_j$ and $\mathbf{w}_{j+1}$ represent real and imaginary parts of a complex pair of vectors $\mathbf{w}_j \pm i\mathbf{w}_{j+1}$ which are to be normalized such that the largest element is unity. If $e + if$ is the element in $\mathbf{w}_j + i\mathbf{w}_{j+1}$ which has the largest modulus, and if the normalized vectors are to be represented by $\mathbf{u}_j^{(k+1)} \pm i\mathbf{u}_{j+1}^{(k+1)}$, then

$$\left.\begin{aligned} \mathbf{u}_j^{(k+1)} &= \frac{e\mathbf{w}_j + f\mathbf{w}_{j+1}}{e^2 + f^2} \\ &\text{and} \\ \mathbf{u}_{j+1}^{(k+1)} &= \frac{-f\mathbf{w}_j + e\mathbf{w}_{j+1}}{e^2 + f^2} \end{aligned}\right\} \qquad (10.80)$$

The modifications to the matrix **P** and the normalization procedure outlined above enable iteration to proceed with real vectors rather than complex vectors without adversely affecting the convergence rate.

Defective interaction matrices

It is possible to obtain a defective interaction matrix even when **A** is not itself defective. This may be illustrated by the following simple example:

$$\text{If} \quad \mathbf{A} = \begin{bmatrix} 1 & 0 & 1 \\ 0 & 1 & 1 \\ 1 & 0 & 1 \end{bmatrix} \quad \text{and} \quad \mathbf{U} = \begin{bmatrix} 0 & 0 \\ 1 & 0 \\ 0 & 1 \end{bmatrix} \quad \text{then} \quad \mathbf{B} = \begin{bmatrix} 1 & 1 \\ 0 & 1 \end{bmatrix} \qquad (10.81)$$

Matrix **A** has eigenvalues 0, 1 and 2 and consequently is not defective, though the interaction matrix **B** is defective. Since rounding error will normally be present, the chance of **B** being precisely defective will be extremely small. It is more probable that **B** will be almost defective. In this case the eigenvectors of **B**, and also the resulting set of reoriented vectors, will exhibit a strong degree of linear dependence. This in turn will cause ill-conditioning or singularity in the matrix **G** during the next iteration cycle. It is advisable to ensure that the procedure does not break down under such circumstances.

Bi-iteration

For unsymmetric matrices it is possible to use a bi-iteration technique in which two sets of trial vectors converge onto the dominant sets of left and right eigenvectors (Jennings and W. J. Stewart, 1975). Eigenvalue estimates will be more accurate than those for lop-sided iteration after the same number of iterations have been performed. However, bi-iteration requires more computation per iteration and also more storage space for computer implementation.

10.10 SIMULTANEOUS ITERATION FOR VIBRATION FREQUENCY ANALYSIS

Dynamical matrix formulation

The frequencies of vibration of structural systems may be obtained as the values of ω which give non-trivial solutions to the equation

$$\omega^2 \mathbf{M}\mathbf{x} = \mathbf{K}\mathbf{x} \tag{10.82}$$

where **M** and **K** are mass and stiffness matrices, both of which are normally symmetric and positive definite (section 7.3). Since this equation corresponds to the linearized eigenvalue problem, properties shown in section 7.4 are applicable. In that section it was noted that the eigenvalues will all be real and positive and consequently will yield valid vibration frequencies.

For the power method or SI it is convenient if the lower frequencies (which are normally the ones required) correspond to the larger λ values. Hence equation (10.82) should be expressed in the form $\mathbf{A}\mathbf{x} = \lambda\mathbf{B}\mathbf{x}$ by equating **M** to **A** and **K** to **B** so that $\lambda = 1/\omega^2$. Premultiplying equation (10.82) by $\lambda\mathbf{K}^{-1}$ then gives

$$\mathbf{K}^{-1}\mathbf{M}\mathbf{x} = \lambda\mathbf{x} \tag{10.83}$$

showing that a dominant set of the eigenvalues of the dynamical matrix $\mathbf{K}^{-1}\mathbf{M}$ is required. The only operation involving this matrix during SI is the premultiplication

$$\mathbf{V} = \mathbf{K}^{-1}\mathbf{M}\mathbf{U} \tag{10.84}$$

This premultiplication may be accomplished by first performing the matrix multiplication

$$\mathbf{X} = \mathbf{M}\mathbf{U} \tag{10.85}$$

and then solving the following set of linear equations

$$\mathbf{K}\mathbf{V} = \mathbf{X} \tag{10.86}$$

Preserving sparseness

If **K** and **M** are sparse matrices, advantage may be taken of sparsity in performing both of the above operations. In order to derive most benefit from the sparsity of the stiffness matrix **K**, it is necessary to use one of the storage schemes for symmetric and positive definite matrices described in Chapter 5. In addition, the variables should be ordered in such a way that the most effective use is made of the particular storage scheme. For instance, a frontal ordering scheme may be used for the variables with a diagonal band or variable bandwidth scheme for the matrix **K**. Because equation (10.86) is repeatedly solved with a different right-hand matrix **X** at each iteration, it is advisable to perform a triangular decomposition of **K** before the outset of the iterative process, so that only forward- and backsubstitution operations need to be executed in solving equation (10.84) within each iteration cycle. The most suitable storage schemes for **M** are ones in which only the non-zero elements are stored. In most cases this will take the form of a packing scheme rather than a band scheme.

Preserving symmetry

The computational efficiency of the method can be further increased by making use of the following transformation (see property (d) of section 7.4). Defining a new set of variables **y** satisfying

$$\mathbf{y} = \mathbf{L}^T\mathbf{x} \tag{10.87}$$

(where **L** is the lower triangular matrix obtained by Choleski decomposition of **K**), equation (10.83) may be transformed to

$$\mathbf{L}^{-1}\mathbf{M}\mathbf{L}^{-T}\mathbf{y} = \lambda\mathbf{y} \tag{10.88}$$

If simultaneous iteration is applied to this transformed system advantage may be taken of the symmetry of $\mathbf{L}^{-1}\mathbf{M}\mathbf{L}^{-T}$. The premultiplication

$$\mathbf{V} = \mathbf{L}^{-1}\mathbf{M}\mathbf{L}^{-T}\mathbf{U} \tag{10.89}$$

can be implemented by performing the following three operations:

$$\left.\begin{array}{lll} \text{(a)} & \text{solve by backsubstitution} & \mathbf{L}^T\mathbf{X} = \mathbf{U} \\ \text{(b)} & \text{matrix multiplication} & \mathbf{Y} = \mathbf{M}\mathbf{X} \\ \text{(c)} & \text{solve by forward-substitution} & \mathbf{L}\mathbf{V} = \mathbf{Y} \end{array}\right\} \tag{10.90}$$

which are essentially the same operations as those required to perform the premultiplication (10.84), except that the order of implementation of the operations has been changed.

Relative numerical efficiencies

If **b** is the average semibandwidth of **K** and **c** is the average number of non-zero elements per row in **M**, then the premultiplication can be executed through the algorithm (10.90) using approximately $nm(2b + c)$ multiplications. Hence, using the symmetric SI procedure shown in Table 10.6 to complete the iteration cycle, the total number of multiplications per iteration is approximately $nm(2b + c + 2m) + (f_s + 4/3)m^3$. In addition it is necessary, before iteration commences, to decompose

K (requiring $nb^2/2$ multiplications) and after convergence has been reached to transform the predicted eigenvectors from the variables **y** to the original variables **x**. The latter operation can be achieved by executing the backsubstitution phase only of equations (10.90) (requiring nbm multiplications). Hence, if a total of k iterations are required to obtain a solution by means of SI, then the total number of multiplications required will be

$$T_{\mathrm{SI}} \simeq \frac{nb^2}{2} + nmk(2b + c + 2m) + (f_s + 4/3)m^3k + nbm \tag{10.91}$$

By comparison, if the same results are obtained by:

(a) computing $\mathbf{L}^{-1}\mathbf{M}\mathbf{L}^{-T}$ explicitly,
(b) performing a Householder tridiagonalization of $\mathbf{L}^{-1}\mathbf{M}\mathbf{L}^{-T}$
(c) obtaining the eigensolution of the tridiagonal matrix and
(d) computing r modes of vibration by transforming r eigenvectors of the tridiagonal matrix,

the total number of multiplications required will be

$$T_{\mathrm{H}} \simeq 2/3n^3 + n^2(b + r) \tag{10.92}$$

Alternatively, if the Sturm sequence method is used as described in section 9.6(c), then, assuming that **M** and **K** have the same band structure, the number of multiplications performed will be

$$T_{\mathrm{SS}} \simeq \frac{3nb^2lr}{2} \tag{10.93}$$

where l is the average number of μ values required to locate one eigenvalue.

Consider, for instance, vibration equations derived from the finite difference form of Laplace's equation with mesh sizes to correspond with those in Table 6.4. The stiffness matrix will be of band form and the mass matrix will be diagonal. Assume that for each problem the lowest five frequencies are required and that six cycles of simultaneous iteration are needed to obtain the required accuracy using eight trial vectors. Assume also that in using the Sturm sequence method it is

Table 10.8 Comparison of methods of obtaining the lowest five frequencies and mode shapes for some vibration problems

Grid size	Number of equations n	Average semibandwidth	Approximate no. of multiplications ($M = 10^6$) by Householder's tridiagonalization	by Sturm sequence	by simultaneous iteration
4 x 8	32	5	0.032M	0.031M	0.050M
12 x 24	288	13	17M	1.9M	0.71M
40 x 80	3,200	41	22,000M	210M	20M
4 x 6 x 12	288	25	18M	6.9M	1.2M
9 x 13 x 26	3,042	118	20,000M	1,600M	62M

necessary to choose, on average, five μ values in order to isolate each eigenvalue to sufficient accuracy. From the comparison of the computational requirements for each of the methods given in Table 10.8 it is evident that SI is particularly efficient for the larger problems.

Relationship of SI with mass condensation

Mass condensation, otherwise known as the eigenvalue economizer method (Irons, 1963), involves defining a reduced set of variables which sufficiently accurately yield the required vibration modes. The condensed vibration equations obtained from the reduced set of variables can be solved much more easily than the original equations can be solved, if the latter are of large order.

If the reduced variables are represented by the $m \times 1$ vector $\mathbf{z}$, then the basic variables $\mathbf{x}$ are assumed to be linearly related to $\mathbf{z}$ according to

$$\mathbf{x} = \mathbf{H}\mathbf{z} \tag{10.94}$$

where $\mathbf{H}$ is an $n \times m$ matrix. It can be shown that equation (10.82) gives rise to the condensed equations

$$\omega^2\bar{\mathbf{M}}\mathbf{x} + \bar{\mathbf{K}}\mathbf{z} = \mathbf{0} \tag{10.95}$$

where $\bar{\mathbf{M}} = \mathbf{H}^T\mathbf{M}\mathbf{H}$ and $\bar{\mathbf{K}} = \mathbf{H}^T\mathbf{K}\mathbf{H}$. The recommended technique for choosing $\mathbf{H}$ is to allocate m master displacements and then to set each column of $\mathbf{H}$ equal to the displacements arising through a unit movement of one of the master displacements (with the other master displacements restrained to zero).

It has been shown by Jennings (1973) that mass condensation is equivalent to the first round of iteration of SI, if $\mathbf{H}$ is used as the set of trial vectors for SI. The computational efficiencies of SI and mass condensation are similar. However, SI is more versatile and reliable because the accuracy of the results can be monitored and controlled and also is not dependent upon an initial choice of the number and position of master displacements.

10.11 SIMULTANEOUS ITERATION MODIFICATIONS WHICH IMPROVE EFFICIENCY

Approximate eigensolution of the interaction matrix

In the early rounds of iteration the error components will be so large that accurate eigensolutions of the interaction matrix will yield results which are only partially meaningful. Furthermore, in the later rounds of iteration the vectors will already be close to the corresponding eigenvectors and hence the interaction matrix eigensolution will only produce a slight adjustment to them. Hence the convergence rate is not usually adversely affected if only an approximate eigensolution of the interaction matrix is computed at each iteration (Jennings, 1967, for symmetric matrices; Clint and Jennings, 1971, for unsymmetric matrices).

Omission of the reorientation process

An alternative to using an approximate eigensolution of the interaction matrix is to include the reorientation process in only a few of the iterations. When **A** is sparse the reduction in the amount of computation will be significant. However, reorientation or orthogonalization should not be omitted from so many consecutive iterations that the components of $\mathbf{q}_1$ in all of the vectors becomes so large as to make the set of trial vectors **U** linearly dependent (Rutishauser, 1969; G. W. Stewart, 1976).

Locking vectors

When the first vector has been computed to the required accuracy it may be *locked.* In the locked states it is ignored during the premultiplication process and is only included in the reorientation to prevent other vectors from converging onto it. After the second and subsequent vectors have been computed to the required accuracy they may also be locked. If **A** is symmetric, the only operation which needs to be performed with the locked vectors is the orthogonalization of the remaining vectors with respect to them (Corr and Jennings, 1976, for symmetric matrices).

Chebyshev acceleration

When **A** is symmetric it is possible to accelerate the rate of convergence by a Chebyshev procedure similar to that for accelerating relaxation methods given in section 6.11 (Rutishauser, 1969).

10.12 LANCZOS' METHOD

Lanczos' (1950) method for the eigensolution of a matrix has similar characteristics to, and indeed is related to, the conjugate gradient method for solving linear equations. Consider the eigensolution of a symmetric matrix **A**. Starting with a single trial vector, the algorithm generates a sequence of mutually orthogonal vectors by means of a process which includes premultiplications by the matrix **A**. Theoretically the sequence of vectors must terminate after n vectors have been generated. The orthogonal vectors combine to produce a transformation matrix which has the effect of transforming **A** to tridiagonal form. The elements of the tridiagonal matrix are generated as the orthogonal vectors are formed and, when the tridiagonal matrix is complete, its eigenvalues may be computed by any suitable technique (e.g. the Sturm sequence, or LR or QR methods).

As in the case of the power method, the premultiplication process amplifies the components of the eigenvectors corresponding to the eigenvalues of largest modulus. Therefore if, when **A** is large, the process of vector generation is terminated after m steps where $m \ll n$, eigensolution of the resulting tridiagonal matrix of order m will yield an approximation to the set of m dominant eigenvalues of **A**. The larger eigenvalues of this tridiagonal matrix give good approximations to the dominant eigenvalues of **A**. This means that the Lanczos algorithm may be used for either full or partial eigensolution of a matrix.

The standard Lanczos algorithm transforms a symmetric matrix into unsymmetric tridiagonal form. The method described below modifies the procedure so that a symmetric tridiagonal matrix is formed. If $\mathbf{Y} = [\mathbf{y}_1 \mathbf{y}_2 \ \cdots \ \mathbf{y}_n]$ is the full compounded set of mutually orthogonal vectors, the transformation to tridiagonal form may be described by

$$\begin{bmatrix} & & \\ & \mathbf{A} & \\ & & \end{bmatrix} \begin{bmatrix} & & \\ \mathbf{y}_1 \mathbf{y}_2 & \cdots & \mathbf{y}_n \\ & & \end{bmatrix} = \begin{bmatrix} & & \\ \mathbf{y}_1 \mathbf{y}_2 & \cdots & \mathbf{y}_n \\ & & \end{bmatrix} \begin{bmatrix} \alpha_1 & \beta_1 & & & \\ \beta_1 & \alpha_2 & \beta_2 & & \\ & \cdot & \cdot & \cdot & \\ & & \cdots & \cdots & \beta_{n-1} \\ & & & \beta_{n-1} & \alpha_n \end{bmatrix} \tag{10.96}$$

This matrix equation may be expanded to give the n vector equations

$$\left.\begin{aligned} \mathbf{A}\mathbf{y}_1 &= \alpha_1 \mathbf{y}_1 + \beta_1 \mathbf{y}_2 \\ \mathbf{A}\mathbf{y}_2 &= \beta_1 \mathbf{y}_1 + \alpha_2 \mathbf{y}_2 + \beta_2 \mathbf{y}_3 \\ &\cdot \quad \cdot \quad \cdot \quad \cdot \quad \cdot \quad \cdot \quad \cdot \\ \mathbf{A}\mathbf{y}_j &= \beta_{j-1} \mathbf{y}_{j-1} + \alpha_j \mathbf{y}_j + \beta_j \mathbf{y}_{j+1} \\ &\cdot \quad \cdot \quad \cdot \quad \cdot \quad \cdot \quad \cdot \quad \cdot \\ \mathbf{A}\mathbf{y}_n &= \beta_{n-1} \mathbf{y}_{n-1} + \alpha_n \mathbf{y}_n \end{aligned}\right\} \tag{10.97}$$

The first vector $\mathbf{y}_1$ is chosen to be an arbitrary non-null vector normalized such that $\mathbf{y}_1^T \mathbf{y}_1 = 1$. If the first of equations (10.97) is premultiplied by $\mathbf{y}_1^T$, it can be established that when $\alpha_1 = \mathbf{y}_1^T \mathbf{A} \mathbf{y}_1$ then $\mathbf{y}_1^T \mathbf{y}_2 = 0$. Thus choosing α_1 in this way ensures that $\mathbf{y}_2$ is orthogonal to $\mathbf{y}_1$. Substitution of this value of α_1 into the first of equations (10.97) yields $\beta_1 \mathbf{y}_2$. Then, since $\mathbf{y}_2^T \mathbf{y}_2 = 1$, $\mathbf{y}_2$ can be obtained from $\beta_1 \mathbf{y}_2$ by Euclidean normalization, $1/\beta_1$ being the normalizing factor. The remaining vectors of the set may be generated in a similar way using the second, third, etc., of equations (10.97), the process being fully described by the following algorithm:

$$\left.\begin{aligned} \mathbf{v}_j &= \mathbf{A}\mathbf{y}_j - \beta_{j-1} \mathbf{y}_{j-1} \quad (\beta_0 = 0) \\ \alpha_j &= \mathbf{y}_j^T \mathbf{v}_j \\ \mathbf{z}_j &= \mathbf{v}_j - \alpha_j \mathbf{y}_j \\ \beta_j &= (\mathbf{z}_j^T \mathbf{z}_j)^{1/2} \\ \mathbf{y}_{j+1} &= (1/\beta_j) \mathbf{z}_j \end{aligned}\right\} \tag{10.98}$$

Theoretically it can be shown that the vectors are all mutually orthogonal and also that the last of equations (10.97) is implicitly satisfied.

Symmetric Lanczos tridiagonalization of the matrix given in Table 8.2 using $\mathbf{y}_1 = \{1 \quad 0 \quad 0 \quad 0\}$ yields

$$\left.\begin{aligned} &\mathbf{y} = \begin{bmatrix} 1 & 0 & 0 & 0 \\ 0 & -0.8018 & -0.4912 & -0.3404 \\ 0 & -0.5345 & 0.3348 & 0.7760 \\ 0 & 0.2673 & -0.8041 & 0.5310 \end{bmatrix} \\ &\alpha_1 = 1 \quad , \quad \beta_1 = 3.7417 \\ &\alpha_2 = 2.7857, \quad \beta_2 = 5.2465 \\ &\alpha_3 = 10.1993, \quad \beta_3 = 4.4796 \\ &\alpha_4 = 1.0150 \end{aligned}\right\} \tag{10.99}$$

(It is interesting to note that the Lanczos method with $\mathbf{y}_1 = \{1 \quad 0 \quad 0 \quad 0\}$ and Householder's method yield the same tridiagonal matrix except for sign changes.)

For this small example the eigenvalues of the tridiagonal matrix are similar to the eigenvalues of the original matrix. However, for larger matrices, particularly where the ratio $|\lambda_1|/|\lambda_n|$ is large, the implicit orthogonality conditions $\mathbf{y}_i^T\mathbf{y}_j = 0$ for $j > i + 1$ will not be satisfied accurately because each step of the process will magnify any rounding errors present. Hence the eigenvalue estimates obtained using the basic algorithm cannot always be guaranteed. This difficulty may be overcome by extending the algorithm (10.98) so that $\mathbf{y}_{j+1}$ is orthogonalized with respect to the vectors $\mathbf{y}_1, \mathbf{y}_2, \ldots, \mathbf{y}_{j-1}$. When modified in this way the algorithm will perform satisfactorily.

For an $n \times n$ matrix $\mathbf{A}$ having an average of c non-zero elements per row, the number of multiplications required to perform the full tridiagonalization by the basic method is approximately $(c + 5)n^2$. However, the extra orthogonalizations require a further n^3 multiplications approximately. Hence the reliable version of the complete tridiagonalization algorithm is less efficient than Householder's tridiagonalization even when the matrix $\mathbf{A}$ is sparse. On the other hand, the algorithm including orthogonalization provides an efficient method for partial eigensolution. If the tridiagonalization is halted when the tridiagonal matrix is of order m then approximately $nm(c + m + 5)$ multiplications will be needed. For $m \ll n$ the eigensolution of the tridiagonal matrix may be considered to involve a negligible amount of computation. However, if the eigenvector $\mathbf{q}$ of $\mathbf{A}$ corresponding to an eigenvector $\mathbf{p}$ of the tridiagonal matrix is required, it needs to be computed from

$$\mathbf{q} = \mathbf{Y}\mathbf{p} \tag{10.100}$$

where $\mathbf{Y} = [\mathbf{y}_1\mathbf{y}_2 \quad \ldots \quad \mathbf{y}_m]$. Hence nmr multiplications are required to compute r of the eigenvectors of $\mathbf{A}$.

The Lanczos algorithm may also be used to compute partial eigensolutions for linearized eigenvalue problems (Ojalvo and Newman, 1970). The efficiency of algorithms for obtaining the set of r dominant eigenvalues depends on the ratio m/r necessary to obtain the required accuracy. Tests by Ojalvo and Newman give ratios, m/r, which makes the Lanczos algorithm generally more efficient than SI (where the criterion of efficiency is taken as the total number of multiplications). However, any possible gain in computational efficiency should be measured against the following possible disadvantages:

(a) The premultiplication $\mathbf{v}_i = \mathbf{A}\mathbf{y}_i$ has to be performed separately for each vector. Thus, if $\mathbf{A}$ is held in backing store, it will have to be transferred to the main store more often than if SI is used. A similar conclusion applies for the solution of $\mathbf{Ax} = \lambda\mathbf{Bx}$ when $\mathbf{A}$ and $\mathbf{B}$ are held in backing store.
(b) It is difficult to assess how far it is necessary to proceed with the tridiagonalization in order to obtain a partial eigensolution of the required accuracy.
(c) There is no indication of the accuracy of any particular solution.
(d) There appears to be a significant chance that an eigenvalue will be omitted from the results because the trial vector contains no component of its eigenvector. (In SI an eigenvalue will only be omitted if all of the trial vectors contain no component of its eigenvector.)

It is possible to overcome the difficulties mentioned in (b) and (c) by computing the eigenvalues of the current tridiagonal matrix within each step of the process. Convergence may be assumed to have taken place when the larger eigenvalues computed from two consecutive tridiagonal matrices are sufficiently close that they satisfy a suitable tolerance criterion.

A Lanczos procedure is also available for transforming an unsymmetric matrix into tridiagonal form. However, this method may be subject to numerical instability.

BIBLIOGRAPHY

Aitken, A. C. (1937). 'The evaluation of the latent roots and vectors of a matrix'. *Proc. Roy. Soc. Edinburgh Sect. A*, **57**, 269–304. (Purification for subdominant eigenvalues.)

Bathe, K.-J., and Wilson, E. L. (1973). 'Solution methods for eigenvalue problems in structural mechanics'. *Int. J. for Num. Methods in Engng.*, **6**, 213–226. (Relative merits of SI.)

Bauer, F. L. (1958). 'On modern matrix iteration processes of Bernoulli and Graeffe type'. *J. Ass. of Comp. Mach.*, **5**, 246–257.

Bodewig, E. (1956). *Matrix Calculus*, North Holland, Amsterdam.

Clint, M., and Jennings, A. (1971). 'A simultaneous iteration method for the unsymmetric eigenvalue problem'. *J. Inst. Maths. Applics.*, **8**, 111–121. (Adopts an approximate eigensolution of the interaction matrix.)

Corr, R. B., and Jennings, A. (1976). 'A simultaneous iteration algorithm for symmetric eigenvalue problems'. *Int. J. Num. Methods in Engng.*, **9**, 647–663.

Crandall, S. H. (1951). 'Iterative procedures related to relaxation methods for eigenvalue problems'. *Proc. Roy. Soc. London Ser. A,* 207, 416–423. (Inverse iteration.)

Dong, S. B., Wolf, J. A., and Peterson, F. E. (1972). 'On a direct-iterative eigensolution technique'. *Int. J. for Num. Methods in Engng.,* 4, 155–161. (An SI procedure.)

Duncan, W. J., and Collar, A. R. (1935). 'Matrices applied to the motions of damped systems'. *Philos. Mag. Ser.* 7, 19, 197–219. (Deflation.)

Fox, L. (1964). *Introduction to Numerical Linear Algebra,* Clarendon Press, Oxford.

Frazer, R. A., Duncan, W. J., and Collar, A. R. (1938). *Elementary Matrices and Some Applications to Dynamics and Differential Equations,* Cambridge University Press, Cambridge. (The power method and deflation.)

Gourlay, A. R., and Watson, G. A. (1973). *Computational Methods for Matrix Eigenproblems,* Wiley, London.

Irons, B. M. (1963). 'Eigenvalue economizers in vibration problems'. *Journal of the Royal Aero. Soc.,* 67, 526–528. (Mass condensation.)

Jennings, A. (1967). 'A direct iteration method of obtaining latent roots and vectors of a symmetric matrix'. *Proc. Camb. Phil. Soc.,* 63, 755–765. (An SI procedure with approximate eigensolution of the interaction matrix.)

Jennings, A. (1973). 'Mass condensation and simultaneous iteration for vibration problems'. *Int. J. for Num. Methods in Engng.,* 6, 543–552.

Jennings, A., and Stewart, W. J. (1975). 'Simultaneous iteration for partial eigensolution of real matrices'. *J. Inst. Maths. Applics.,* 15, 351–361.

Lanczos, C. (1950). 'An iteration method for the solution of the eigenvalue problem of linear differential and integral operators'. *J. Res. Nat. Bur. Stand.,* 45, 255–282.

Ojalvo, I. U., and Newman, M. (1970). 'Vibration modes of large structures by an automatic matrix-reduction method'. *AIAA Journal,* 8, 1234–1239. (Lanczos' method for partial eigensolutions.)

Ramsden, J. N., and Stoker, J. R. (1969). 'Mass condensation: a semi-automatic method for reducing the size of vibration problems'. *Int. J. for Num. Methods in Engng.,* 1, 333–349.

Rutishauser, H. (1969). 'Computational aspects of F. L. Bauer's simultaneous iteration method'. *Numer. Math.,* 13, 4–13. (A method for symmetric matrices which includes Chebyshev acceleration.)

Stewart, G. W. (1976). 'Simultaneous iteration for computing invariant subspaces of non-Hermitian matrices'. *Numer. Math.,* 25, 123–136.

Stiefel, E. (1958). 'Kernel polynomials in linear algebra and their numerical applications'. *Nat. Bur. Standards Appl. Math. Ser.,* 49, 1–22. (Connects the method of conjugate gradients and Lanczos' method.)

Wilkinson, J. H. (1965). *The Algebraic Eigenvalue Problem,* Clarendon Press, Oxford. (Bauer's SI method and inverse iteration for complex conjugate pairs of eigenvalues.)

Wilkinson, J. H., and Reinsch, C. (1971). *Handbook for Automatic Computation Vol. II, Linear Algebra,* Springer-Verlag, Berlin. (An algorithm by H. Rutishauser on simultaneous iteration for symmetric matrices and an algorithm by G. Peters and J. H. Wilkinson on the calculation of specified eigenvectors by inverse iteration.)

Appendix A
Checklist for Program Layout

Nine tenths of wisdom is being wise in time
Theodore Roosevelt

General questions

Q1. Are alternative simpler or more economical matrix formulations available for the problem (e.g. section 2.2 for electrical resistance networks)?

Q2. Can the basic matrices be constructed automatically from simple input data (see section 2.3 for electrical resistance networks)?

Q3. Do any of the matrices have useful properties, e.g. symmetry or positive definiteness?

Q4. Where the algebraic formulation includes the inverse of a matrix, can the computation of the inverse be avoided by using, instead, an equation-solving procedure (see sections 1.13 and 4.5)?

Q5. Where a multiple matrix product is to be evaluated, what is the most efficient sequence for carrying out the multiplications (see section 1.3)?

Q6. Should special storage schemes be used to take advantage of sparseness, symmetry or other properties that any of the matrices possess (see section 3.8 and following sections for sparse matrices)?

Q7. Where large matrices need to be stored, is it necessary to use backing store, and does this affect the choice of procedure for the matrix operations (see section 3.7 for matrix multiplication and section 5.10 for sparse elimination procedures)?

Q8. Are suitable standard or library procedures available for the required matrix operations?

Q9. Has a simple numerical problem been processed by hand to ensure the viability of the method?

Q10. Can automatic checks be incorporated in the algorithm to check the accuracy or the validity of the results (e.g. by substituting the solution back into a set of linear equations to determine the residuals (section 4.13) or by checking known physical requirements that the results must satisfy)?

Questions on equation solving

Q11. Is an elimination or iterative procedure to be preferred (see Table 6.5)?

Q12. Could the equations be so ill conditioned that special measures are required to counteract ill-conditioning (see section 4.11 and following sections)?

Q13. If elimination is to be used, is it necessary to select pivots (see section 4.6)?

Q14. If elimination is to be used for equations which have a sparse coefficient matrix, can the variables be easily ordered in such a way that the matrix has a narrow diagonal band or variable bandwidth structure (see sections 5.5 and 5.6)?

Questions on eigensolution

Q15. Is the matrix requiring eigensolution symmetric or, alternatively, is the problem one which transforms to the eigensolution of a symmetric matrix (see section 7.4)?

Q16. What is known about the nature and range of the eigenvalues (e.g. are they real and positive)?

Q17. Is a complete or partial eigensolution required?

Q18. If a partial eigensolution is sought, can the eigenvalue formulation be so arranged that the required eigenvalues are the dominant set (see sections 10.3 and 10.10)?

Q19. Are good initial estimates available for the required eigenvalues which may be useful in the Sturm sequence or inverse iteration methods?

Q20. Are good initial estimates available for the required eigenvectors which may be useful in the power method or simultaneous iteration?

Q21. Could the presence of equal eigenvalues, eigenvalues of equal modulus or non-linear elementary divisors cause difficulty with the eigensolution (see section 10.2 for the power method)?

Appendix B
Checklist for Program Preparation

Q1. Is the program composed of well-defined sections such that the correct functioning of each can be easily checked?

Q2. Has the action of the program been simulated by hand using a simple numerical example?

Q3. If similar sets of instructions occur in different parts of the program, can the duplication be avoided by the use of subroutines?

Q4. Wherever a division is carried out, is there any possibility of the divisor being zero?

Q5. Wherever a square root is obtained, could the argument be negative; and, if the argument is definitely positive, is the positive or negative root required?

Q6. Are the parts of the program which will be most heavily utilized programmed in a particularly efficient manner?

Q7. If many zeros occur in matrices which are not packed, can the efficiency be improved by avoiding trivial computations?

Q8. Where sparse matrix storage schemes are used (particularly packing schemes), is there a safeguard to ensure that array subscripts do not go out of bounds during execution?

Q9. Does the program form a closed loop such that several sets of input data can be processed in one computer run?

Q10. Is the input data in as simple a form as possible so that they can be easily prepared and checked?

Q11. Are there any parameters which are set within the program which would be better to be read in as input parameters (e.g. the tolerance for iterative methods)?

Q12. Can the output data be easily interpreted without reference to the input data (since these two documents may be separated)?

Q13. Are failure conditions properly monitored in the output?

Q14. If a failure condition is encountered, can the program execution be continued (e.g. by proceeding to the next problem contained in the input data)?

Q15. If an iterative sequence is present in a program, is there a facility to exit from the loop if iteration continues well beyond the expected number of cycles? (A non-convergent iteration may occur through a fault in either the program, the input data or the logic of the method).

Appendix C
Checklist for Program Verification

Q1. Can a variety of test problems be obtained from published literature? (A useful reference is R. T. Gregory and D. L. Karney, *A Collection of Matrices for Testing Computational Algorithms*, Wiley, New York, 1969.)

Q2. Do the test problems verify all parts of the program?

Q3. Does the set of test problems include examples which are most likely to give numerical instability or cause ill-conditioning in linear equations?

Q4. Can the accuracy of the solutions to the test problems be verified by independent checks (e.g. if the eigenvectors of a symmetric matrix have been obtained, they should satisfy the orthogonality condition given in section 1.20)?

Q5. Does the convergence rate for an iterative method conform with any available theoretical predictions (since a mistake in an iterative procedure may cause a slowing down in the convergence rate, although it may still converge to the correct solution)?

Index